AAPS Introductions in the Pharmaceutical Sciences

Founding Editor
Robin Zavod

Volume 2

Series Editor
Claudio Salomon, National University of Rosario, Rosario, Argentina

The *AAPS Introductions in the Pharmaceutical Sciences* book series is designed to support pharmaceutical scientists at the point of knowledge transition. Springer and the American Association of Pharmaceutical Scientists (AAPS) have partnered again to produce a second series that juxtaposes the *AAPS Advances in the Pharmaceutical Sciences* series. Whether shifting between positions, business models, research project objectives, or at a crossroad in professional development, scientists need to retool to meet the needs of the new scientific challenges ahead of them. These educational pivot points require the learner to develop new vocabulary in order to effectively communicate across disciplines, appreciate historical evolution within the knowledge area with the aim of appreciating the current limitations and potential for growth, learn new skills and evaluation metrics so that project planning and subsequent evolution are evidence-based, as well as to simply "dust the rust off" content learned in previous educational or employment settings, or utilized during former scientific explorations. The *Introductions* book series will meet these needs and serve as a quick and easy-to-digest resource for contemporary science.

S. Narasimha Murthy
Editor

Topical Products and Dermal Drug Delivery

Springer

aaps

Editor
S. Narasimha Murthy
Topical Products Testing LLC
Batesville, MS, USA

ISSN 2522-834X ISSN 2522-8358 (electronic)
AAPS Introductions in the Pharmaceutical Sciences
ISBN 978-3-031-80524-0 ISBN 978-3-031-80525-7 (eBook)
https://doi.org/10.1007/978-3-031-80525-7

This Springer imprint is published by the registered company Springer Nature Switzerland AG
The registered company address is: Gewerbestrasse 11, 6330 Cham, Switzerland

Preface

The book *Topical Formulations and Dermal Drug Delivery* addresses a wide range of topics pertinent to professionals in industry, academia, and clinical research. The motivation to compile and edit this book arose from the extensive research carried out by investigators globally over the past decade, much of which has been supported by research grants from the United States Food and Drug Administration (USFDA). Consequently, readers will notice a regulatory perspective that runs through many of the chapters. One of the book's distinguishing features is that it presents some topics in unprecedented detail, marking the first time such in-depth coverage appears in a book chapter.

Focusing on critical quality attributes (CQAs) from both stability and performance perspectives is crucial in developing topical products. As such, the opening chapter of this book thoroughly explores the fundamentals and importance of CQAs in topical formulations, providing specific examples. The second chapter on metamorphosis delves into how the evaporation of solvents at the application site alters CQAs, affecting the overall performance of the product. A dedicated chapter on rheological characterization offers readers essential knowledge and practical insights. Furthermore, the book introduces cutting-edge microscopic tools and techniques for examining microstructure (Q3 characteristics), active pharmaceutical ingredient (API) distribution, and drug absorption into the skin layers—valuable resources for both researchers and practitioners.

The book includes an in-depth chapter on in vitro release testing (IVRT) of topical products, featuring recent case studies particularly relevant to the generic pharmaceutical industry. In vitro permeation testing (IVPT) serves as a highly effective method for evaluating the performance of topical formulations. Regulatory guidance outlines the specific experimental conditions required for conducting IVPT. The chapter dedicated to in vitro permeation testing of topical products addresses the practical challenges investigators encounter, offering potential solutions to overcome them.

The book highlights innovative approaches to dermal drug delivery, including techniques for investigating in vivo cutaneous pharmacokinetics using dialysis-based methods. It also presents case studies of in vivo experiments that underscore the importance of these techniques in comparing generic topical products with reference-listed drugs (RLD). In a dedicated chapter, emphasis is placed on the critical role of statistical data analysis in decision-making processes. Another standout chapter explores the role of sensory characteristics in pharmaceutical semisolids and their impact on the clinical outcomes of therapy. The final chapter is particularly noteworthy, as it provides a detailed history and implementation of the FDA guidance, the Maximum Usage Trial (MuST).

Overall, the authors are confident that readers will find this book to be a unique, informative, and valuable resource for activities related to the development, characterization, and evaluation of topical products.

Batesville, MS, USA S. Narasimha Murthy

Contents

About the Editor

S. Narasimha Murthy is the Chief Scientific Officer of Topical Products Testing (TPT) LLC, USA (www.TopicalProductsTesting.com). Dr. Murthy is a pioneer in topical and transdermal drug delivery technologies. His research programs are funded by National Institutes of Health (NIH), United States Food and Drug Administration (USFDA), and Pharmaceutical companies. Formerly, Dr. Murthy served as a Professor of Pharmaceutics at the University of Mississippi School of Pharmacy for over 16 years. Dr. Murthy has published over 150 research papers and presented over 250 scientific posters at national and international scientific meetings. He has authored 3 books and over 20 book chapters.

Dr. Murthy has received several awards, such as the Cumberland Researcher of the Year from the University of Mississippi, the Global Indus Technovator award from MIT, Endowed Chair for Research at the Ohio Northern University, and he was inducted as a Fellow of the American Association of Pharmaceutical Scientists in 2017. He was bestowed with the Distinguished Scientist award by the American Association of Indian Pharmaceutical Scientists in 2018. Dr. Murthy was honored with the prestigious 2022 American Association of Pharmaceutical Scientists (AAPS) Pharmaceutical Global Health Award. In 2023, the Academy of Pharmaceutical Sciences, UK, awarded him a Fellowship, the highest award given to very few selected members recognizing their research accomplishments in any sector of pharmaceutical sciences.

Dr. Murthy is also one of the co-founders of a non-profit research organization in India, "The Institute for Drug Delivery and Biomedical Research (IDBR)" (www.IDBResearch.com), founded in 2013. IDBR's primary goals are to foster research in drug delivery and biomedical sciences to find solutions to unmet medical needs and help aspiring scientists and students. The organization also plays a significant role in several community service activities.

Chapter 1
Topical Formulations and Critical Quality Attributes

H. N. Shivakumar, Avichal Kumar, Ashana Puri, and S. Narasimha Murthy

1.1 Introduction

The quality attributes determine the release and absorption of Active Pharmaceutical Ingredients (APIs) from the topical products. The quality attributes of a topical product depend on several factors, such as the type of dosage form, composition, and manufacturing process. From the regulatory perspective, the quality attributes may include physicochemical characteristics and biological or microbiological attributes. The critical quality attributes should be within an appropriate limit, range, or distribution to ensure the product has the desired quality. Some of the quality attributes of topical products are listed in Table 1.1.

The impact of any quality attributes on the formulation performance is not universal but depends on the API's physicochemical nature, formulation type, and indication being treated [1]. For instance, the drug partitioning is determined by the pH of the formulation and the pK_a of the API [2]. However, the pH may not affect the partitioning into the skin of a highly nonpolar API. Similarly, water activity, which is the amount of thermodynamically active water in the formulation, may not be relevant in the case of an anhydrous and hydrocarbon-base ointment.

The nature of excipients and concentration in topical products influence the API solubility, thermodynamic activity, drug release, and drug penetration into or

H. N. Shivakumar · A. Kumar
Department of Pharmaceutics, KLE College of Pharmacy, Bangalore, Karnataka, India

A. Puri
Department of Pharmaceutical Sciences, Bill Gatton College of Pharmacy, East Tennessee State University, Johnson City, TN, USA

S. N. Murthy (✉)
Topical Products Testing LLC, Batesville, MS, USA

S. N. Murthy (ed.), *Topical Products and Dermal Drug Delivery*,
AAPS Introductions in the Pharmaceutical Sciences 2,
https://doi.org/10.1007/978-3-031-80525-7_1

Table 1.1 Quality attributes and their importance in topical products

Quality attribute	Importance of topical products
pH	Determines the formulation stability, skin irritancy, and drug absorption.
Phase distribution of drug	The API dissolved in the cream partitions between the immiscible phases depending on its relative affinity and formulation characteristics.
Particle shape and size	Particle size influences the dissolution of API, release, and permeation from topical products.
Polymorphic form	It determines the solubility, release, absorption, and stability of the drug substance in the product.
Globule size	The globule size distribution is critical in determining the stability of semisolids, sensorial characteristics, and the rate and extent of drug absorption.
Water activity	Impacts the formulation's drying rate, drug release, skin hydration, and skin permeability.
Rheological characteristics	The viscosity is also known to impact the stability of creams and determine the residence time and performance. It also influences the sensorial characteristics of the product.
Apparent density	Determines the volume of the formulation and the dose applicable per cm^2.
Work of adhesion	It is a function of the extent of affinity and interaction of the formulation with the skin.

permeation across the skin. Thus, each functional and structural ingredient is titrated during product development to favor optimal drug absorption and therapeutic efficacy. The impact of different quality attributes on the formulation's performance warrants systematic investigation so that optimal characteristics are embodied in the formulation to maximize its effectiveness. Such carefully developed and clinically tested novel FDA-approved products are called Innovator Drug Products or Reference-Listed Drug (RLD) products.

Generic products are new products that are compared with the RLD to demonstrate that they are bioequivalent (i.e., performing in the same manner as the innovator drug). Once the Abbreviated New Drug Application (ANDA) is approved, the generic product may be manufactured and marketed to provide a safe, effective, low-cost alternative. Thus, the primary formulation goal in generic drug product development is to ensure that it has the same components (Q1 sameness) at the same concentration as the RLD (Q2 sameness). The generic product formulator needs to ensure that the excipients are of the same grade, as different grades would considerably impact the critical quality attributes of the topical drug products. Generic product formulators perform reverse engineering of the RLD using validated analytical methods to establish Q2 sameness [3]. Unlike in the case of solutions and simple homogenous products, the dispersed systems, a generic semisolid product, despite the Q1/Q2 sameness criteria being met, might exhibit differences in microstructure (arrangement of matter) compared to the innovator product (Q3 sameness), owing to the plausible differences in the manufacturing process variables. The deviation in the quality attributes of the generic product could significantly impact its performance, potentially rendering it non-bioequivalent [4]. A similar rheological profile, globule size, and solid-state characteristics of the drug product are some essential aspects in demonstrating Q3 equivalence.

1.2 Critical Quality Attributes

The fundamentals of some of the critical quality attributes of topical products are discussed in the following sections.

1.2.1 pH of the Formulation

The pH, one of the most essential characteristics of a topical formulation, is adjusted to a specific value for one or more of the following reasons:

API Centric

- The API is most stable within a pH range.
- The API needs to exist in an ionic or non-ionic form.
- To modulate the interaction of API with the excipients.

Skin Centric

- To modulate the skin permeability for the API
- To minimize skin irritation

Formulation Centric

- To induce pH-sensitive gelation
- To maximize the preservative efficacy
- To modulate the excipient interactions

The formulation pH could affect the drug solubility, the extent of ionization, the polymorphic state, the fraction of the dissolved API, and the relative amount of drug distributed between nonpolar and polar phases in a dispersion [5]. From the physiological compliance perspective, the preferred pH of a topical formulation is in the range of 4–8. An extreme deviation from the skin's pH is known to impair the skin barrier and potentially irritate the skin [6–8].

1.2.1.1 pH and Drug Delivery

The permeation of a drug through the skin depends on the extent of ionization, solubility in the formulation, and the extent of partitioning of the drug into the skin.

In most cases, the ionized species of the drug is known to possess a lower partition coefficient and, thus, a low permeability coefficient. The total flux of the permeant through the skin is the sum of the transport of the ionized and the unionized fractions as per Eq. 1.1.

$$J_{\text{flux}} = K_p^{\text{union}} \cdot C_{\text{union}} + K_p^{\text{ion}} \cdot C_{\text{ion}} \tag{1.1}$$

where J_{flux} is the total flux through the skin, K_p^{union} and K_p^{ion}. The permeability coefficients of the unionized and ionized species, whereas C_{union} and C_{ion} would be the concentrations of the unionized and ionized fractions of the drug, respectively.

Most dermatological therapeutic agents are weak acids or weak bases. The pH of the topical dosage form and the pKa value of these molecules will determine the extent of ionization of the active moiety. Equations 1.2 and 1.3 represent the fraction of ionized and unionized species of weakly acidic and weakly basic therapeutic agents, respectively:

$$\text{pH} = pK_a + \log \frac{[\text{I}]}{[\text{UI}]} \tag{1.2}$$

$$\text{pH} = pK_a + \log \frac{[\text{UI}]}{[\text{I}]} \tag{1.3}$$

where [I] and [UI] refer to the concentration of ionized (I) and unionized (UI) species of the drug substance at the pH of the formulation, while pK_a would be the dissociation constant of the drug substance. The pH of the formulation is adjusted to a specific value to increase the fraction of the unionized species to favor the permeation of the API. For instance, topical formulations of nonsteroidal anti-inflammatory drugs such as salicylic acid and ibuprofen have pK_a ranging from 4.0 to 4.5. An increase in the donor solution pH from 5 to 8 reduced the fraction of unionized species, eventually decreasing the permeability across the human epidermis [9].

On the other hand, the permeability coefficient of three basic drugs was found to increase when the pH of the donor solution was increased from 5.8 to 8.0, owing to an increased unionized fraction of the API. The skin permeation of lignocaine, sufentanil, and fentanyl, which possess a pK_a of 8, 8.5, and 9, respectively, was enhanced by increasing the pH from 3 to 7.4. [9] The effects of pH on the permeability of acidic and basic drugs are collectively captured in Fig. 1.1.

Most often, the ionized species commonly have a high aqueous solubility but a low K_p^{ion}, while the unionized fraction has a low solubility but a high K_p^{union}. When the relative solubilities of the ionized and unionized forms differ significantly, the high solubility of the ionized species might compensate for its low permeability, eventually enhancing the API's overall flux [10].

In some instances, a cosolvent can significantly alter the solvent polarity, and, therefore, the influence of pH would affect the solubility and the permeation of drugs

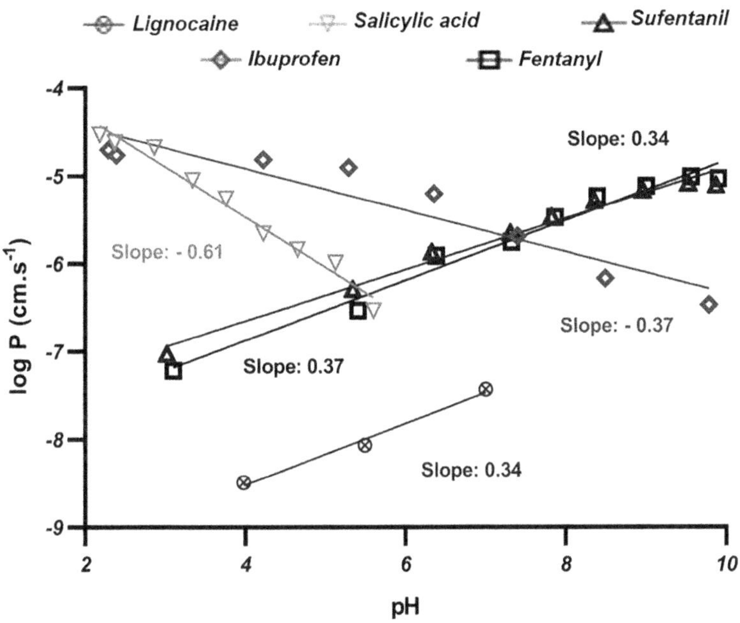

Fig. 1.1 Effect of pH on the permeability coefficients of weakly acidic and basic drugs. The permeability coefficient of acids was found to decrease, whereas the permeability coefficient of bases was found to increase with the increase in pH. (Digitized and replotted from [9])

from a topical product. The changes in pH would determine the extent of ionization depending on the polarity of the vehicle. The flux could increase owing to increased solubilization of the ionized species in the cosolvent used in the formulation, leading to increased thermodynamic activity. The increased thermodynamic activity of the ionic species substantially favors permeation, counteracting the negative impact of ionization in some cases [10]. Thus, the flux can be improved when the solubility of the ionized form in the formulation is substantially higher than the unionized form to maintain high thermodynamic activity.

Methotrexate is an antineoplastic agent used to treat psoriasis and severe rheumatoid arthritis. The drug is a weak acid with a pKa value of 4.3 (-carboxyl) and 5.5 (-carboxyl). Due to its ionic nature at physiological pH, it exhibits permeability issues. In vitro permeation of methotrexate from a saturated solution in propylene glycol (50%) across human cadaver skin was found to be pH-dependent [11].

The saturation solubility of the drug in 50% propylene glycol was found to be high at a pH <2 (1.4 mg/ml), while it decreased at pH 3.87 (0.46 mg/ml). As the pH exceeded 4, there was a sharp increase in drug solubility in the same solvent system. A set of five saturated solutions of the drug in propylene glycol (50%) that had a pH of 1.98, 2.98, 3.87, 5.29, and 6.34 were used for the studies. A saturated drug solution at pH 4.12 in the same solvent system was used as a control. When permeation studies were performed with the saturated solutions, an increase in the

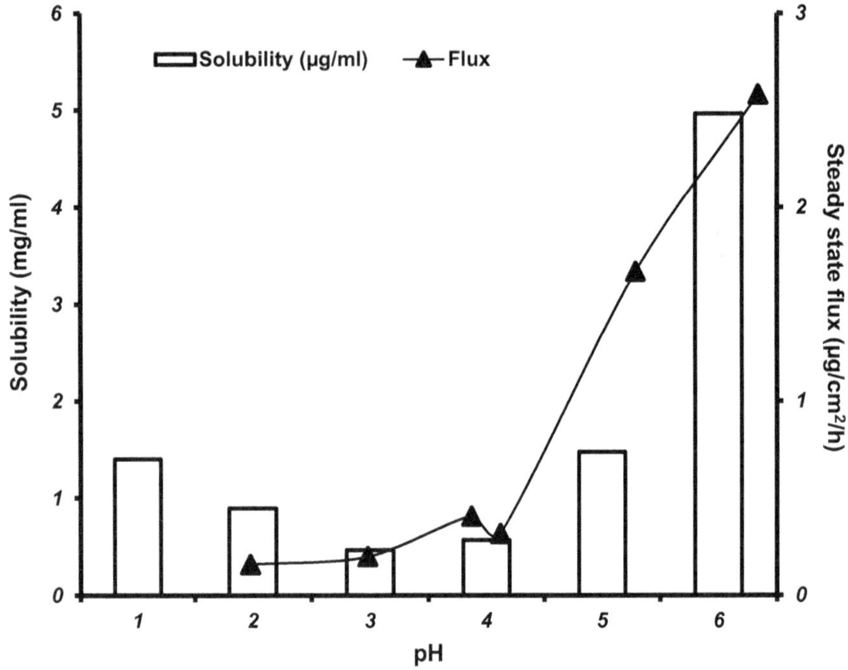

Fig. 1.2 pH-dependent solubility and permeability profiles of methotrexate. (Plotted using the data from [11])

permeation flux and the extent of ionization of the API was observed (Fig. 1.2). The permeation rate increased due to the increased solubility of the ionized species in the binary solvent mixture. Thus, a combined approach of modulation of pH and the use of cosolvent improved the solubility of ionized species and, thereby, the transdermal drug permeation rate of methotrexate.

1.2.1.2 pH and Drug Stability

One of the primary factors to consider when developing topical products is the chemical stability of the API and excipients in the formulation. In the case of APIs, wherein stability is sensitive to pH changes, the pH needs to be adjusted to the optimal stability range to mitigate the degradation of the drug substance [12].

Fluocinolone acetonide cream is used to treat several skin conditions, such as eczema, dermatitis, allergies, and rashes. A study was conducted to elucidate the effect of the cream's pH (ranging from 2.3 to 6.0) on drug stability [13]. Figure 1.3 shows the degradation rate at different pH values at a temperature of 50 °C. The studies conclusively indicate that the maximum stability of fluocinolone acetonide was about pH 4.1. The studies at different temperatures lead to the same outcome.

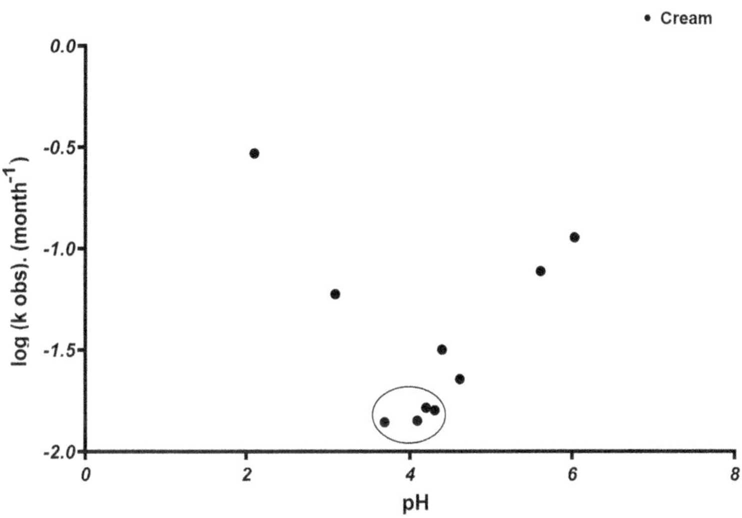

Fig. 1.3 pH-log (rate constant) profiles depicting degradation of fluocinolone acetonide in creams at 50 °C. (Digitized and replotted from [13])

1.2.1.3 pH-Induced Gelation

The pH is known to influence the viscosity or the consistency of certain topical drug products. Carbomers are acrylic acid polymers cross-linked with polyalkenyl polyethers commonly used as gelling agents to manufacture topical semisolids. The widely used pH-sensitive carbomers in pharmaceutical applications include Carbopol® 934P and Carbopol® 940. The two differ in molecular weight and generate semisolids of different viscosity and rigidity [14]. Aqueous dispersions of carbomers are acidic (pH<3.5). Conventionally, carbomers are used as a gelling agent in the 0.1–1% concentration range. The aqueous solutions are free-flowing liquids at lower pH values, whereas they form viscous, transparent gels at neutral pH values. The pH of the aqueous solution of carbomers is often neutralized using an alkali to transform them into gels. The ionization of the carboxyl group above pH 7.0 is known to result in the uncoiling of the polymer chains and forming a rigid gel, as depicted in Fig. 1.4. The pH affects the viscosity of pH-sensitive polymers like Carbopol® 934P [15]. For example, the viscosity of methotrexate gels composed of Carbopol® 934P (2%) was found to increase from 40,500 cps at pH 5.1 to 44,800 cps at pH 7.1, demonstrating the pH-sensitive nature of the polymer. The diffusion studies indicated that the diffusion coefficient of the drug from 2% Carbopol® gel decreased from 3.58×10^{-6} to 1.41×10^{-6} when the pH was increased from 5.1 to 7.1, respectively. This is attributed to a decrease in the diffusivity of the API with an increase in the gel viscosity.

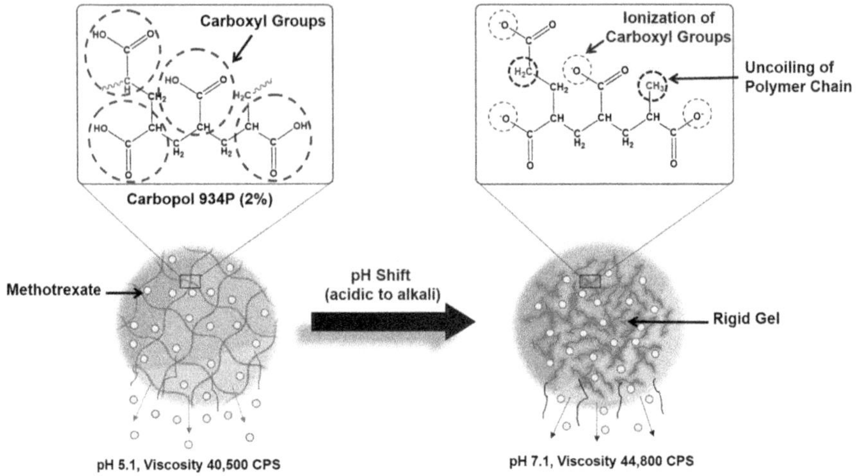

Fig. 1.4 pH-dependent gelation of Carbopol® 934. The ionization of the carboxyl group above pH 7.0 results in the uncoiling of polymer chains, leading to the formation of a rigid gel that determines the drug release

1.2.1.4 pH and the Antimicrobial Efficacy

The pH of the topical product would dictate the choice of the preservative to be incorporated. Generally, the preservatives are more effective when present in the unionized form. For instance, organic acids employed as preservatives are highly effective in the pH range of 3–5. Likewise, commonly used preservatives like parabens are known to be effective in a pH range of 4–8. On the contrary, quaternary compounds are more effective at a pH above 6 [16]. The pH of the formulation affects the extent of ionization of the preservative, as well as its interaction with and partitioning into the microorganism's cell wall [3].

Chlorhexidine is a biguanide cationic molecule employed as an antimicrobial in topical products. Chlorhexidine, at antiseptic concentrations (0.5–4%) is known to possess both bactericidal and bacteriostatic activity against gram-negative and gram-positive bacteria. The antimicrobial activity of chlorhexidine is pH-dependent, with an optimal activity in the range of 5.5–7.0 [17].

1.2.2 Degree of Saturation

The drug concentration in the formulation is crucial to driving the drug across the biological barriers (Fig. 1.5a, b). Achieving a high degree of saturation in a topical product is needed to maximize drug absorption across the skin [18].

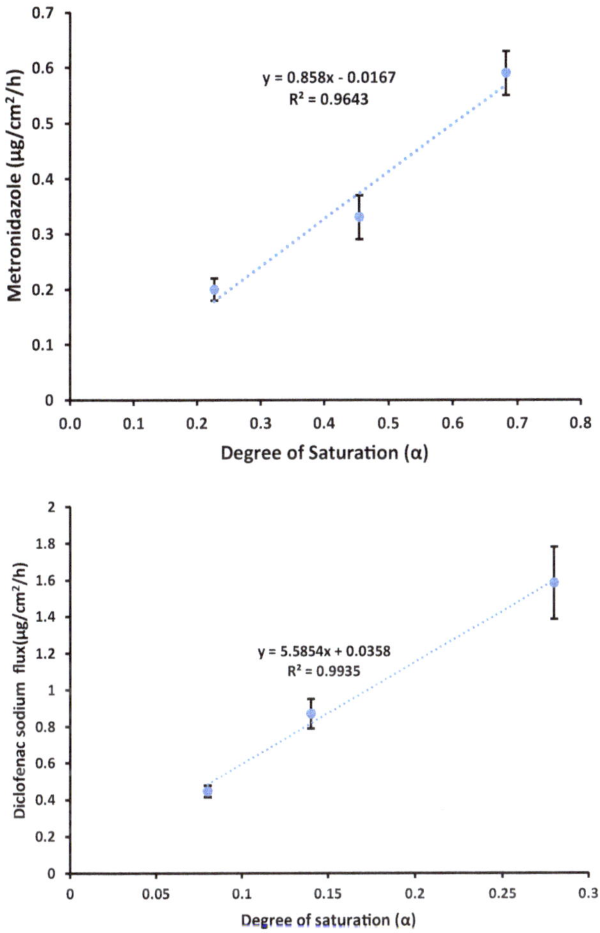

Fig. 1.5 (**a**) The effect of the degree of saturation on the permeation flux of metronidazole. The steady-state transdermal flux of metronidazole increased linearly with the degree of saturation. The gels were prepared with hydroxyethyl cellulose (HEC; 5%w/w), PEG 200 (30%w/w), sod. EDTA, and water. (**b**) The effect of the degree of saturation on the permeation flux of diclofenac. The steady-state transdermal flux of diclofenac was found to increase linearly with an increase in the degree of saturation. The gels were prepared with hydroxyethyl cellulose (HEC 5%w/w), PEG 200 (30%w/w), sod. EDTA, and water

Therefore, several approaches are employed to enhance the solubility of the API in the formulation vehicle during product development.

Generally, if the API is nonpolar or insensitive to pH changes, cosolvency would be one of the common approaches to enhance the drug solubility. Cosolvents such as glycerin, propylene glycol, polyethylene glycol, ethanol, dimethyl isosorbide (DMI), dimethyl formamide (DMF), and N-methyl pyrrolidone (NMP) are commonly used as cosolvents to improve the solubility of the drug in topical products.

Table 1.2 Improvement of solubility of silymarin constituents in topical products by complexation with sulfobutyl ether-beta-cyclodextrin

Silymarin constituents	Intrinsic solubility (mg/mL)	Solubility of the complex (mg/mL)
Taxifolin	0.135 ± 0.003	2.62 ± 0.31
Silychristin	0.145 ± 0.01	5.25 ± 0.28
Silydianin	0.056 ± 0.002	7.71 ± 0.20
Silybin A	0.003 ± 0.001	2.24 ± 0.12
Silybin B	0.011 ± 0.002	4.40 ± 0.18
Isosilybin A	0.007 ± 0.002	2.75 ± 0.08
Isosilybin B	0.002 ± 0.001	0.79 ± 0.03

The use of surfactant is another approach in which the drug is solubilized by embedding the API in the micelles. Non-ionic surfactants such as sorbitan esters of fatty acids and poly-oxyethylene esters of these acids are the most preferred. Modifying API to a salt or ester form would be another strategy to enhance permeation. Some examples of such derivatives would be diclofenac sodium or diclofenac diethylammonium, currently available as commercial topical products. Some formulations utilize complexing agents such as cyclodextrins to enhance the solubility of drugs in topical products. For example, in some countries, topical products containing piroxicam-cyclodextrin complex are available for topical use. The solubility of anandamide, an endocannabinoid, was improved by 25000-fold in water by complexation with hydroxypropyl beta-cyclodextrin [19]. In another study, complexation with (1:1) sulfobutyl ether-beta-cyclodextrin enhanced the solubility of silymarin constituents several folds, as shown in Table 1.2.

One has to ensure that the topical product is formulated within the solubility limits that are less prone to cause the crystallization of the drug owing to minor unanticipated changes in temperature, batch-to-batch variation in excipients, and toward the end of the shelf-life due to evaporation of the solvent [20]. There is a need to identify and avoid potential excipient impurities, such as multivalent ions, likely to induce drug precipitation.

1.2.3 Particle Size

In a topical suspension product, the vehicle is saturated with the API, and an additional amount of API particles are dispersed in the product. The drug particles are intended to serve as a reservoir to ensure drug saturation in the vehicle as the dissolved drug gets absorbed into the skin. Therefore, the amount of the API to be dispersed in a formulation is determined by the dissolution rate of the drug and the rate of absorption of the API across the skin. For example, in the case of 5% w/w acyclovir cream, the amount of API dissolved in the vehicle is less than 1% of the total API, and the rest is suspended. Acyclovir has a very poor dissolution rate and must be present in exceedingly larger amounts to maintain saturation of the API in the continuous phase.

The particle size determines the rate of dissolution of the API. Therefore, some products contain micronized API to enhance the dissolution rate in the continuous phase. The micronized form of API would act as a reservoir to maintain saturation spontaneously. Some azelaic acid and progesterone topical semisolid products contain micronized API. In these cases, particle size is known to impact the permeation of a drug from the topical formulation. Due to fluctuations in the storage conditions, the ratio of the dissolved to the undissolved drug in the topical product might change temporarily. When the formulation returns to equilibrium, the drug can potentially deposit on the larger particles, which can lead to the enlargement of particles due to Ostwald's ripening. The API that gets precipitated will now display a different distribution of particles that might affect the permeation of the drug from topical products [12].

In an infinite-dose study that investigated the effect of particle size on skin permeation and retention of piroxicam (PXC) in aqueous suspension, the cumulative permeation was inversely proportional to the size of the particles (Figs. 1.6a, b). The authors placed an aqueous suspension of piroxicam prepared in acetate buffer, in the donor compartment and performed permeation studies across hairless mouse skin with PBS; pH 7.4) in the receiver compartment. The solubility of piroxicam in the donor medium was about 29 μg/ml. In this case, the smaller particles were able to replenish the donor compartment more rapidly than the larger particles, owing to the faster dissolution of the smaller particles. A similar particle size-dependent effect was observed with skin retention too [21]. A very limited number of studies convincingly demonstrated the passive penetration of drug particles intact into the skin [22, 23].

In another infinite-dose permeation study, similar results were observed with nitrofurazone topical gel. The drug's particle sizes in the gels were 299, 611, and 906 nm, respectively [24]. The dissolution and skin permeation of nitrofurazone from the gel-containing nanoparticles depended on the particle size. The extent of permeation achieved from the gel formulations showed an inverse relationship with the particle size (Fig. 1.6c).

1.2.4 Polymorphic Form of the API

The API in different polymorphic forms may differ in their physicochemical characteristics. The polymorphic form, having a lower melting point and a higher log P, is likely to be absorbed well compared to the other polymorphic forms. Therefore, choosing the proper polymorphic form is key to product development. In some cases, the polymorphic form also determines the extent of solubility of the API in the vehicle. The change in the polymorphic form in the formulation could affect the overall product performance [3]. Therefore, it is necessary to select a stable polymorph during topical product development. The polymorphic forms of

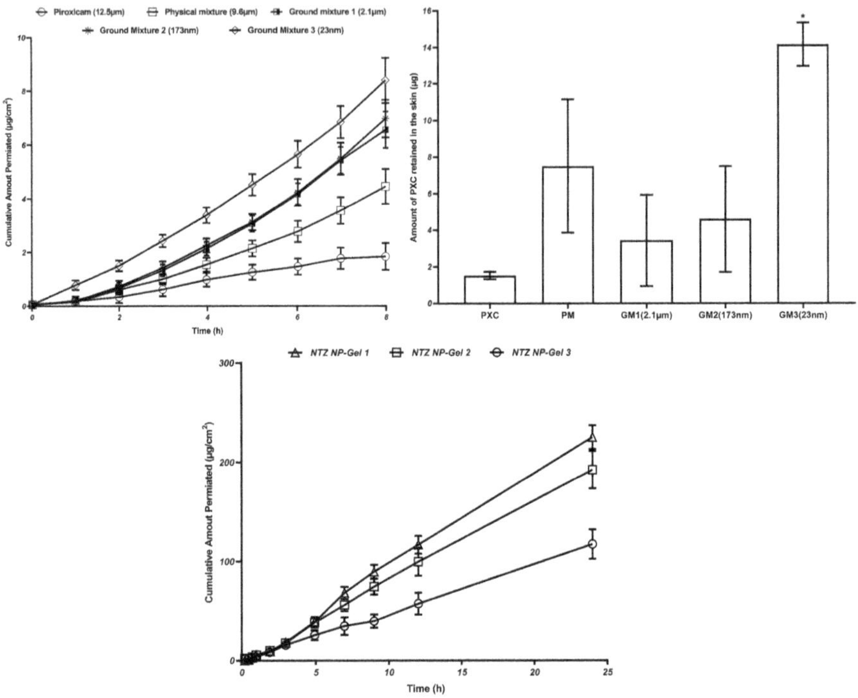

Fig. 1.6 (**a**) Effect of particle size on permeation of piroxicam from aqueous suspensions (digitized and replotted from [21]). Permeation was found to increase in the order: ground mixture 3 (23 nm) *[unfilled diamonds]* was more than ground mixture 2 (173 nm) *[filled circles]* was more than a ground mixture 1 (2.1 μm) [filled square] was more than physical mixture (9.65 μm) *[unfilled square]* was more than piroxicam (12.5 μm) *[unfilled circles]*. (**b**) Effect of particle size on skin retention of piroxicam (PXC) from aqueous suspensions (digitized and replotted from [21]). Piroxicam (12.5 μm), Physical mixture (PM: 9.6 μm), ground mixture 1 (GM1: 2.1 μm); ground mixture 2 (GM2: 173 nm); ground mixture 3 (GM3: 23 nm). (**c**) Effect of particle size on nitrofurazone permeated from gels (digitized and replotted from [24]). The permeation was highest for gels composed of smaller particles (299nm) *[triangle]*, followed by intermediate-size particles (611 nm) [squares], and lowest for large particles (906 nm) *[circles]*

the drug substance can be identified by techniques such as X-ray diffractometry, thermal analysis, microscopic techniques, and confocal Raman spectroscopy [25]. Even when one develops generic products of a topical suspension, it is necessary to characterize the polymorphic form besides other solid-state characteristics, such as particle size distribution and shape, to demonstrate the Q3 sameness with the innovator product. Often, slight differences in polymorphic nature could also contribute to differences in the performance between generic and innovator products [26]. For example, the polymorphic form of acyclovir was the same in the innovator product (Zovirax) and a generic product approved in Austria (Acyclovir 1A). It is evident from Fig. 1.7 that acyclovir in both products resulted in a similar

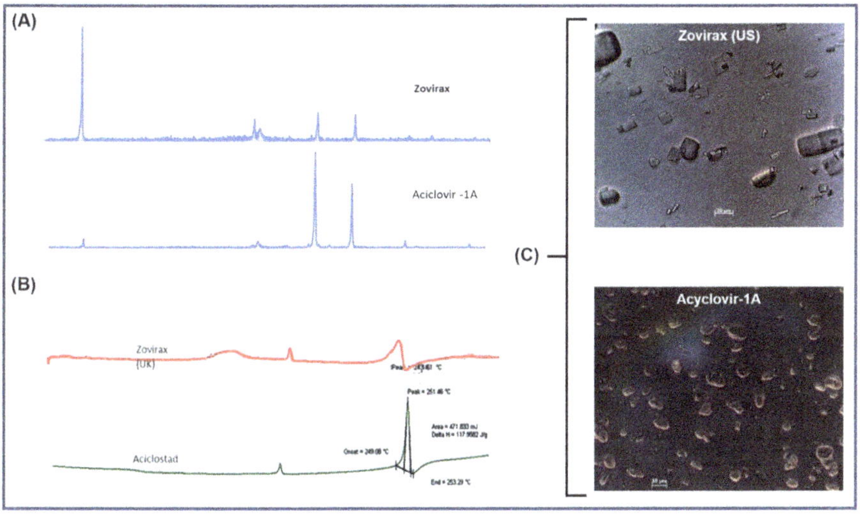

Fig. 1.7 The X-ray diffraction spectrum (**a**), differential scanning spectroscopic pattern (**b**), and microscopic pictures (**c**) of acyclovir present in the RLD, Zovirax, and a generic product approved in Austria

X-ray diffraction pattern and differential scanning thermograms. However, the microscopic appearance and size of the particle were significantly different.

1.2.5 Water Activity

Solvent systems constitute a significant component of a topical product. The thermodynamic activity of the solvent system can impact drug absorption. For example, the water activity in a formulation is the actual amount of thermodynamically available water present in a formulation:

$$a_w = \frac{\rho}{\rho_0}$$

where a_w represents the water activity of the topical product, while ρ and ρ_0 represent the partial vapor pressure of free water and pure water, respectively, at the equilibrium temperature. Water activity does not depend on absolute water content in a system but is the measure of the thermodynamic energy status of water [27]. Water can interact with chemical species like salts, sugars, amino acids, glycols, and polymers via surface interaction (dipole-dipole), ionic bonds, hydrogen bonds, and Van der Waals forces. The binding or association of water with other formulation components could impact the water activity. Water activity can range between 0 (completely dry) and 1.0 (pure water).

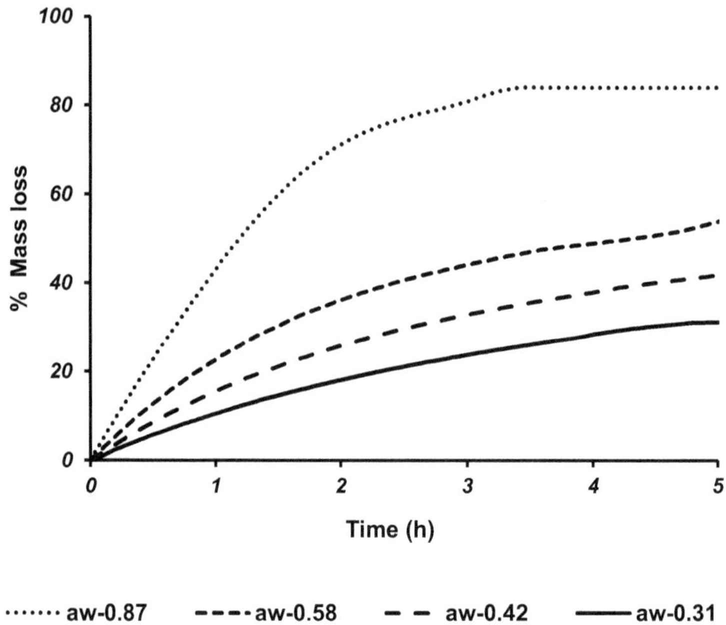

Fig. 1.8 Effect of water activity of topical products on the percent moisture loss. Products with higher water activity tend to dry relatively quickly compared to those with lower water activity

In pharmaceutical applications, water activity has been found to influence the microbiological/physical/chemical stability of the product, drug solubility, dissolution, and processability (powder flow and compaction) of pharmaceutical systems [28, 29]. Water activity measurements have emerged as a useful in vitro comparative test to characterize the generic and reference-listed drugs in product-specific guidance [30].

The drying rate is a vital formulation characteristic that water activity would influence. Formulations with higher water activity would dry relatively faster compared to formulations with lower water activity, as indicated in Fig. 1.8.

The formulations that are the same in composition could have significantly different water activity due to the difference in the microstructure owing to the manufacturing variables. The formulations with the same amount of water could differ in their water activity due to the difference in the extent of hydration of the polymer or the nature and molecular weight of the polymers. Glycols that replace water in the formulations would substantially decrease the water activity of the formulations due to their association with water.

In a study by Kerdudo and co-workers, different methods of mixing 16%w/w glycerin in water were found to impact the water activity of the binary solvent mixture [31]. They found that incorporating glycerin dropwise with continuous mixing reduced the water activity more than combining all at once or adding water dropwise to the glycerin. The authors imply that the dropwise incorporation and mixing of glycerin led to an increase in the contacting surface and molecular interactions between the solvents. Similar results were observed with other glycols as well. The mixing duration also impacted the water activity of the binary mixture of glycols and water.

In a study by Angamuthu and others, the manufacturing process was found to affect water activity [32]. A w/o cream product manufactured using different homogenization protocols led to the formation of creams with significantly different water activity values. The amount of water content measured in both creams was found to be the same. The w/o type of cream displayed water activity of 0.93 when processed at a lower homogenization speed (3500 rpm) for 10 min. However, increasing the mixing speed and the mixing time (7500 rpm for 25 min) reduced the water activity to 0.83. The difference in water activity was attributed to differences in the microstructure and quality attributes of the products.

Angamuthu and Murthy extensively investigated the role of water activity on the permeation of drugs across the skin. The water activity of the formulations was found to play a pivotal role in determining the release of the drug from the formulation and impacting skin hydration and permeability. The transport of drugs across synthetic membranes decreased with a decrease in the water activity of the formulations. Likewise, in vitro, the permeation of caffeine across the porcine epidermis decreased with a reduction of water activity (Fig. 1.9). Formulations with low water activity-induced structural changes in the epidermis, such as stratum corneum detachment due to dehydration. However, including hygroscopic agents like propylene glycol restored the epidermis's hydration levels and structural integrity.

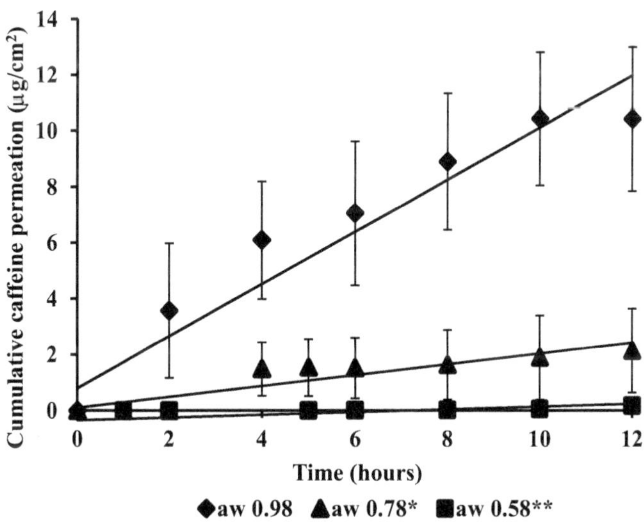

Fig. 1.9 Effect of water activity of topical vehicle on caffeine permeation across porcine epidermis ($n = 6$)

Bjorkland and co-workers worked extensively on the impact of water activity on skin hydration and permeability [33]. The authors state that the skin barrier property is susceptible to the water activity of the formulation applied. At low water activity values, the skin is dehydrated, and mobility of the lipids is affected, thus affecting the skin permeability. The barrier properties, skin hydration, and permeability, even at low water activity, can be promoted by incorporating humectants such as urea and glycols in the formulation.

1.2.6 Phase Distribution of API

Multiple phases could coexist in complex topical products such as creams and lotions. Depending on its partition coefficient, the API dissolved in the system partitions between the immiscible phases.

Generally, the API would localize in the aqueous phase if the drug is highly water-soluble. If the dispersed phase is an o/w system, the API would exist in the continuous phase with a negligible amount in the oil phase.

If the API is moderate to extremely lipophilic, it would be predominantly partitioned in the oil phase. Drug release/permeation would occur mainly from the phase in which the API is predominantly localized. If the drugs are moderately lipophilic, the release/permeation would occur from both phases directly, as well as the drug would be released from the dispersed phase into the continuous phase (Fig. 1.10).

For instance, Zovirax cream was approved by the USFDA for treating herpes labialis (cold sores) in adults and adolescents. The topical aqueous cream is an o/w emulsion containing acyclovir (5%w/w). The drug concentration in Zovirax cream is 5%w/w, which exceeded the equilibrium solubility of acyclovir in the pH range of 2–9. The drug dissolved in the aqueous phase contributes to the thermodynamic activity determining drug absorption. The drug in the lipid phase was twice the amount in the aqueous phase [34]. However, when the oil phase was separated and subjected to permeation studies, the permeation did not occur from the lipid phase for 24 h. In contrast, the permeation occurred from the aqueous phase. Therefore, it

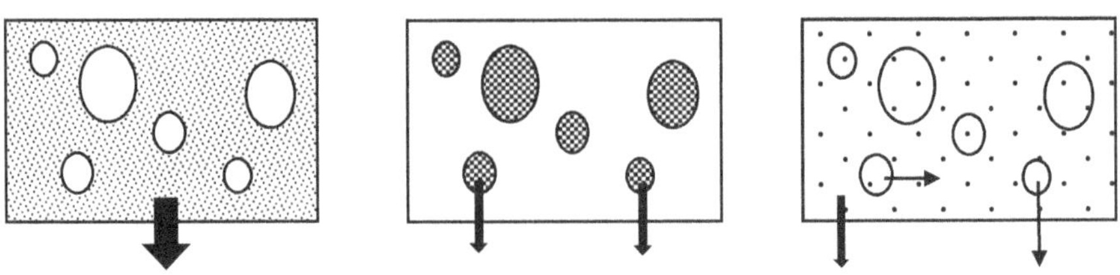

Fig. 1.10 A pictorial representation of pathways of drug release from a topical formulation that has (**a**) extremely hydrophilic drug, (**b**) extremely lipophilic drug, (**c**) moderately lipophilic or hydrophilic (soluble in both phases in significant amounts)

is likely that the API is partitioned into lipids that also act as a reservoir and maintain the saturation levels of API in the aqueous phase.

Likewise, in the case of an FDA-approved metronidazole cream (Metrocream® 0.75% w/w) the amount of the drug in the aqueous phase was about twice the amount in the dispersed phase. When separated and subjected to permeation studies, the permeation occurred both from the lipid and the aqueous phases. The drug likely gets absorbed directly from both phases. The mutual exchange of drugs between the two phases may also be happening while absorption is in progress.

In a custom-made cream product, Nile red was incorporated as a permeant. Due to the extremely lipophilic nature of the dye, most of it is localized in the oil phase (log P: 5) [35]. The permeation was found to occur only from the dispersed phase of an o/w cream. The permeation flux was, therefore, dependent on the average globule size (discussed under the globule size distribution section).

1.2.7 Viscosity of the Formulation

Viscosifying agents, gelling agents, polymers, particles, and dispersed phases impart the viscosity of a formulation. The microstructure of the product (arrangement of matter in the formulation) is one of the phenomena that can contribute to the system's consistency. There are multiple dimensions to viscosity's influence on the performance of the topical products [36].

The viscosity of the formulation determines its stability, especially in the case of dispersions. Further, the system's viscosity influences the extrudability of the topical product from the collapsible tubes and sensorial characteristics such as texture, feel, and spreadability. The viscosity of the formulations changes due to the shear applied during application and the pseudoplastic nature of the formulations. The extent of structural breakdown depends on the magnitude of shear (frequency and time) applied to the system. The formulation may completely regain its original viscosity if the structural breakdown is minimal. However, the recovery would be partial if the structural breakdown is substantial. The viscosity of the cream formulation was found to increase due to a decrease in the droplet size. The reduction in the droplet size, in turn, increases the number and the interfacial surface area, increasing the cream viscosity [37]. The rheological properties of semisolids can be altered by modulating the viscosity of the continuous phase and volume fraction of the dispersed phase and by varying the emulsifying agent concentration [38].

The drug release decreased with an increase in viscosity due to the increased dense gel network of Carbopol® 980. The viscosity of the diclofenac sodium gel was found to increase from 50,000 to 1,50,000 cps when the concentration of Carbopol was increased from 1.2% to 2% w/w. The release studies performed using Franz diffusion cells showed that the extent of drug release from a gel with a viscosity of 40975 cps was found to be 72.89%, whereas the gel that had a viscosity of 184000 cps (2%w/w) was only 46.32% [39]. The hindrance to diffusion of the permeant as the viscosities of gels increased was found to be the rate-limiting factor for the release of diclofenac sodium from Carbopol gels.

Attempts were made in the past to elucidate the effect of formulation viscosity on the human epidermal flux and epidermal retention [40]. Four sunscreen cream formulations designated as 1, 2, 3, and 4 of benzophenone-3, having viscosities of 68,000, 1,80,000, 4,30,000, and 16,00,000 cps, respectively, were developed. Formulation 1, containing cetomacrogol, served as a control, while formulation 2 was prepared with Tego Care 450, a lipid emulsifier. Formulations 3 and 4 comprised Carbopol 980 at a concentration of 0.2 and 0.5%, respectively. Following infinite dosing, the steady-state flux and the epidermal retention decreased with an increase in viscosity, as depicted in Fig. 1.11. Likewise, the amount of API retained in the epidermis in creams with high viscosity was significantly lesser compared to lower viscosity products.

The rheological characterization of topical products provides a comprehensive understanding of their flow behavior and microstructure. The next chapter will cover the fundamentals of rheology and the rheological characterization of topical products, along with some case studies.

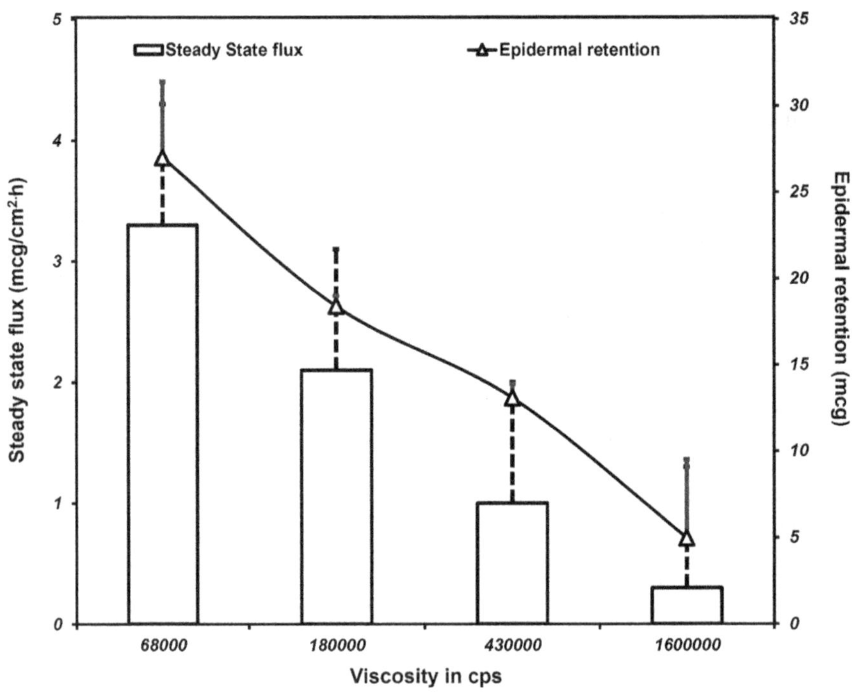

Fig. 1.11 Effect of formulation viscosity on steady-state flux and drug retention of Oxybenzone. [Created using the data from [40]]

1.2.8 Globule Size Distribution

The globule size distribution is a critical characteristic determining the stability of semisolid creams and lotions. Poor control of globule size often leads to creaming, phase separation, or cracking of a topical cream. Differences in the globule size between a prospective generic and the reference product often affect the quantity of drug entrapped in the globule, its subsequent partitioning into the aqueous phase, and eventually the drug release and partitioning into the skin [4]. Different techniques, such as dynamic light scattering, laser diffraction, Raman spectroscopy, and optical microscopy, have been employed to determine the globule size distribution. Thus, globule size would be a critical quality attribute that determines the performance of the topical product.

If the drug is predominantly located in the dispersed phase, the globule size generally affects the permeation inversely [30, 41, 42]. However, suppose the drug is predominantly located in the continuous phase; the globule size distribution is likely to have a negligible impact on the drug absorption unless the phase volume ratio of the creams is high enough to affect the active diffusion area (area of contact between the continuous phase carrying the drug and the skin surface as applicable to this case).

In an in-house study by the investigators, a set of custom-made creams with the same composition (Table 1.3) prepared using different homogenization protocols were found to have significantly different properties. The appearance of the products was significantly different due to the differences in their globule size distribution (Table 1.4) (Fig. 1.12).

The globule size differences also impacted the steady-state permeation flux of Nile red. When the globule size and permeation flux were correlated (Fig. 1.13), an inverse trend was observed.

In another study, the release of lidocaine from o/w type emulgels was inversely related to the globule size distribution. The drug release rate from an emulgel with a globule size distribution of $d_{10} = 1.1$, $d_{50} = 2.2$, $d_{90} = 3.6$ μm (Product 1) was 146.18 ± 10.13 μg/cm^2, whereas the release rate from the emulgel with the same composition but different globule size distribution ($d_{10} = 1.1$, $d_{50} = 3.1$ and $d_{90} = 7.5$ was 61.61 ± 8.97 ug/cm^2 (Product 2) (Fig. 1.14). The increase in the release can be attributed to an increase in the interfacial surface area with a decrease in the globule size [43].

Table 1.3 Composition of the custom-made creams prepared using different manufacturing protocols

Ingredients (o/w)	%w/w
Nile red	1
Cetostearyl alcohol	7
Cremophor A6	1.5
Cremophor A25	1.5
Mineral oil	12
Propylene glycol	9
Water	Q.s

Table 1.4 Homogenization speeds and duration employed and globule size for the different o/w cream products

Formulation code	Homogenization protocol	Globule size (μm)
F1	500 rpm – 20 min	11.37 ± 7.03
F2	1000 rpm – 20 min	7.41 ± 2.19
F3	3000 rpm – 20 min	2.98 ± 1.25
F4	5000 rpm – 20 min	1.71 ± 0.41
F5	3000 rpm – 10 min	4.30 ± 1.33
F6	3000 rpm – 40 min	4.36 ± 0.88

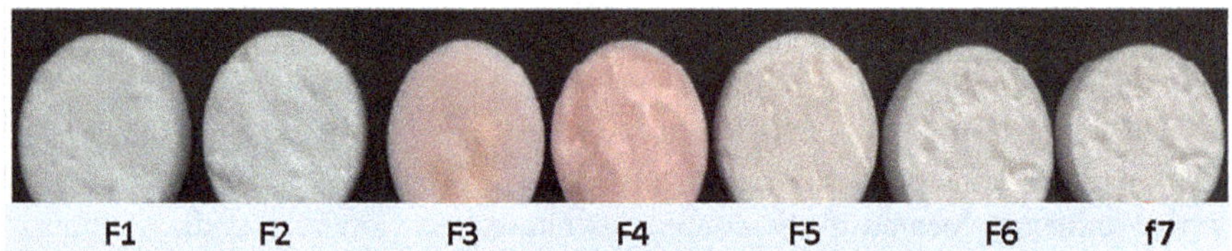

Fig. 1.12 The variation in the color intensity of the o/w cream products, which were of the same composition (Q1/Q2 same) but were different in their globule size

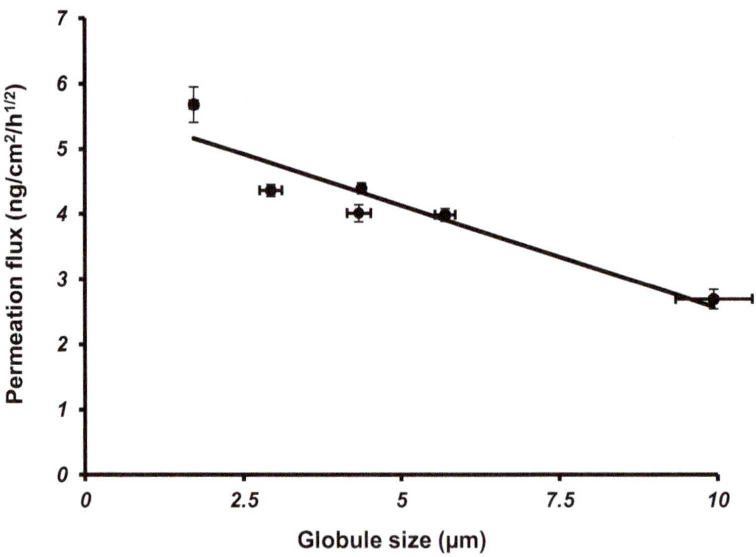

Fig. 1.13 The effect of the globule size of o/w cream on the steady-state flux of Nile red. The permeation flux decreased as the globule size increased

1.2.9 Work of Adhesion

The topical product must interact with the skin surface and adhere well for the drug absorption to be effective [44]. Therefore, there is a need to improve the

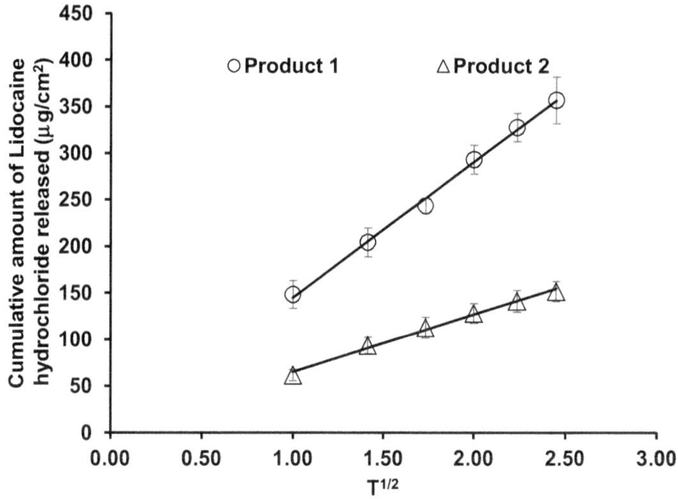

Fig. 1.14 The in vitro drug release profile from two formulations with the same composition (Q1 and Q2) and the globule size was different. The rest of the quality attributes were similar between the products

adhesive properties to optimize drug absorption from topical formulations [45, 46]. Work of adhesion is a parameter that measures the extent of interaction of the formulation with the skin. A texture analyzer can be used to characterize the work of adhesion. Adhesion is a sensitive parameter that varies with even slight composition alternations, microstructure changes, and quality attributes. The probes used in texture analyzers are constructed using acrylates, which may differ in surface-free energy from the skin surface. Nevertheless, using the analyzer to characterize topical products is a useful strategy. Systematic investigations are warranted to understand the relationship between globule size, viscosity, and other quality attributes with the work of adhesion. There is scope for exploring a biorelevant probe (material and design) that mimics the surface properties of biological membranes.

1.2.10 Apparent Density of the Topical Products

The apparent density of the products (ρ') is the weight-to-volume ratio of the product. The apparent density could be calculated for homogeneous solvent mixtures using the density of the individual component (ρ) and its mole fraction. However, predicting the density in a heterogeneous mixture where multiple phases coexist is challenging. The product's microstructural characteristics, such as micelles, aggregates, globule size distribution, hydrate forms, polymeric fibers, and suspended API, can enhance the complexity of the product, making it even more difficult to predict the apparent density. Therefore, it is practically determined using a typical gravimetric method or, more precisely, a gas pycnometer.

The apparent density of the topical product dictates the dose withdrawn and applied [47]. A high-apparent density formulation will result in a low volume. On the other hand, low-density formulations, such as creams with a high-volume fraction of oil and foams, would result in a larger formulation volume. In the former case, it would be challenging to spread the formulation across the affected area in vivo (for clinical evaluation of the product) or across the active diffusion area, in vitro to perform permeation studies. Confining the formulation to the desired surface area will be challenging in the latter case. Therefore, the apparent density should be monitored during product development so as not to let it become unreasonably high or low. The density of the formulation is modulated by the change in the particle size of API, the globule size of the dispersed phase, the volume fraction of the oil, and the dehydration of polymers and suspended particles due to the incorporation of solvents like alcohol.

1.3 Conclusion

The release and absorption of drugs from the topical products are significantly influenced by the critical quality attributes. Critical quality attributes of topical products such as pH, phase distribution of drug, polymorphic form, globule size, water activity, viscosity, density, adhesivity, particle size, and size distribution have been found to influence the performance of the topical products. The pH is a crucial factor that determines the solubility and stability of the drug substance. The degree of saturation is a dominant factor that governs drug permeation. The drug's particle size contributes to the dissolved drug's thermodynamic activity. Similarly, the globule size of a topical product tends to increase the specific interfacial area and, therefore, the absorption from topical products. The critical quality attributes can be tweaked in a formulation to achieve optimal drug absorption.

Acknowledgment The authors acknowledge the USFDA's funding support for the research on the acyclovir products presented in this chapter.

References

1. Yacobi A., Shah V.P., Bashaw E.D., Benfeldt E., Davit B., Ganes D., Ghosh T., Kanfer I., Kasting G.B., Lindsey K, Robert L., Guang W.L, Howard I.M, Lynn K.P, Russell J.R, Andre R, Chinmay G.S, Kailas T, Nathalie W, Elizabeta Z, Majella E.L. Current challenges in Bioequivalence, quality and novel assessment technologies, for topical products. Pharm. Res. 31, 2014, 837–46.
2. Valenta C, Hadgraft J. pH, pKa and dermal delivery. Int J of Pharm. 2000; 200: 243–247.
3. Chang R-W, Raw A., Lionberger R., Yu Lawrance. Generic Development of Topical Dermatologic Products: Formulation Development, Process Development, and Testing of Topical Dermatologic Products. The AAPS Journal, Vol. 15, No. 1, January 2013a, 41–52.

4. Ilic, T., Pantelic, I., Savic, S. The Implications of Regulatory Framework for Topical Semisolid Drug Products: From Critical Quality and Performance Attributes towards Establishing Bioequivalence.Pharmaceutics 2021, 13, 710. https://doi.org/10.3390/pharmaceutics13050710
5. Shanley, A. Topical Formulation: Moving from Art to Science. APIs, Excipients, and Manufacturing 2016, Supplement to Pharmaceutical Technology 40 (9)
6. Brown M.B., Turner R and Lim S.T. Topical Product Formulation Development. In Transermal and topical Drug Delivery. Theory and Practice. Benson H.A.E. Watkinson A.C. (Eds.). 2012, John Wiley & Co. New Jersey, U.S.A
7. Ali SM., Yosipovitch G. Skin pH: From basic science to basic skin care. Acta Derm. Venereol. 2013: 93(3), 261–67.
8. Irena A.-F. Irritants and Skin Barrier Function. In Agner T (ed): Skin Barrier Function. Curr Probl Dermatol. Basel, Karger, 2016, vol 49, pp 80–89 (DOI: https://doi.org/10.1159/000441547)
9. Chantasart D, Chootanasoontorn S, Suksiriworapong J, Li SK: Investigation of the pH influence on skin permeation behavior of weak acids using nonsteroidal anti-inflammatory drugs. J Pharm Sci 2015; 104: 3459–3470.
10. Hadgraft J, Valenta C. (2000). pH, pK(a) and dermal delivery. Int J Pharm, 200:243–247
11. Vaidhyanathan R., Chaubal M.G., Vasavada R.C. Effect of pH and solubility on the in vitro skin penetration of methotrexate from a 50% propylene glycol-water vehicle. Int J Pharm., 1985, 25, 85–93.
12. Chang R-W, Raw A., Lionberger R., Yu Lawrance. Generic Development of Topical Dermatologic Products, Part II: Quality by Design for topical semisolid products. The AAPS Journal, Vol. 15, No. 3, July 2013b, 674–83.
13. Kenley R.A., Lee M.O., Sukumar L., Powell M.F. Temperature and pH dependence of Flucinolone acetonide degradation in topical creams. Pharm Res. 1989, 4 (4), 342–47.
14. Srivastava P. Excipients for Pharmaceutical semisolids In Excipient development for Pharmaceuticals, Biotechnology and Drug Delivery Systems. Katdare A, Chaubal M.V. (Eds.) 2006, Informa Healthcare USA, Inc. New York, pp 251–70.
15. Lu G., Jun W. Diffusion studies of methotrexate in Carbopol and Poloxamer gels. Int J Pharm. 1998, 160, 1–9.
16. Malik S. Microbiological stability for skin care products In Handbook of formulating Dermal applications Dayan N (Ed), 2017, John Wiley & Sons, Inc. Hoboken, New Jersey, U.S.A pp 485–513.
17. Jennings M.C., Minbiole K.P.C., Wuest W.M. Quaternary Ammonium Compounds: An antimicrobial platform for innovation to address bacterial resistance. ACS Infectious Diseases. 2015, 1, 7, 288–303.
18. Yoshida, H.; Tamura, S.; Toyoda, T.; Kado, K.; Ohnishi, N.; Ibuki, R. In vitro release of Tacrolimus from Tacrolimus ointment and its speculated mechanism. Int. J. Pharm. 2004, 270, 55–64.
19. Police A, Shankar VK, Pandey P., Rangappa S, Doerksen RJ, Murthy SN. Novel topical anandamide formulation for alleviating peripheral neuropathic pain. Int J Pharm. 2023, 641, 123085 (https://doi.org/10.1016/j.ijpharm.2023.123085).
20. Osborne D.W. Impact of Quality by Design on Topical Product Excipient Suppliers, Part II: Reasonable Expectations. Pharm Tech 2016b, 40(11), 30–33.
21. Morbi K., Shibata M., Furuishi T., Higashi K., Tomono K., Yamamoto K. Effect of particle size on skin permeation and retention of Pioxicam in aqueous suspension. Chem Pharm Bull 2010, 58 (8), 1096–99.
22. Shim J, Kang H.S, Park W-S., Han S-H, Kim J, Chang I-S. Transdermal delivery of mixnoxidil with block copolymer nanoparticles. J Cont. Rel 2004: 97(3), 477–484. https://doi.org/10.1016/j.jconrel.2004.03.028
23. Vogt A, Combadiere B, Sabrina Hadam S, Stieler KM., Lademann J, Schaefer H, Autran B, Sterry W, Blume-Peytavi U. 40 nm, but not 750 or 1,500 nm, Nanoparticles Enter Epidermal CD1a+ Cells after Transcutaneous Application on Human Skin. J Invest Dermatol 2006; 125(6), 1316–22. https://doi.org/10.1038/sj.jid.5700226

24. Liu X., Shen B., Shen C., Zhong R., Wang X., Yaun H. Nanoparticle-loaded gelsfor topical delivery of nitrofurazone: Effect of skin permeation and retention. Journal of Drug Deliv. Sci. Tech. 2018a, 45, 367–72
25. Jung N., Namjoshi S, Mohammad Y., Griece J.E., Benson H.A.E., Raney S.G., Roberts M.S., Windbergs M. Application of confocal Raman Microscopy for characterization of topical semisolid formulations and the penetration into human skin ex vivo. Pharm Res.2022, 39, 935–48.
26. Garcia-Arieta, A., Gordon J., Gwaza, L., Merino, V. Mangas-Sanjuan, V. Regulatory Requirements for the Development of Second-Entry Semisolid Topical Products in the European Union. Pharmaceutics 2023, 15, 601. https://doi.org/10.3390/pharmaceutics15020601
27. Nardi-Ricart A, Linares M.J, Villca-Pozo F, PeÂrez-Lozano P, Sune-Negre J.M, Bachs-deMiquelL, Roig-Carreras M, Sune-Pou M, Nofrerias-Roig I, Garcia-Montoya E Plus One Sept 2018, https://doi.org/10.1371/journal.pone.0201643
28. Friedel R. Pharmacopeial forum, 1999, pp 8974–8981.
29. Varvaresou A, Papageorgiou S, Tsirivas E, Protopapa E, Kintziou H., Kefala V, Demetzos C. Self preserving cosmetics. Int. J Cosmetic. Sci. 2009; 31, 163–75.
30. Jin X., Imran M., Mohammed, Y. Topical Semisolid Products—Understanding the Impact of Metamorphosis on Skin Penetration and Physicochemical Properties. Pharmaceutics 2022, 14, 2487.
31. Kerdudo A., Fontaine-Vive F., Faure C. Fernandez X. Optimization of Cosmetic preservation: Water activity reduction. Int. J Cosmetic Sci. 2015, 37(1), 31–40.
32. Angamuthu M, Shankar V.K, Murthy S.N. Water activity and its significance in Topical dosage forms. J.Pharm Sci 2018, 107 (6), 1656–66. https://doi.org/10.1016/j.xphs.2018.02.013
33. Bjorkland S., Engblam J, Thuresson K, Sparr E. Glycerol and urea can be used to increase skin permeability in reduced hydrated conditions. Eur. J Pharm Sci., 2013, 50 (5), 638–45.
34. Krishnaiah Y.S.R, Xu X., Rahman Z., Yang Y., Katragadda U., Lionberger R., Peters J.R., Uhl K., Khan M.A. Development of performance matrix for generic product equivalence of acyclovir topical creams. Int. J. Pharmaceutics 475 (1 -2) 110–122 (2014)
35. Rashid-Doubell F and R. W. Horobin, Selection of Florescent Golgi Complex probes using structure-activity relationship. In Bach PH, Reynolds CH, Clark JM, Mottley J, Polle PL. (Eds.) Biotechnology Applications of Microinjection, Microscopic Imaging, and Fluorescence, Springer, London, 1993, pp. 73–78.
36. Binder, L.; Mazál, J.; Petz, R.; Klang, V.; Valenta, C. The Role of Viscosity on Skin Penetration from Cellulose Ether-Based Hydrogels. Ski. Res. Technol. 2019, 25, 725–734
37. Simos A., Veiga F, Vitorino C. Figueiras A. A tutorial for developing cream formulations by quality by design approach. J.Pharm. Sci. 2018, 107, 2653–62.
38. Crowley M. Solutions, emulsions, suspensions, and extracts. In: Felton L, ed. Remington: Essentials of Pharmaceutics. London, United Kingdom: Pharmaceutical Press; 2012:435–468.
39. Khanolkar A., Thorat V., Raut P., Samanta G. Application of Quality by Design: Development to Manufacturing of Diclofenac Sodium Topical Gel AAPS PharmSciTech, 2017, https://doi.org/10.1208/s12249-017-0755-8.
40. Cross, S.E.; Roberts, M.S.; Jiang, R.; Benson, H.A.E. Can Increasing the Viscosity of Formulations Be Used to Reduce the Human Skin Penetration of the Sunscreen Oxybenzone? J. Investig. Dermatol. 2001, 117, 147–150.
41. Patel, H.K.; Barot, B.S.; Parejiya, P.B.; Shelat, P.K.; Shukla, A. Topical Delivery of Clobetasol Propionate Loaded Microemulsion Based Gel for Effective Treatment of Vitiligo: Ex Vivo Permeation and Skin Irritation Studies. Colloids Surf. B Biointerfaces 2013, 102, 86–94.
42. Simoes, A.; Veiga, F.; Vitorino, C. Progressing Towards the Sustainable Development of Cream Formulations. Pharmaceutics 2020, 12, 647
43. Echanur V.A, Anusha V.M,·Pragathi S.G., Sarasija S., Thean Y, Badruddoza A.Z, Jaymin Shah J,·Kulkarni V, Ajjarap S, Reena N.M, Shivakumar H.N, Narasimha Murthy S.N Continuous Manufacturing of Oil in Water (O/W) Emulgel by Extrusion Process. AAPS PharmSciTech (2023) 24:76.

44. Gudin J., Webster L.R., Greuber E., Vought K., Patel K., Kuritzky L. Open-label adhesion performance studies of a new lidocaine topical system 1.8% versus lidocaine patch 5% and lidocaine medicated plaster in Healthy subjects J. Pain Res. 2021: 14, 513–26.
45. Wokovich AM, Prodduturi S, Doub WH, et al. Transdermal drug delivery system (TDDS) adhesion as a critical safety, efficacy and quality attribute. Eur J Pharm Biopharm. 2006;64(1):1–8. https://doi.org/10.1016/j.ejpb.2006.03.009
46. Wang C., Stasinger C., Shen M., Tsong Y. Statistical consideration in assessing in vivo adhesion with transdermal and topical delivery systems for new drug application. AAPSPharmSciTech 2020: 20, 137.
47. Murthy, N.S. Critical Quality Attributes (Q3 Characterization) of Topical Semisolid Products. 2016. Available online: http://www.pharmtech.com/pharmtech-webcasts

Chapter 2
Rheological Behavior of Topical Pharmaceutical Formulations

Santanu Kundu, Kyle Callahan, and S. Narasimha Murthy

2.1 Introduction

Topical dermatological formulations are meant to be applied to the skin. These formulations should have the right consistency for proper dosing and application. Highly viscous formulations can be difficult to apply, while low-viscosity formulations can run down the skin. Topical pharmaceutical formulations or topical formulations come in different types, such as gels, lotions, and creams, and cannot be characterized by a single viscosity value due to their complex microstructure. These formulations exhibit both viscous and elastic behavior and can be classified as soft viscoelastic materials. Understanding the viscoelastic behavior of these formulations is crucial for their packaging, stability, delivery, and applications. Investigating their viscoelastic behavior through rheometry is a continuously evolving research area [1–8]. There are challenges to overcome while developing a rheological protocol for the given product. For example, performing rheological studies on formulations that dry out rapidly or liquefy at higher shear is difficult. Formulations like topical foams that break down rapidly with time are challenging to study. In this work, we present our efforts on rheological investigations of pharmaceutical formulations, including some approaches to method development.

S. Kundu (✉) · K. Callahan
Dave C Swalm School of Chemical Engineering, Mississippi State University, Mississippi State, MS, USA
e-mail: santanukundu@che.msstate.edu

S. N. Murthy
Topical Products Testing LLC, Batesville, MS, USA

S. N. Murthy (ed.), *Topical Products and Dermal Drug Delivery*,
AAPS Introductions in the Pharmaceutical Sciences 2,
https://doi.org/10.1007/978-3-031-80525-7_2

2.2 Basic Rheometry Concepts

The subject of rheology studies how matter deforms and flows when subjected to mechanical deformation in terms of mechanical strain and stress. Perfectly elastic solids obey Hooke's Law, in which the resultant stress of a solid due to applied mechanical deformation is linearly proportional to the applied strain, as $\sigma = G\gamma$. Here, σ and γ are shear stress and strain, respectively. The proportionality constant G is the shear modulus. On the opposite side of the spectrum are Newtonian fluids, where shear stress is directly proportional to the rate of deformation, not the extent of deformation. This behavior is represented as $\sigma = \eta\dot{\gamma}$. Here $\dot{\gamma}$ is the shear rate or rate of deformation. The proportionality constant, η, is defined as the viscosity of a material. Viscosity also quantifies the internal resistance to flow. Note that the viscosity is independent of the shear rate for Newtonian fluids and remains constant over the shear-rate range.

In contrast, pharmaceutical formulations such as gels, ointments, and emulsions cannot be characterized entirely as solids or Newtonian fluids. These soft materials behave somewhat in between and are defined as viscoelastic materials. For example, although the gels behave more like solids, the viscous behavior is still present somewhat. On the other hand, in ointments, the viscous behavior is higher than the elasticity of these samples. Further, unlike Newtonian fluids, the viscosity of these specimens depends on the shear rate. Most commonly, the viscosity decreases with increasing shear rate, and these specimens are categorized as shear-thinning fluids. The opposite occurs in shear-thickening behavior, where the viscosity increases with increasing shear rate, but examples of such behavior are not common in pharmaceutical formulations.

The packaging, stability, delivery, and applications of topical pharmaceutical formulations required a detailed investigation of their flow behavior subjected to mechanical stress and strain. Rheometers are the tools used toward that, and different experimental protocols and frameworks can be used to determine the pertinent functions for characterizing these fluids [9, 10].

Shear or rotational rheometers are the most commonly used instruments for characterizing viscoelastic materials. Various geometries can be used in these instruments, and temperature can be controlled using a Peltier device. The most common experimental geometries are parallel plates, cone plate, and cup-bobs (Fig. 2.1). The sample volume for both parallel plates and cone plate is ~1 mL (depending on the geometry size), whereas for the cup-bob system, the sample volume could be a few mL. All these systems can be equipped with solvent traps necessary to restrict the drying of specimens during rheological investigation. As discussed below, in cone-plate geometry, a sample is subjected to a constant shear rate, whereas in parallel-plate geometry, the shear rate varies across the geometry. In that sense, cone-plate geometry is beneficial. However, the characterization of pharmaceutical formulations often requires rough surfaces to avoid slippage at the geometry and sample interface. Toward that, either textured geometries or adhesive-

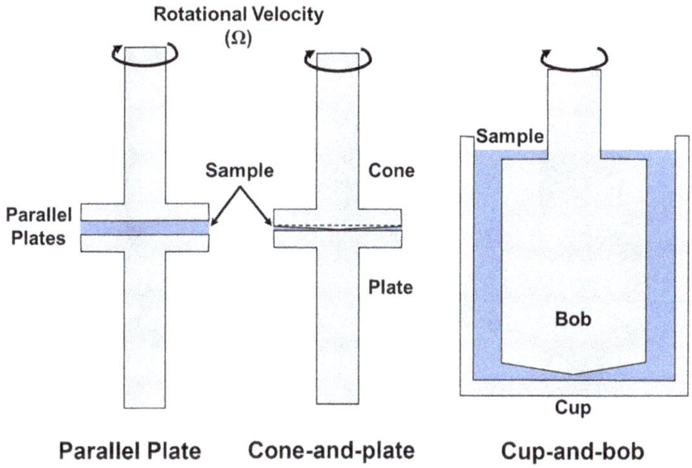

Fig. 2.1 Three different types of geometries for rotational rheometers that can be used to characterize topical pharmaceutical formulations

backed sandpapers are added to the parallel-plate geometries. Such is difficult to achieve in cone-plate geometries.

For both parallel-plate and cone-plate systems, a small volume of sample is loaded between two plates or cone and plates. Depending upon the instrument, either the top or bottom geometry is rotated continuously at a set rotational velocity (Ω) or oscillated at a given frequency (ω). The sample resists the motion, and the torque necessary to overcome that is measured.

Both parallel-plate and cone-and-plate geometries are capable of handling viscous fluids but are not suitable for lower viscosity samples, as the specimen can flow out of the system and may not produce adequate torque. Increasing the surface area, that is, using larger diameter geometries, may remedy this problem by providing a larger surface area to generate higher torque values. However, a cup-and-bob geometry may be used if the fluid viscosity is too low. Here, a bob is submerged into the sample fluid and held in a cup. A small gap is maintained between the bob and cup surfaces. The larger surface area generates sufficient torque values for the instrument to register.

For a cone-plate geometry, with a cone angle of θ_0, the shear rate across the geometry is constant and can be presented as $\dot{\gamma} = \Omega/\theta_0$. Note that this formulation considers the cone angle to be very small, ~1–4°. During the experiments, the instrument measures the torque (T),which is then converted to shear stress τ as $\tau = \frac{3T}{2\pi R^3}$, where R is the radius of the cone and plate. The shear viscosity is then, $\eta(\dot{\gamma}) = \tau\left(\dot{\gamma}\right)/\dot{\gamma}$. Here, $\eta(\dot{\gamma})$and $\tau(\dot{\gamma})$ are both functions of $\dot{\gamma}$, typical of pharmaceutical formulations, as these formulations do not behave like Newtonian fluids. For parallel-plate systems, the shear rate varies across the geometry as $\dot{\gamma} = \frac{r\Omega}{H}$, where r is the radius from the center of the geometry, and H is the gap

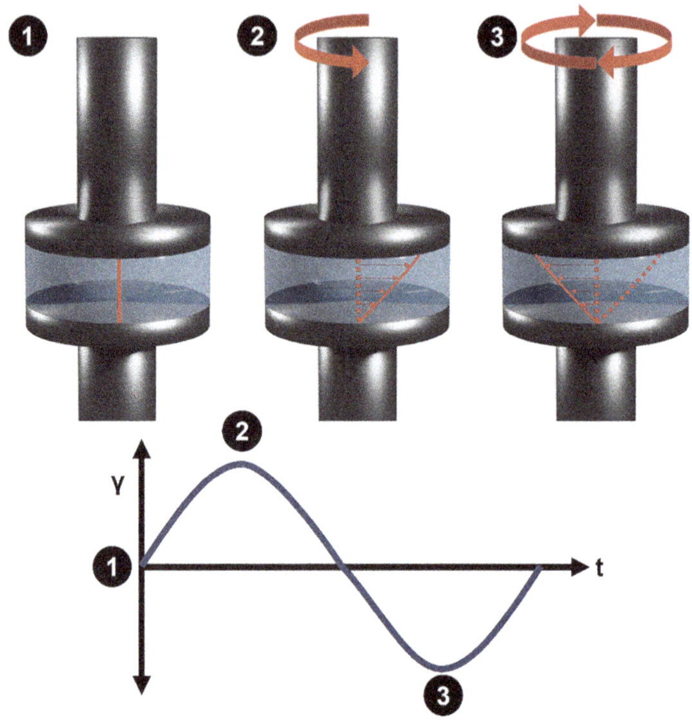

Fig. 2.2 Schematic representation of applied strain and corresponding sample deformation in a rotational rheometer

between two plates. As r changes from 0 (at the center) to R (at the edge), $\dot{\gamma}$ also varies. The shear viscosity at $r = R$ is $\eta_R(\dot{\gamma}) = \frac{2T(\dot{\gamma})}{\pi R^3 \dot{\gamma}_R} R$.

For the cup-and-bob geometry, $\dot{\gamma} = \frac{\kappa \Omega}{1-\kappa}$ where κ = radius of bob (inner cylinder)/radius of cup (outer cylinder). Note that the shear rate is uniform throughout the gap, but this is applicable when the gap is small. The corresponding shear viscosity can be estimated as $\eta_R\ (\dot{\gamma}) = \frac{T(\dot{\gamma})\ (\kappa-1)}{2\pi R^2 L \kappa^3 \Omega}$.

The discussion above has focused on steady-flow experiments. Although these experiments provide shear viscosity of samples, additional experiments need to be conducted to obtain more information, particularly related to the viscoelastic behavior of the specimens. Toward that, dynamic rheology experiments can be conducted. Here, a sinusoidal, time-dependent strain is applied to the sample as $\gamma(t) = \gamma_0 \sin\ (\omega t)$, as shown in Fig. 2.2 for a parallel-plate geometry. Here, γ_0 is strain amplitude, and ω is oscillation frequency. The stress response depends on the nature of the material. For elastic solids yields, $\tau(t) = G\gamma_0 \sin\ (\omega t)$.

As discussed, the stress response is directly proportional to the applied strain for elastic solids. The maximum stress occurs when the strain reaches its maximum. Therefore, stress and strain are in-phase (see Fig. 2.3). For Newtonian fluids, stress

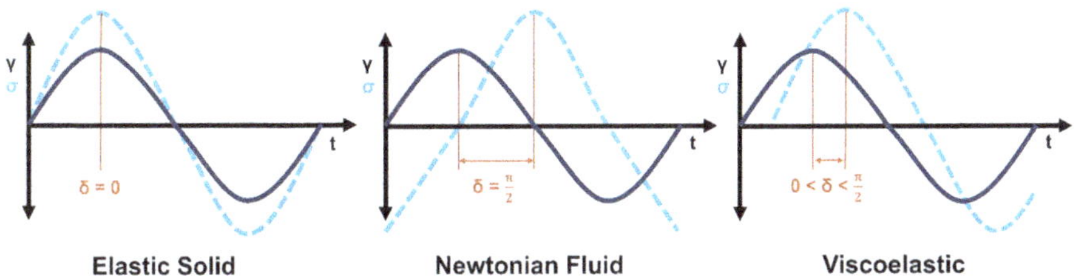

Fig. 2.3 Applied strain and corresponding stress responses for elastic solids, Newtonian fluids, and viscoelastic soft materials

is proportional to the strain rate, not the strain. As strain rate is $d\gamma/dt$, sinusoidal strain causes the strain rate to be zero at maximum strain. The stress response is then, $\tau\,(\mathrm{t}) = \eta\frac{\mathrm{d}\gamma}{\mathrm{dt}} = \eta\gamma_0 \sin\left(\omega \mathrm{t} + \frac{\pi}{2}\right)$. Stress and strain are, therefore, out-of-phase for Newtonian fluids.

Soft materials such as topical formulations are viscoelastic, meaning their characteristics exist somewhere between elastic solids and Newtonian fluids. The extent of how out-of-phase the material is often represented by a quantity called the phase angle (δ). When a material is in-phase, $\delta = 0$, and when it is out-of-phase, $\delta = \frac{\pi}{2}$. Viscoelastic soft materials have a phase angle between 0 and $\frac{\pi}{2}$. Stress for viscoelastic materials during oscillatory strain is $\tau(t) = \sigma_0 \sin\,(\omega t + \delta)$.

To further measure the in-phase and out-of-phase components, the shear stress can be written as $\tau(t) = \gamma\ _0G'\ \sin\,(\omega t) + \gamma_0 G''\ \cos\,(\omega t)$. The G' is the storage modulus and is defined as $G'\ = \tau_0/\gamma_0 cos\delta$. The loss modulus is $G''\ = \tau_0/\gamma_0 sin\delta$. Several other auxiliary parameters can also be defined, such as complex modulus (G^*) and complex viscosity (η^*). The complex modulus incorporates both G' and G'', and is defined as $G^* = G'\ + iG''$, whereas complex viscosity is $\eta^* = \left(G'^2 + G''^2\right)^{\frac{1}{2}}/\omega$. G^* represents the total resistance of a sample against an applied strain. Further, for many samples, achieving high shear rates is difficult because of the geometry's limitations. In those cases, instead of shear viscosity (η), complex viscosity (η^*) can be used. For rheologically simple fluids, the Cox-Mertz rule is applicable, which indicates $\eta \approx \eta^*$, for $\dot{\gamma} = \omega$ [9, 10]. An additional term useful in characterizing soft materials is *tan* δ, defined as $G''/\ G'$. Specimens closer to solids will have $G'\ \gg G''$, or tan$\delta < 1$ [11–13]. Fluid-like behavior is prevalent when $G'\ \ll G''$, or tan $\delta\ > 1$. For a viscoelastic sample, G′ is in the same order of magnitude as G''.

Although modern rheometers can perform various tests, the most common tests that are performed on topical pharmaceutical formulations are time sweep, amplitude sweep, frequency sweep, temperature sweep, and steady-shear rate sweep. The time, amplitude, and frequency sweep experiments are conducted in oscillation mode. Whereas in rate sweep experiments, the rotational speed (Ω) is increased in steps, and for each step, the steady-shear viscosity (η) is obtained after the torque (T)/shear stress (τ) reaches a steady state. Temperature sweep experiments can be conducted in both oscillation and steady-shear modes; however, rheological measurements are mostly conducted at skin temperature for topical pharmaceutical formulations. In typical time-sweep experiments, the samples are subjected to constant strain and frequency at a constant temperature over a selected time period, and the changes of G', G'', and tanδ, as a function of time, are captured. These experiments are useful in monitoring the time-dependent properties of samples, such as sample stability over long periods of time, gelation, or decomposition behavior, and sample relaxation after a mechanical deformation is applied.

In amplitude sweep experiments, the applied strain amplitude is increased in steps while keeping frequency and temperature constant. These experiments are typically used to determine the linear viscoelastic (LVE) regime, where sample microstructure does not change significantly with applied strain and both G' and G'' remain constant with strain. Other strain-dependent properties, such as strain hardening and strain softening, most common for topical pharmaceutical formulations, can be captured. As discussed below, yield stress can be determined from these experiments. Frequency sweeps alter the oscillation frequency while keeping the strain amplitude and temperature constant. The strain rate dependence and other viscoelastic properties are determined through these experiments.

2.3 Rheological Parameters and Measurements Relevant to Topical Pharmaceutical Formulations

2.3.1 *Yield Stress*

Yield stress is an important parameter for topical pharmaceutical formulations. It indicates a critical stress value below which the material does not flow. Knowing the yield stress value is important for dispensing a sample from a packaging tube and during the initial spreading process. Although it is an important physical property, determining yield stress is challenging. Many techniques can be found in the literature. The yield stress values may vary depending on the approach used to determine them. We have determined yield stress using the tangent analysis method (Fig. 2.4). Here, G' from the amplitude sweep experiments is plotted against the stress amplitude values. A line is drawn parallel to the constant G' values (LVE region), and another is drawn parallel to the strain-softening regime. The stress value corresponding to the intercept of these two lines is defined as yield stress.

2.3.2 *Typical Shear Rates for the Spreading of Topical Products on the Skin*

The approximate shear rates relevant to clinical conditions can be estimated (Fig. 2.5). A hypothetical case is discussed here. During application, for example, the

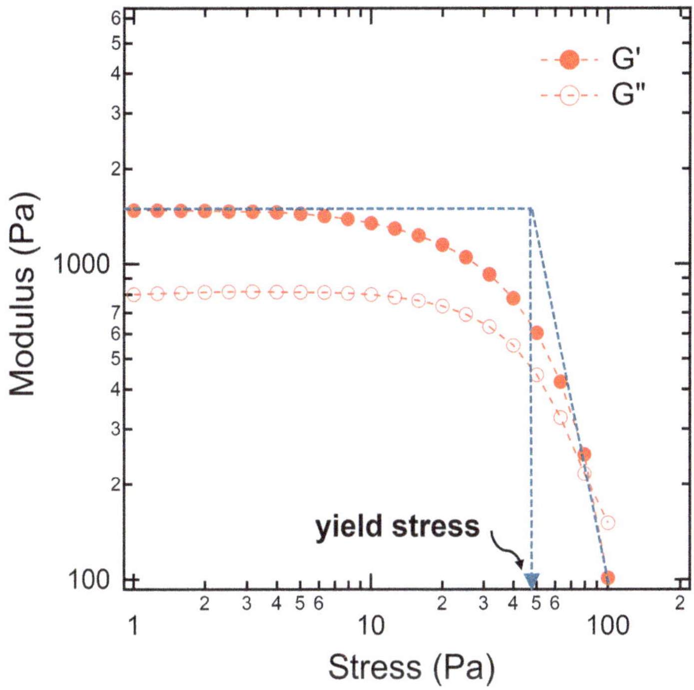

Fig. 2.4 Storage (G') and loss (G'') moduli as a function of shear stress. Yield stress estimation using the tangent method is noted

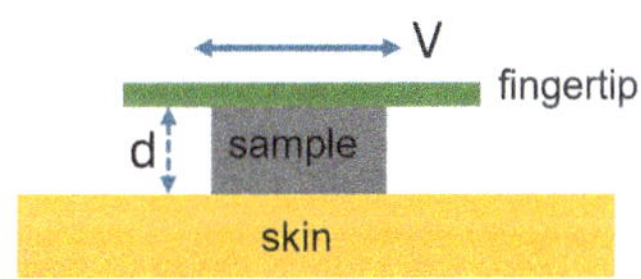

Fig. 2.5 Schematic representing the application of tropical product on skin and corresponding shear rate estimation

initial (dispensed sample) thickness of the formulation applied on the skin is approximately 5 mm, and the sample is spread at 2 cycles/s. If the application area considered is 1 square inch, then it signifies that the fingertip needs to travel 1 inch in length in 250 milliseconds. Thus, the estimated fingertip velocity is $\approx$0.1 m/s. Considering simple shear flow conditions, the estimated shear rate is ~velocity (*V*)/sample thickness (*d*) is $\approx$20 s^{-1}. A larger application area will further reduce the shear rate, as the fingertip must travel a longer distance and the corresponding velocity will be lower. At the later stage of application, the sample thickness decreases significantly. For example, if the sample thickness decreases to about 30 μm with the same spreading rate, the estimated shear rate would be $\approx$3333 s^{-1}. The shear rate becomes zero once the sample is applied to the skin and the fingertip velocity is zero. Therefore, to capture the specimen behavior, we determined the shear viscosity over the shear rate range of 0.004 s^{-1} to 3750 s^{-1}, as shown below.

2.3.3 *Estimation of Drug Diffusivity Using Zero Shear Viscosity Values*

The diffusivity of the drug (Active Pharmaceutical Ingredient [API]) will depend on the microstructure of the sample. The drug diffusivity (D) through the formulation can be predicted using the Stokes–Einstein equation as $D = \frac{k_B T}{6\pi \eta r}$, where k_B is the Boltzmann constant, T is temperature, r is the radius of the active pharmaceutical ingredient, and η is the viscosity. The diffusivity is inversely proportional to the sample viscosity, and it can be hypothesized that drug diffusion will be more efficient in low-viscosity samples. Since diffusion mainly occurs when the formulation is on the skin at the rest condition, that is, the shear rate of ~0. In some cases, when the corresponding zero shear viscosity was difficult to determine, the lowest possible shear rate achieved for topical formulations using DHR-2 rheometer, ~0.004 s^{-1}, was used instead.

2.3.4 *Thixotropic Loop*

Recovery of the microstructure of a sheared sample, as the shear flow is stopped, is an important property. Flow-sweep loop or thixotropic loop tests are conducted to investigate this property of a product. In these experiments, the samples are subjected to increasing shear rate (denoted as forward sweep or F) up to a given shear rate to disrupt the microstructure of the samples. Subsequently, the shear rate is decreased (backward sweep or B) to observe if the product would recover its microstructure, partially or completely. As shown for a typical pharmaceutical formulation (Fig. 2.6), the low-shear viscosity in the backward sweep has been found to be lower than that of the forward sweep. This indicates microstructure recovery was not complete. The area within the loop, called the hysteresis loop, provides an estimate of the energy required to break down the microstructure. This technique can be used to compare different formulations.

2.4 Rheological Characterization of Pharmaceutical Formulations

In this section, we present the results for different tropical pharmaceutical formulations. Several strategies have been implemented to overcome the challenges while working on different types of complex dosage forms. Unless mentioned, the rheological data presented in this chapter were obtained using DHR-2 rheometer (TA Instruments Inc., Delaware, USA). The experiments were conducted at a constant temperature of 32 °C (skin surface temperature), controlled with a Peltier system. Adhesive-backed sandpapers (grit number# 600) were attached to the

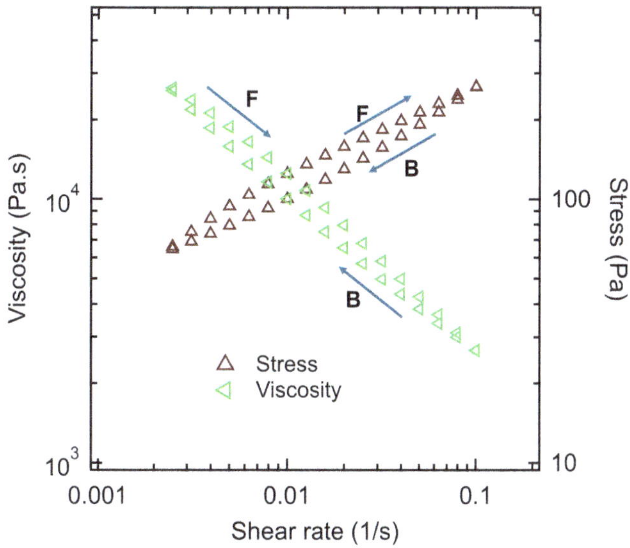

Fig. 2.6 Typical thixotropic loop for a pharmaceutical formulation

upper and lower plates to minimize sample slippage at the sample-rheometer plate interface. A 25 mm parallel-plate geometry was used to incorporate the sandpapers, which is not feasible in a cone-plate geometry. Although a gap of 1 mm is commonly used for 25-mm parallel-plate geometry, a smaller gap of 500 mm for the topical pharmaceutical formulation was selected, as it allows one to use a smaller quantity of samples. A solvent trap was used to avoid drying the sample during the experiment. Drying of pharmaceutical formulations during rheological experiments is a significant challenge, and we paid special attention to minimizing that. Specifically, the results from the time-sweep experiments were tracked to ensure that there was not so significant drying during the prolonged experiments.

For a typical test for a sample, the following experiments were conducted in series: (1) time-sweep (time, t = 300 s, strain amplitude, $\gamma_0 = 0.1\%$, frequency, $\omega = 1$ Hz), (2) strain amplitude sweep ($\gamma_0 = 0.05\%-50\%$, $\omega = 1$ Hz), (3) time-sweep (t = 300 s, $\gamma_0 = 0.1\%$, ω = 1 Hz), and (4) thixotropic loop (forward flow sweep with $\dot{\gamma}$ = 0.004 s^{-1}–3750 s^{-1} and backward flow sweep with $\dot{\gamma}$ = 3750 s^{-1}–0.004 s^{-1}). The time-sweep tests were conducted multiple times at different sequences to allow the samples to relax after deformation. Both storage (G') and loss (G'') were tracked as a function of time, and the sample relaxation led to a plateau of G' and G''. As discussed above, these experiments are also used to capture sample drying, as drying will lead to a continuous increase of G' and G'' with time. Note that in some cases, fresh samples were used for each test mentioned above, and the above series of experiments were beneficial as only a small amount of sample was necessary for each test.

2.4.1 Rheological Characterization of Acyclovir Topical Cream Products

Acyclovir is a nucleoside analog indicated for the topical treatment of viral infections like cold sores (HSV-1) and herpes (HSV-2) [14]. Acyclovir (5% w/w dose strength) is available as a topical dermatological product in both cream and ointment dosage forms. The rheological properties (one of the quality attributes) were evaluated for five model acyclovir cream 5% drug products currently in clinical use. Those are Zovirax® cream 5% w/w (GlaxoSmithKline-US), Zovirax® cream 5% w/w (GlaxoSmithKline-UK and Austria), Aciclovir-1A cream 5% w/w (1A Pharma, Austria), and Aciclostad cream 5% w/w (Stada, Austria). The Zovirax creams of different countries and the Austrian generic creams differed in composition. Hence, the objective was to investigate the differences in the rheological characteristics across the five creams.

The steady-shear experiments were conducted to evaluate the viscosity change as a function of shear rate. Figure 2.7 displays the results for five samples. The samples did not show a plateau at the lowest achievable shear rate in the rheometer, and the viscosity decreased with the increase in shear rates. However, the decrease in viscosity was not monotonous, and a change of slope with the viscosity reduction was observed at an approximate shear rate of 0.5 s^{-1}. The viscosity of Aciclostad and Aciclovir-1A samples decreased rapidly with increasing shear rates and displayed lower viscosity than that observed for Zovirax® (US) above the shear rate of 1 s^{-1} (Fig. 2.7, Table 2.1). On the other hand, the viscosity of Zovirax® (UK) and Zovirax® (AUT) did not change significantly and cannot be investigated at higher shear rates, as the samples were ejected out of the rheometer.

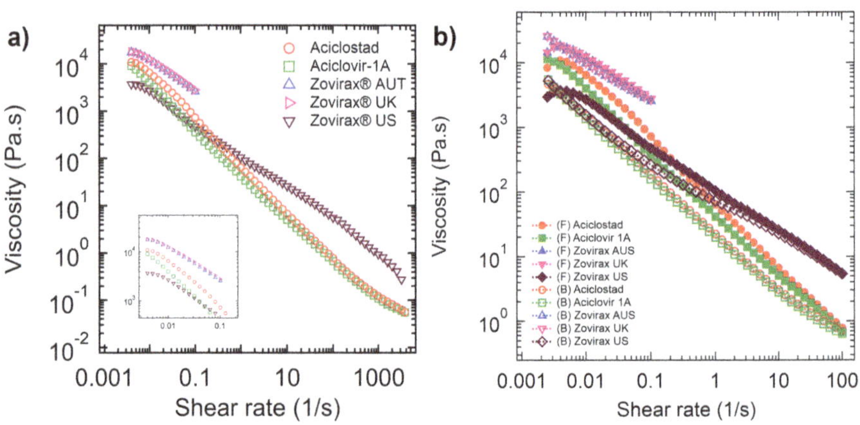

Fig. 2.7 (**a**) Viscosity as a function of shear rate for different creams at 32°C. The low-shear viscosity values are shown in inset. (**b**) Thixotropic behavior of different creams. Viscosity as a function of shear rate is shown. F represents the increasing, and B represents decreasing shear rates

Table 2.1 Viscosity of acyclovir creams under different shear rates at 32°C. N/A indicates that experimental results are not available in that condition

	Viscosity, Pa. s		
Products	@ shear rate: 0.004 s^{-1}	@ shear rate: 20 s^{-1}	@ shear rate: 3162 s^{-1}
Zovirax® (US)	3,697 ± 857	17 ± 0.2	0.28
Zovirax® (UK)	17,872 ± 1,345	N/A	N/A
Zovirax® (AUT)	17,611 ± 594	N/A	N/A
Aciclostad	10,873 ± 1,153	3.4 ± 0.1	0.06
Aciclovir-1A	9,089 ± 3,595	2.7 ± 0.3	0.06

The samples did not show a plateau at the lowest achievable shear rate in the rheometer, that is, the zero shear viscosity could not be obtained. Therefore, the viscosity values at the lowest shear rate measured were used for the diffusivity estimation. As discussed in Sect. 2.3.2, the viscosity at a shear rate of 20 s^{-1} is essential to mimic the application of topical formulations to the skin.

Recovery of the microstructure of a sheared sample, as the shear is removed, is an important property. That is particularly important for packaging where the sample is subjected to mechanical deformation. However, the microstructure recovery will be beneficial to maintain the product's stability and precise dose application and would impact the product's performance when applied to the skin. Deformation could also occur when the formulation is applied to the skin. The extent of deformation and the recovery of the microstructure lost due to the application depends on the magnitude of the shear. To investigate microstructural recovery after mechanical deformation, flow-sweep loop or thixotropic loop tests were conducted to characterize the thixotropic behavior of the cream formulations. In these experiments, the samples were subjected to an increasing shear rate (denoted as forward sweep or F) up to a given shear rate, and it is anticipated that the microstructure of the samples to be disrupted during this process. Subsequently, the shear rate decreased (backward sweep or B), and partial or complete microstructure recovery was expected. As displayed in Fig. 2.7, Zovirax® (US), Zovirax® (UK), and Zovirax® (AUT) returned to their initial low-shear viscosity in the thixotropic loop experiments. In contrast, Aciclostad and Aciclovir-1A samples did not show the same behavior, i.e., the low-shear viscosity in the backward sweep is lower than that of the forward sweep. This indicates that microstructure recovery was incomplete for the Aciclostad and Aciclovir-1A samples. The area within the loop, called the hysteresis loop, provides an estimate of the energy required to break down the microstructure.

The yield stress values were determined using the tangent method. G' and G'' values from all experiments are shown in Table 2.2. From these results, G′ as a function of stress is summarized in Fig. 2.8. As discussed above, a tangent is drawn to the plateau region of the G′ vs. stress plot, and another tangent is drawn to the decreasing portion of the modulus. The intersection point of these two tangents, i.e., the stress value at which the deviation from the plateau region takes place, is considered yield stress for the material. The results indicate that Zovirax® (US) and

Table 2.2 Plateau modulus and yield stress of acyclovir creams

Products	Plateau modulus		Yield stress (Pa)
	G′(Pa)	G'' (Pa)	
Zovirax® (US)	586 ± 112	277 ± 52	≈50
Zovirax® (UK)	3,560 ± 401	3,141 ± 373	≈260
Zovirax® (AUT)	3,200 ± 162	2,878 ± 8	≈235
Aciclostad	1,574 ± 183	934 ± 80	≈70
Aciclovir-1A	517 ± 234	183 ± 106	≈40

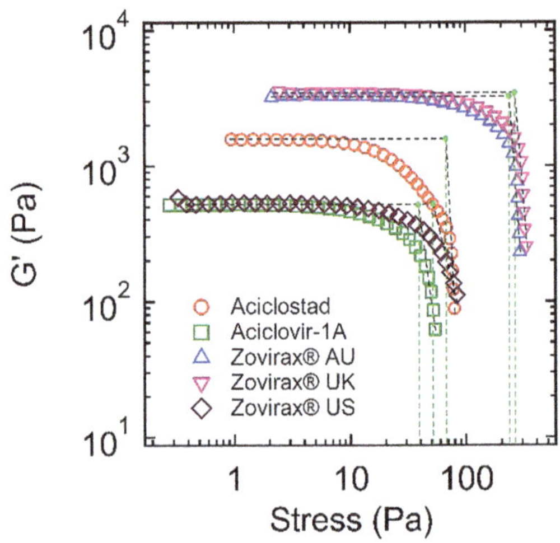

Fig. 2.8 Yield stress determined for the acyclovir creams

Aciclovir-1A have the lowest yield stress, and Zovirax® (UK) and Zovirax® (AUT) have the highest yield stress values (Table 2.2).

2.4.2 Rheological Characterization of Metronidazole Topical Products

Unlike in the acyclovir topical products case study, in the case of metronidazole topical products, a USFDA-approved RLD cream and gel and respective generic products were available to work with and validate the methodology. The manufacturer's name is used in the description and figures for the convenience of the readers.

G' and G'' as a function of strain amplitudes for six samples are shown in Fig. 2.9. The results for gels, creams, and lotions were plotted separately, and differences in their behavior can be seen. G' and G'' initially remained unchanged (independent of strain) but then decreased with increasing strain (strain softening). A decrease in moduli with applied strain indicates breaking down or dissociation of structure with applied strain or stress, as expected during the application of these products. For

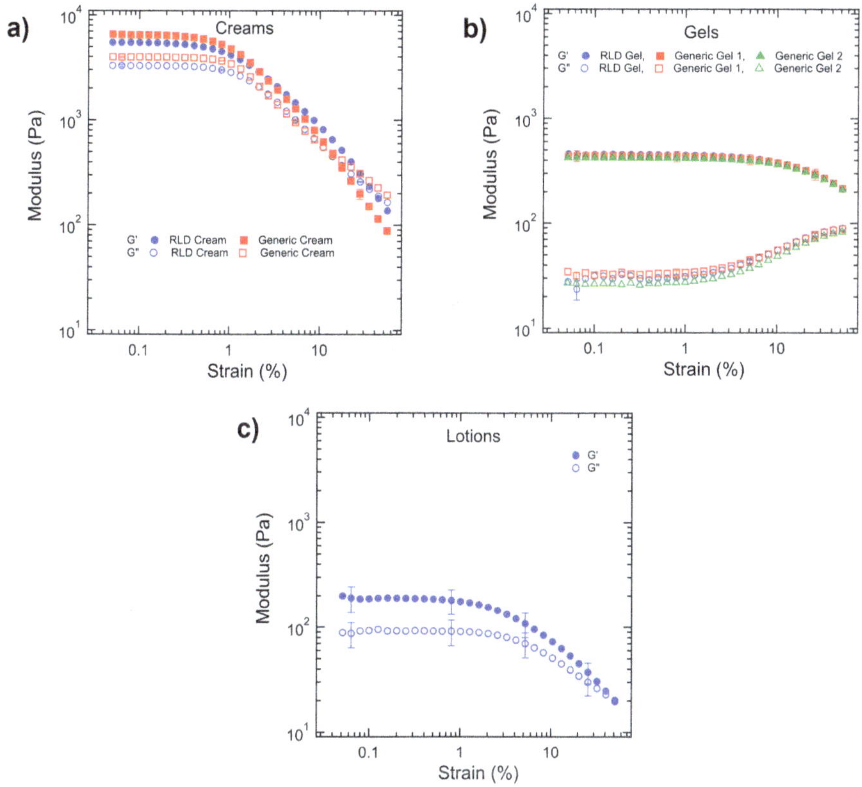

Fig. 2.9 Storage (G') and loss (G'') moduli as a function of strain for metronidazole semi-solid topical formulations: (**a**) creams, (**b**) gels, and (**c**) lotions

the cream and lotion samples, a crossover between G' and G'' has been observed at higher strain, and the viscous response dominates the behavior beyond that strain value. Such crossover has not been observed for the gel samples. Note that the strain value initiating the strain-softening responses and the crossover strain significantly differed for the cream and the lotion samples.

Interestingly, the moduli values for creams at low strain were the highest, and those for the lotions were the lowest. As expected, for the gel samples, $G'/G'' > 10$. For the creams, G' and G'' are of the same order of magnitude, indicating the sample's viscoelastic nature. It is important to note that the RLD and generic samples for each type display similar behavior, suggesting that they are rheologically similar.

Yield stress values measured using the tangent method are summarized in Table 2.4, indicating that the cream and gel samples have the highest yield stress on the order of 100 Pa. In contrast, the lotions have the lowest yield stress values. Statistical analysis (Mann–Whitney U-test for the unpaired data) was performed to evaluate whether the generic and RLD products are similar. For the cream samples, we

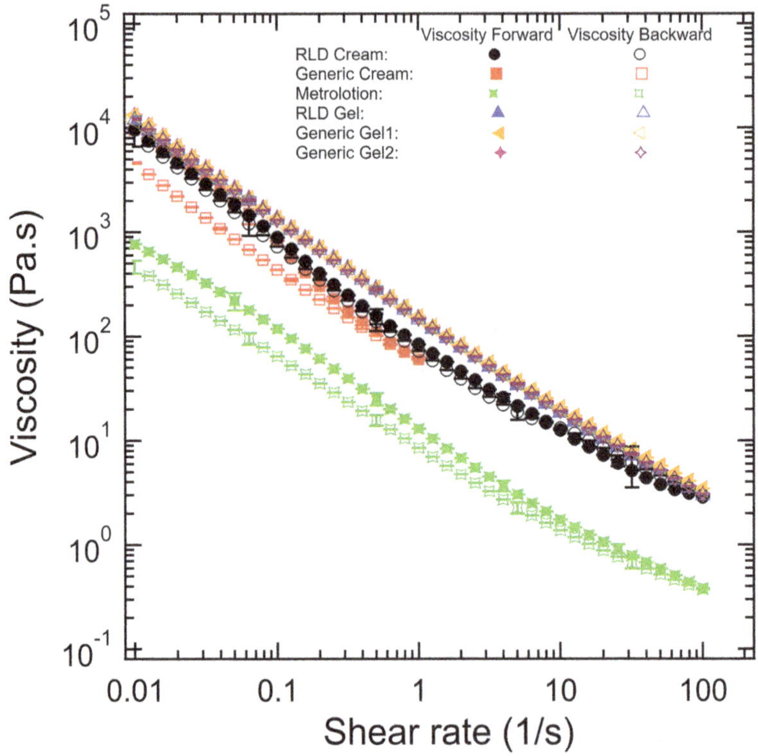

Fig. 2.10 Viscosity as a function shear rate for six metronidazole semi-solid topical formulations

obtained a p-value equal to 0.063 (with a 95% confidence interval); indicating that the null hypothesis is false, i.e., the yield stress values are not statistically different. Similarly, RLD and generic gel 1 gels are similar (p-value of 0.637), and RLD and generic gels are similar (p-value of 0.637).

Thixotropic loop data for metronidazole semisolid topical formulations are shown in Fig. 2.10. During an increase in the shear rate, viscosity decreases, that is, the samples display shear-thinning behavior, which helps the samples spread. Also, the "lotion" sample displays overall low viscosity compared to the other samples. For the "lotion," the forward and backward viscosity values were not the same, indicating that the structure deformed during the increasing shear rate and did not recover when the shear rate was decreased. On the contrary, all other samples did not show such behavior. We can infer that the deformed structure recovered during the decrease in the shear rate. We also compared the low-shear viscosity values, at the shear rate of 0.01 s^{-1}, between the RLD and generic products. The generic cream and all gels have shown similar behavior to their respective RLDs. Therefore, the generic and RLD products are considered rheologically similar (Table 2.4).

2.4.3 Topical Semisolid Creams: Manufacturing Variables, Microstructure, and Rheological Attributes

The microstructure of products could differ even when they are compositionally identical (Q1 and Q2 identical), depending on the variables implemented during cream manufacturing [15–21]. The authors evaluated the impact of manufacturing variables on the globule size distribution in products with identical composition (Q1 and Q2 are the same) in an in-house study. The composition and the manufacturing variables are given in Tables 2.3 and 2.4, respectively, in Chap. 1.

Figure 2.11 displays G' and G'' for six samples obtained from amplitude sweep tests. The samples display behavior similar to that observed for other formulations described above. At low strain, G' was higher than G'', and both G' and G'' were found to be almost independent of applied strain. At higher strain, a decrease in both G' and G'' and a crossover have been observed. Beyond crossover, the samples behaved like a viscous liquid, indicating deformation and breakdown of the structure. Although these samples are emulsions, higher G' compared to G'' indicates the specimens have some elasticity. Yield stress can be estimated from the amplitude sweep data using the tangent analysis method, as shown in Fig. 2.12. The F6 sample has the highest yield stress of ~160 Pa, and the F1 sample has the lowest yield stress of ~20 Pa.

The samples were also subjected to thixotropic loop tests. With increasing shear rate (forward), a drop in viscosity (shear-thinning behavior) was observed, whereas viscosity gradually increased with the decreasing shear rate (backward). All these samples, as shown in Fig. 2.13, displayed a hysteresis loop, which indicates that the

Table 2.3 Products considered for the study

	Creams		Gels			Lotion
Product and company	Metrocream (Galderma)	Fougera	Tolmar	Taro	Prasco	Metrolotion* (Galderma)
Nomenclature used in this chapter	RLD cream	Generic cream	RLD gel	Generic gel 1	Generic gel 2	Lotion

*Lotion has a different composition and microstructure from the two groups (Cream and Gels); hence, it was used as an experimental control

Table 2.4 Yield stress and viscosity values at a shear rate of 0.01 s^{-1} for metronidazole semisolid topical formulations

Products	Yield stress (Pa)	Low-shear viscosity (Pa. s @0.01 s^{-1})
RLD cream	94 ± 0	9,822 ± 283
Generic cream	70 ± 3	8,162 ± 16
RLD gel 1	50 ± 0	10,534 ± 263
Generic gel 1	49 ± 5.2	12,489 ± 1692
Generic gel 2	50 ± 4	12,779 ± 1215
Lotion	7.3 ± 2.3	768 ± 175

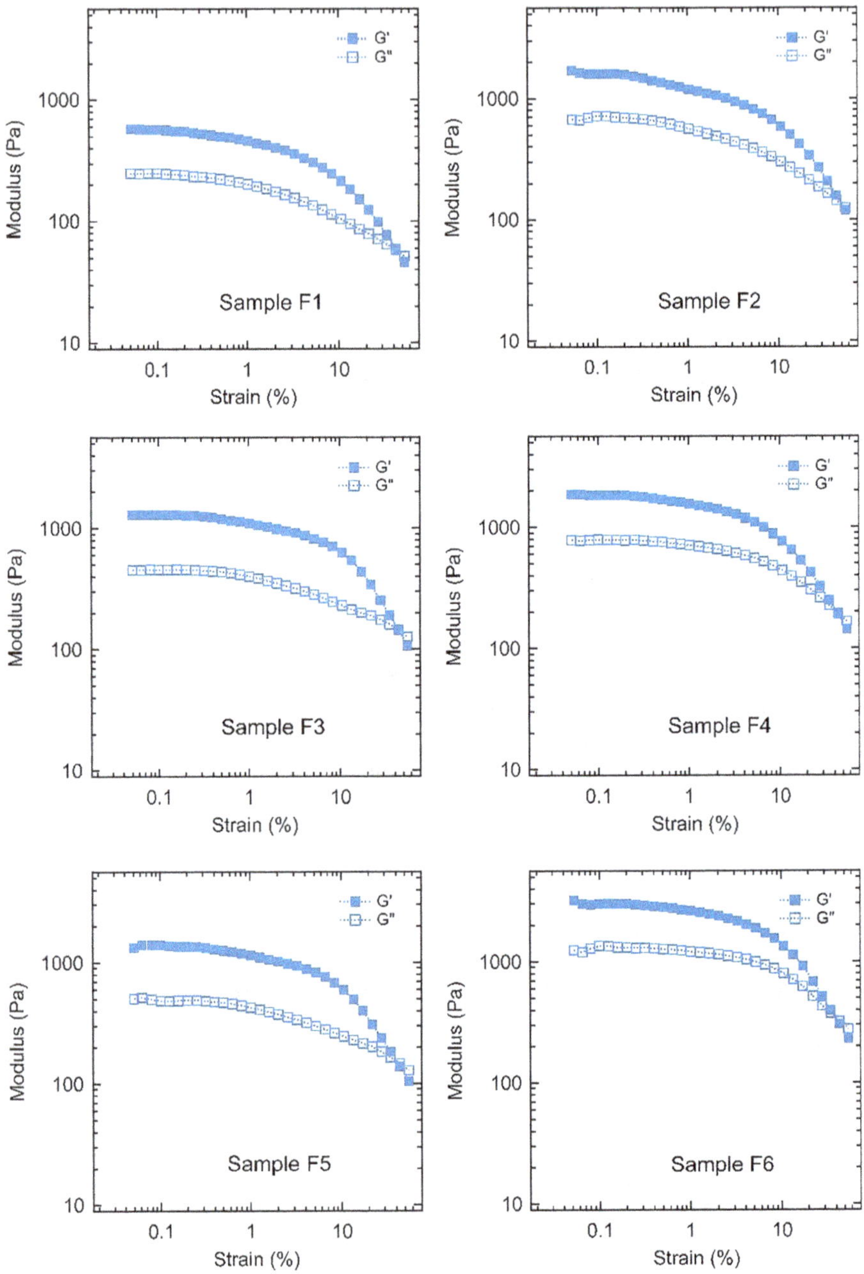

Fig. 2.11 Rheological behavior of the creams as a function of strain amplitude obtained in oscillatory shear experiments at a frequency of 1 Hz

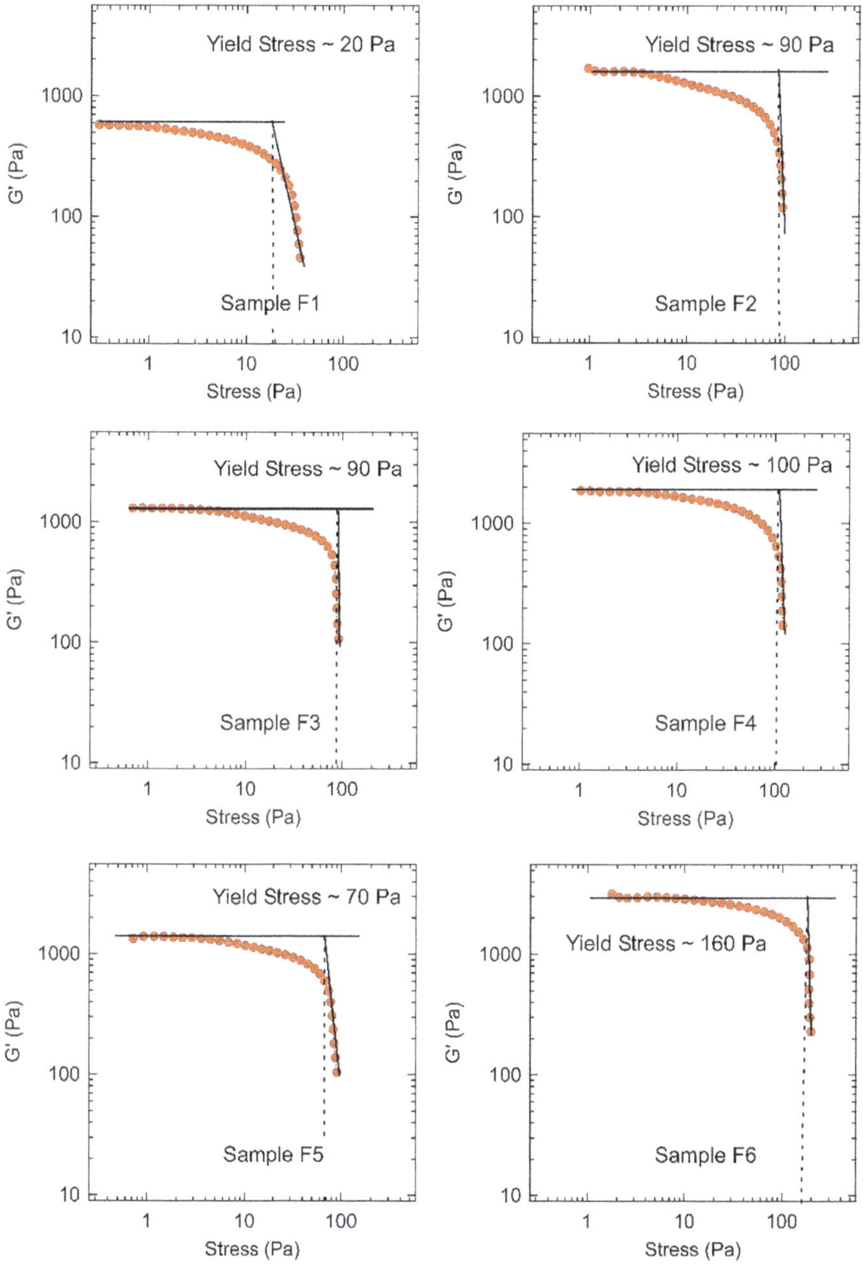

Fig. 2.12 Yield stress of the creams obtained from tangent analysis in an oscillatory strain sweep experiment at a frequency of 1 Hz

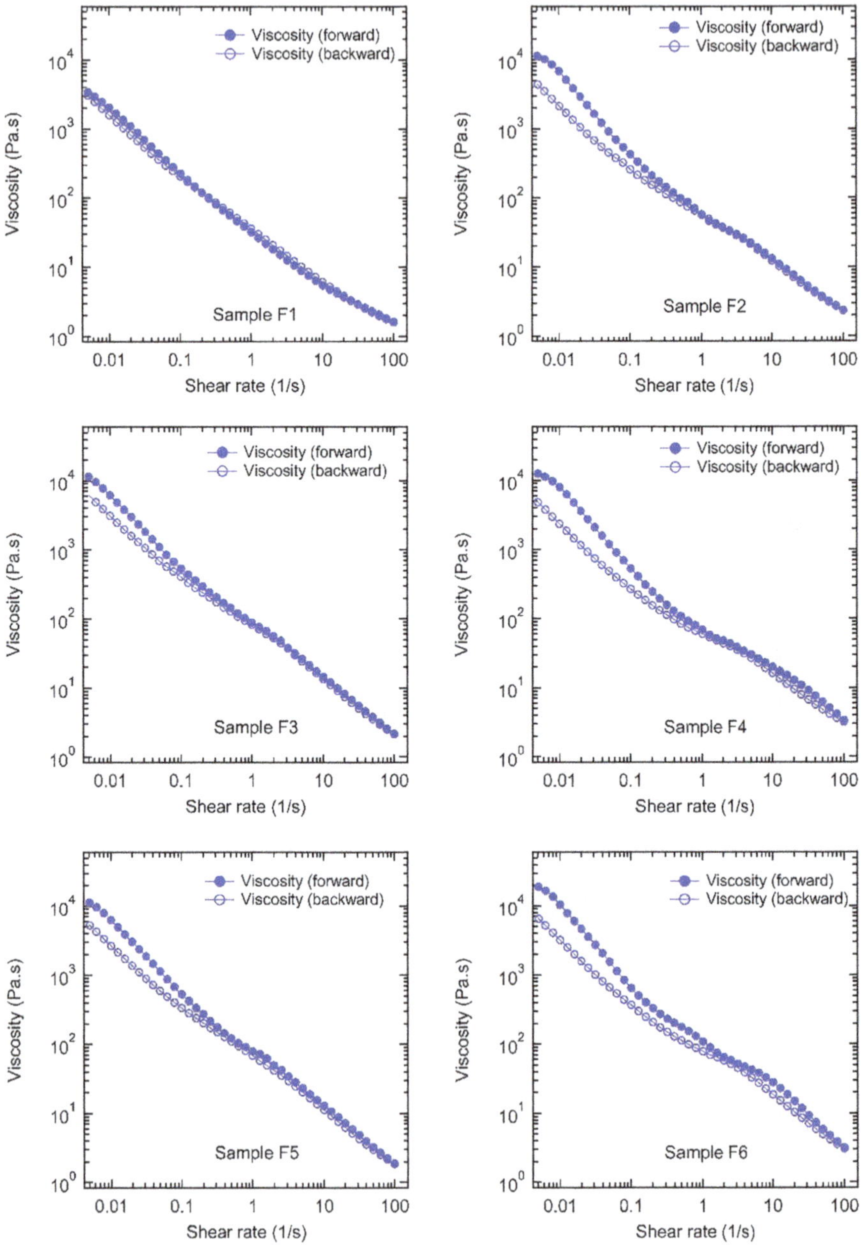

Fig. 2.13 Hysteresis loop associated with thixotropic behavior of the creams in flow sweep experiments. The data obtained in two steps, shearing the creams from low to high shear rate followed by shearing in reverse process from high to low shear rate

deformed structure in the forward loop did not recover completely in the backward direction.

This study showed that creams' rheological characteristics depend on the composition and microstructure of the semisolid products. The manufacturing variables can significantly impact the characteristics of the cream. The creams' rheological properties were found to depend on the globule size distribution in a cream product.

2.5 Concluding Remarks

The rheological characteristics of a product could change with the composition, manufacturing process, and microstructure. Rheological studies not only reveal the flow behavior of the product but also provide information about the product's robustness, the extent of structural breakdown when subjected to shear flow, and subsequent recovery when shear flow is stopped. Acyclovir creams presented in this chapter differed in their rheological behavior due to differences in their compositions. The rheological studies of metronidazole topical products demonstrated the similarity of generic creams and gels with the respective innovator products. Custom-made cream products using different manufacturing variables differed in their globule size distribution, resulting in differences in the rheological attributes. Thus, rheological studies are an essential tool in understanding the behavior, microstructure, and performance of topical products.

Acknowledgments The authors acknowledge the USFDA's funding support for the research on the acyclovir and metronidazole products presented in this chapter.

References

1. Mendonsa, N. S.; Murthy, S. N.; Hashemnejad, S. M.; Kundu, S.; Zhang, F.; Repka, M. A. Development of Poloxamer Gel Formulations via Hot-Melt Extrusion Technology. *Int. J. Pharm.***2018**, *537* (1), 122–131. https://doi.org/10.1016/j.ijpharm.2017.12.008.
2. Mendonsa, N. S.; Pradhan, A.; Sharma, P.; Prado, R. M. B.; Murthy, S. N.; Kundu, S.; Repka, M. A. A Quality by Design Approach to Develop Topical Creams via Hot-Melt Extrusion Technology. *Eur. J. Pharm. Sci.***2019**, *136*, 104948. https://doi.org/10.1016/j.ejps.2019.06.002.
3. Manda, P.; Kushwaha, A. S.; Kundu, S.; Shivakumar, H. N.; Jo, S. B.; Murthy, S. N. Delivery of Ziconotide to Cerebrospinal Fluid via Intranasal Pathway for the Treatment of Chronic Pain. *J. Controlled Release***2016**, *224*, 69–76. https://doi.org/10.1016/j.jconrel.2015.12.044.
4. Bhagurkar, A. M.; Angamuthu, M.; Patil, H.; Tiwari, R. V.; Maurya, A.; Hashemnejad, S. M.; Kundu, S.; Murthy, S. N.; Repka, M. A. Development of an Ointment Formulation Using Hot-Melt Extrusion Technology. *AAPS PharmSciTech***2016**, *17* (1), 158–166. https://doi.org/10.1208/s12249-015-0453-3.
5. Bhide, S.; Panda, A.; Rangappa, S.; Shettar, A.; Repka, M. A.; Prado, R. M. B.; Kundu, S.; Richardson, N.; Murthy, S. N. Evaluating the Impact of Fatty Alcohols on Permeation of Clotrimazole from Topical Creams. 8.

6. Beard, M. C.; Cobb, L. H.; Grant, C. S.; Varadarajan, A.; Henry, T.; Swanson, E. A.; Kundu, S.; Priddy, L. B. Autoclaving of Poloxamer 407 Hydrogel and Its Use as a Drug Delivery Vehicle. *J. Biomed. Mater. Res. B Appl. Biomater. n/a* (n/a). https://doi.org/10.1002/jbm.b.34703.
7. Affairs, O. of R. Topical Drug Products (7/94). *FDA***2022**.
8. Rawat, A.; Gupta, S. S.; Kalluri, H.; Lowenborg, M.; Bhatia, K.; Warner, K. Rheological Characterization in the Development of Topical Drug Products. In *The Role of Microstructure in Topical Drug Product Development*; Langley, N., Michniak-Kohn, B., Osborne, D. W., Eds.; Springer International Publishing: Cham, 2019; pp. 3–45. https://doi.org/10.1007/978-3-030-17355-5_1.
9. Faith, M. *Understanding Rheology*; Oxford University Press: New York, 2001.
10. Macosko, C. W. *Rheology: Principles, Measurements, and Applications*; VCH, 1994.
11. Mishra, S.; Badani Prado, R. M.; Kundu, S. Concentration-Dependent Mechanical Behavior of Physically Assembled Triblock Copolymer Gels. *ACS Appl. Polym. Mater.***2020**, *2* (12), 5388–5397. https://doi.org/10.1021/acsapm.0c00583.
12. Varadarajan, A.; Kearney, L. T.; Keum, J. K.; Naskar, A. K.; Kundu, S. Effects of Salt on Phase Behavior and Rheological Properties of Alginate–Chitosan Polyelectrolyte Complexes. *Biomacromolecules***2023**, *24* (6), 2730–2740. https://doi.org/10.1021/acs.biomac.3c00171.
13. Hashemnejad, S. M.; Kundu, S. Nonlinear Elasticity and Cavitation of a Triblock Copolymer Gel. *Soft Matter***2015**. https://doi.org/10.1039/C5SM00330J.
14. Kimberlin, D. W.; Whitley, R. J. Antiviral Therapy of HSV-1 and -2. In *Human Herpesviruses: Biology, Therapy, and Immunoprophylaxis*; Arvin, A., Campadelli-Fiume, G., Mocarski, E., Moore, P. S., Roizman, B., Whitley, R., Yamanishi, K., Eds.; Cambridge University Press: Cambridge, 2007.
15. Tamburic, S.; Craig, D. Q. M.; Vuleta, G.; Milic, J. A Comparison of Electrical and Rheological Techniques for the Characterisation of Creams. *Int. J. Pharm.***1996**, *137* (2), 243–248. https://doi.org/10.1016/0378-5173(96)04528-0.
16. Tamburic, S.; Craig, D. Q.; Vuleta, G.; Milic, J. An Investigation into the Use of Thermorheology and Texture Analysis in the Evaluation of W/O Creams Stabilized with a Silicone Emulsifier. *Pharm. Dev. Technol.***1996**, *1* (3), 299–306. https://doi.org/10.3109/10837459609022599.
17. Lashmar, U. T.; Richardson, J. P.; Erbod, A. Correlation of Physical Parameters of an Oil in Water Emulsion with Manufacturing Procedures and Stability. *Int. J. Pharm.***1995**, *125* (2), 315–325. https://doi.org/10.1016/0378-5173(95)00149-D.
18. Chang, R.-K.; Raw, A.; Lionberger, R.; Yu, L. Generic Development of Topical Dermatologic Products, Part II: Quality by Design for Topical Semisolid Products. *AAPS J.***2013**, *15* (3), 674–683. https://doi.org/10.1208/s12248-013-9472-8.
19. Chang, R.-K.; Raw, A.; Lionberger, R.; Yu, L. Generic Development of Topical Dermatologic Products: Formulation Development, Process Development, and Testing of Topical Dermatologic Products. *AAPS J.***2012**, *15* (1), 41–52. https://doi.org/10.1208/s12248-012-9411-0.
20. Murthy, R.; Rangappa, S.; Repka, M. A.; Vanaja, K.; Shivakumar, H. N.; Murthy, S. N. Infrared Thermal Measurement Method to Evaluate the Skin Cooling Effect of Topical Products and the Impact of Microstructure of Creams. *J. Drug Deliv. Sci. Technol.***2017**, *39*, 296–299. https://doi.org/10.1016/j.jddst.2017.04.015.
21. Raney, S. G.; Franz, T. J.; Lehman, P. A.; Lionberger, R.; Chen, M.-L. Pharmacokinetics-Based Approaches for Bioequivalence Evaluation of Topical Dermatological Drug Products. *Clin. Pharmacokinet.***2015**, *54* (11), 1095–1106. https://doi.org/10.1007/s40262-015-0292-0.

Chapter 3
Evaporative Metamorphosis in Topical Products

S. Narasimha Murthy

3.1 Introduction

Topical products are available in various forms, such as solutions, semisolids, and aerosols. Most formulations contain a considerable solvent fraction. The microstructure and quality attributes of the product are least affected when the solvent fraction remains unchanged in the formulation. However, when applied to the skin in clinically relevant doses, the solvents and volatile components are lost gradually, leading to changes in the formulation's composition, quality attributes, and microstructure [1]. The phenomenon of changes in the formulation brought about due to the loss of solvents is termed "evaporative metamorphosis."

When topical products are subjected to in vitro permeation testing (IVPT), a specific finite dose (clinically relevant dose) of the formulation is applied uniformly on the skin surface, and the donor compartment is unoccluded to allow the evaporation of contents, mimicking the real-world clinical situation. The evaporative metamorphosis plays a significant role in determining the product's performance.

In the infinite dose permeation study, an adequately larger dose is applied to the skin, and the donor compartment is occluded to prevent the loss of contents from the formulation. An infinite dose permeation study is considered when the aim is to investigate the permeation of new chemical entities or compare the permeation profile of formulations that vary in their composition or quality attributes. Infinite dose permeation studies are also used to investigate the effect of permeation enhancers in the formulation. The donor compartment of the diffusion cell is occluded to mitigate any solvent loss so that the observed results can be attributed solely to the variable factor under evaluation.

S. N. Murthy (✉)
Topical Products Testing LLC, Batesville, MS, USA

S. N. Murthy (ed.), *Topical Products and Dermal Drug Delivery*,
AAPS Introductions in the Pharmaceutical Sciences 2,
https://doi.org/10.1007/978-3-031-80525-7_3

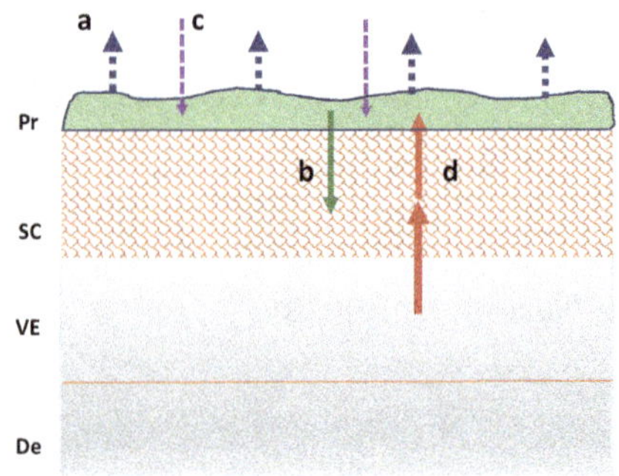

Fig. 3.1 Hydrodynamics in the topical formulation on the skin surface. *Pr* Topical product, *SC* Stratum corneum, *VE* Viable epidermis, and *De* Dermis

When developing a new topical product, it's essential to thoroughly study the impact of quality attributes to achieve the target drug release and absorption. It is also important to account for any changes in the quality attributes that may occur as a result of solvent evaporation and the associated hydrodynamics on the skin surface (Fig. 3.1).

Primarily, the solvents leave the formulation due to evaporation and dermal absorption. The formulations containing a significant fraction of hygroscopic materials could absorb atmospheric moisture. Another important mechanism that promotes continued interaction between the skin and formulation is trans-epidermal water loss, the outward water transport in the skin. This chapter discusses the potential changes in the quality attributes when applied to the skin and the impact of such changes on the product performance.

a) Gradual evaporation of solvents occurs from the formulation
b) Solvents can get absorbed into the skin layers
c) The formulation containing hygroscopic substances absorbs moisture from the atmosphere
d) The formulation uptake water due to the trans-epidermal water transport.

The formulation composition, microstructure, region of application, dose, and environmental variables determine the evaporation rate of solvents (Fig. 3.2a–d). Formulations with hydroalcoholic solvents dry out more rapidly than those with only water, oils, and water-miscible higher boiling point solvents. The drying profile of a formulation can be determined by gravimetry. The formulation applied on a substrate or skin shall be placed at 32°C and ambient humidity conditions, and the sample's weight shall be recorded at different time points. Cross, Benson, and Roberts measured the evaporation rate of solvents from sunscreen formulations. They calculated the evaporation index by fitting the curve to pseudo-first-order kinetics.

The loss of solvent contents from a semisolid formulation applied to the skin typically happens in two phases. Initially, there is a phase of rapid loss of solvents from the formulation, followed by a slower loss. The slower loss of solvent in the second phase is due to the increased viscosity of the formulation and the formation of a surface crust due to the loss of solvent from the superficial layers during the

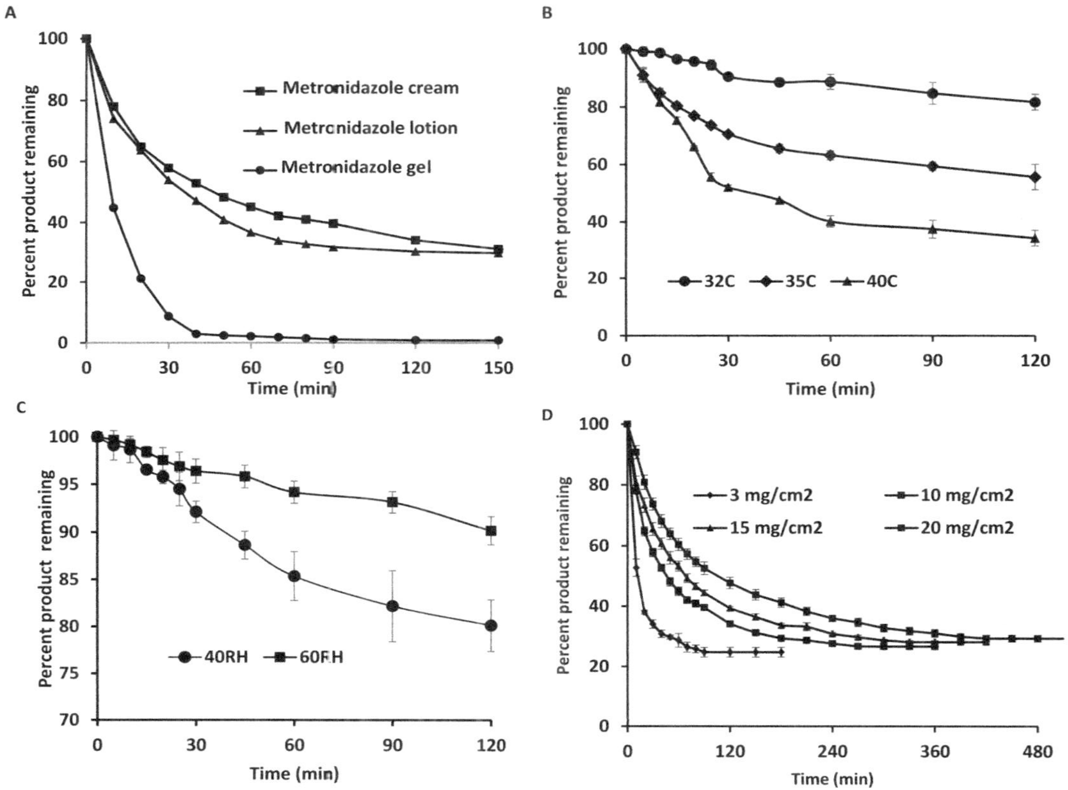

Fig. 3.2 (**a**) The drying rate varies with the type of formulation. In this example, the gel dried out more rapidly than the cream and lotion. (**b**) The drying rate varies directly with the temperature. (**c**) The relative humidity influenced the drying rate of formulations (at a constant temperature). (**d**) The rate of drying varies inversely with the dose

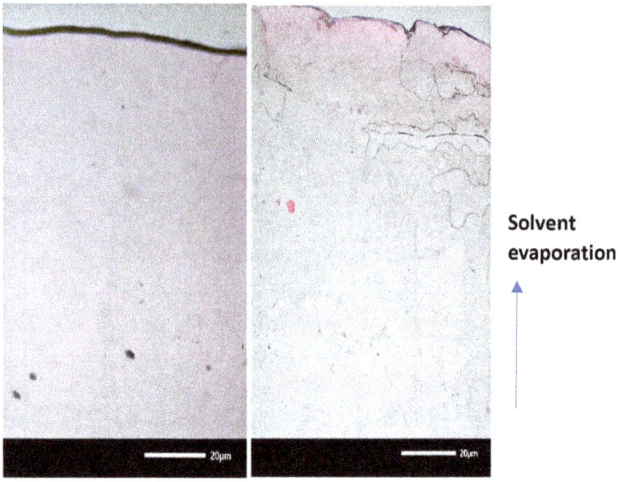

Fig. 3.3 The section of gel formulation incorporated with a water-soluble dye subjected to drying on a glass plate. The left side picture shows a uniform dye distribution in the gel before solvent evaporation. The right-side picture shows the higher intensity of the dye toward the surface. (The microscopic image is the gel section after rapid partial evaporation of the solvent)

initial phase. During the two-phase drying process, more so in the rapid phase, active pharmaceutical ingredients (APIs) could migrate from the bulk of the formulation toward the air formulation interphase. When the solute has relatively more affinity to the solvent than to the excipients in the formulation, the evaporation of the solvent can carry the solute to the surface, forming a drug-rich zone away from the skin surface (Fig. 3.3). This might impact the release and absorption of the API from the product.

As mentioned earlier, the evaporation of the solvent system in the formulation would significantly change the weight ratio of ingredients, microstructure, and quality attributes (Q3 characteristics) [2]. The following sections discuss the quality attributes of the products that are potentially affected by solvent evaporation.

3.2 Impact of Solvent Evaporation on the Quality Attributes

3.2.1 Concentration of the API

The concentration of the API in the formulation vehicle is one of the main driving factors for its permeation across the skin. The concentration of API is proportional to the chemical potential of the drug. The API could exist in multiple modes in nonemulsion-type formulations. In the case of emulsion formulations (creams, lotions, and aerosols), the partitioning of API between the two immiscible phases

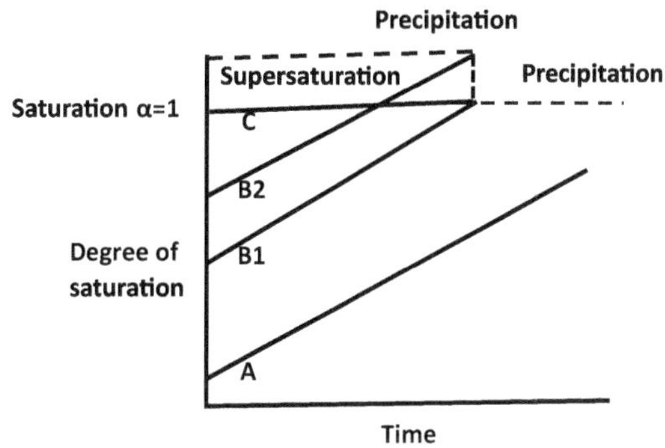

Fig. 3.4 The time course of the degree of saturation of API in the formulation due to solvent evaporation

would add another dimension to the complexity of the API distribution in the formulation. Typically, the existence of APIs in non-emulsion topical products can be categorized into the following scenarios (Fig. 3.4).

A. The API is completely dissolved, and its degree of saturation is significantly lesser than its saturation solubility in the vehicle.
B. The API is completely dissolved in the vehicle, and its degree of saturation is near saturation solubility in the vehicle.
C. API is dissolved (the vehicle is saturated), and excess of the drug is suspended in the formulation vehicle.

In a formulation containing the API wholly dissolved in the vehicle, with a degree of saturation significantly lesser than the saturation solubility, the solvent evaporation will increase the API concentration at the application site. If the formulation primarily contains only one solvent, then the degree of saturation is likely to increase continuously with time (type A in Fig. 3.4). On the other hand, when the formulation vehicle has two or more solvents, the degree of saturation will depend on the real-time solubility of the API in the remnant solvent system on the skin. In such a case, the degree of saturation may increase or decrease randomly.

Here is a case study: a hydrogel was prepared with a binary solvent system, polyethylene glycol (PEG 400), hydroxyethyl cellulose (HEC), and water. Hydroxyethyl cellulose (5%w/w) was used to create a viscous gel. The gels contained 1%, 10%, and 20% PEG 400. Due to the high viscosity imparted by the HEC polymer, the viscosity differences caused by different amounts of PEG 400 in the three formulations were insignificant. The in vitro permeation studies were performed in a semi-infinite setup, with a 300 mg/cm^2 dose in Franz diffusion cells (FDC) with the donor compartment open to the atmosphere. The formulation in the donor was sampled at different time points for six hours, and the amount of PEG 400, water, and API was analyzed. The time course of change in concentration of PEG 400, water, and drug is shown in Fig. 3.5. Water evaporates rapidly; hence, the concentration of PEG 400 increased with time. The degree of saturation (DS) calculated as concentration of the API in the formulation at different time points

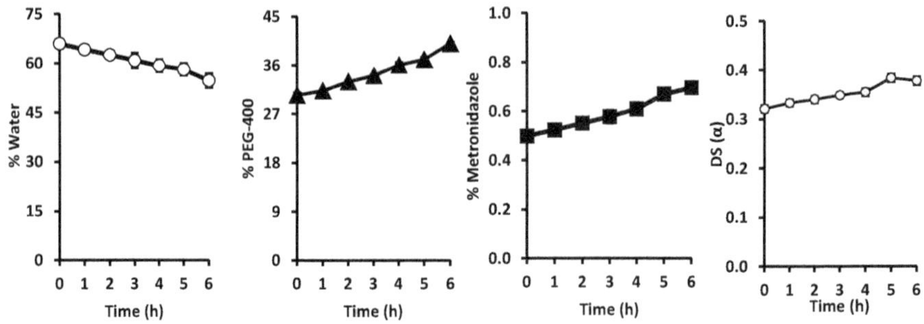

Fig. 3.5 The time course of water, PEG 400, and the API in a gel formulation are presented, respectively. The evaporation of water leads to an increase in the concentration of PEG 400 and a decrease in the overall volume of the solvent system. Therefore, the concentration of the API increased gradually

over the solubility of the drug in the real-time solvent composition is shown in Fig. 3.5d. A solubility curve of the API was constructed at different concentrations of PEG 400 in water to calculate the degree of saturation using the real-time drug concentration at different time points during the drying of the formulation. The degree of saturation (α) in the gel with PEG 400 increased up to 5 h and did not increase further. Whereas, in the case of the gel prepared with only water, the degree of saturation increased linearly with time.

The second case scenario (B), shown in Fig. 3.4, represents a formulation in which the API is completely dissolved in the vehicle, and its degree of saturation value is closer to one. The solvent evaporation would soon lead to saturation of API in the remnant vehicle and precipitation of the drug (B1). Some APIs can hold their molecular dispersion state beyond saturation solubility, resulting in a supersaturated system (B2). The API that precipitates beyond saturation/supersaturation thresholds will not be available for absorption. However, it can act as a reservoir to replenish the drug in the remnant fluid when the dissolved drug continues to get absorbed.

The third scenario (C) in Fig. 3.4 is relevant to formulations in which the vehicle is saturated with the drug and, in addition, contains an excess amount of drug dispersed in it. When applied to the skin, solvent evaporation changes the dissolved/undissolved drug ratio with time. Despite the changes, the degree of saturation is maintained at unity until an adequate amount of solvent is available on the skin surface.

3.2.2 *pH of the Formulation*

A formulation's inherent pH can be attributed to its composition and, to some extent, its microstructure. As discussed in Chapter One, the pH and pKa of an API determine the extent of ionization of weakly acidic or basic APIs. Therefore, the pH of the formulation is adjusted to maintain most of the API in the desired ionic state that favors its absorption. However, the formulation's pH could change for multiple reasons when applied to the skin.

- The evaporation of solvents from the formulation will change its composition, which might lead to a change in the pH.
- The formulation can dissolve atmospheric gases, which might change the pH too.
- The acidic milieu on the skin surface can change the formulation pH due to the incorporation of endogenous ingredients.
- Another crucial process that could change the formulation's pH is the skin's pH-homeostasis mechanism, which interacts with the formulation to transform it to a physiologically compatible pH range gradually.

An ideal formulation can sustain a pH favorable to enable drug absorption at the affected site. Even a unit change in the pH of the formulation due to the above-listed factors can substantially change the ionization of the API. The shift in the ionized-unionized fraction of the API could impact the formulation's performance. The impact of ionization on drug absorption could be negative or positive depending on the nature of the API (acid or base), the relative permeability of ionized and unionized species, the solubility of the ionic and nonionic species, and the predominant pathway of absorption across the skin.

In an in-house study by the authors, a few custom-made gels and creams were prepared. The composition of the creams is given in Annexure 1. The initial pH of the formulation was noted using a microelectrode. The formulations were then dried on an inert substrate at 32° C. The remnant formulation was recovered, and the pH was measured. Table 3.1 shows the change in the formulation pH when it lost 50% of the solvent.

The formulation should be developed robust enough to counteract any pH change. The topical formulations are a mixture of several ingredients, such as fatty acids, esters, surfactants, preservatives, salts, polymers, and solvents [3]. The acidic/basic nature of the excipients enables the formulation to have some buffering ability. The inherent ability of a formulation to resist any significant

Table 3.1 Impact of solvent evaporation on the pH of the formulation: The drying process takes about 4–8 h until 50% of the solvent is evaporated ($n = 3 \pm$ SD)

Formulation	Initial pH	Change in pH after evaporation of about 50% solvent (in vitro) ($n = 3 \pm$ SD)	Increase (↑)/decrease (↓)
Cream 1	6.98	0.29 ± 0.06	↓
Cream 2	4.36	0.93 ± 0.02	↑
Carboxymethyl cellulose, sodium gel	7.00	0.85 ± 0.01	↑
Hydroxyethyl cellulose gel	5.11	0.15 ± 0.04	↑

change in its pH is termed the "*Inherent Buffering Ability of the Formulation*" (IBF). What magnitude pH change is a "significant change"? The answer to this question is subjective. It depends on the API's susceptibility to pH-induced effects like degradation, polymorphic change, solubility, charge transformation, and ionic state. Therefore, for all practical purposes, the IBF was parameterized as the amount of an acid or an alkali to be incorporated to change the formulation pH by one unit. This is analogous to how the term buffer capacity (β) is defined in the case of buffer solutions. The formulations are highly heterogeneous in composition, unlike standard buffers with balanced amounts of acid or base with their respective salts. Therefore, the formulations might show different tolerance levels to acidic and alkaline challenging materials (ACMs). Thus, there is a need to express IBF as IBF_a and IBF_b to differentiate the IBF when the nature of the challenging factor is acidic and alkaline, respectively.

Practically, one can calculate IBF by titrating a known quantity of the formulation with a concentrated ACM and measuring the change in pH. One should choose the concentration of the ACM rationally. A low concentration of the ACM solution can dilute the formulation, leading to a loss of formulation characteristics. A highly concentrated ACM solution could be harsh and perturb the microstructure of the formulation. One can prepare the ACM solution for titration in the vehicle/continuous phase of the formulation to minimize its impact on the microstructure and composition of the formulation.

The IBF could be calculated as follows:

$$IBF_a = \frac{\text{Amount of the acidic challenging material incorporated}}{\text{Change in pH (delta pH)}}$$

$$IBF_b = \frac{\text{Amount of the alkaline challenging material incorporated}}{\text{Change in pH (delta pH)}}$$

When topical products are developed, it is a common practice in the industry to adjust the formulation's pH using appropriate acidic or alkaline pH modulators. Then, the pH modulators becomes an integral part of the composition. The pH modulators maintain a consistent pH throughout the product's shelf-life. However, when applied to the skin, solvent evaporation might lead to a more drastic change in the formulation pH due to an increase in the concentration of the pH modulator.

The formulations in Table 3.2 were adjusted to different pH levels using acidic and alkaline pH modulators commonly used in topical formulations. The pH of the formulations was measured after evaporating 50% of the solvent following the same procedure discussed earlier. The pH of the formulation changed more than that of the formulations prepared without using the pH modulator (Table 3.2).

Table 3.2 The pH change after 50% solvent loss in formulations incorporated with pH modulators. The pH change was significantly higher than in formulations without the pH modulator

Formulation	Initial pH	Adjusted pH	pH modulator used	Increase (↑)/ decrease (↓)	Delta pH due to 50% solvent evaporation
Cream 1	6.98	4	HCl	↓	1.00 ± 0.02
		5	HCl	↓	0.9 ± 0.02
		9	TEA	↑	1.32 ± 0.01
Cream 2	4.36	4	HCL	↓	1.00 ± 0.01
		9	NaOH	↑	1.21 ± 0.02
Carboxymethyl cellulose, sodium gel	7.00	9	NaOH	↑	1.22 ± 0.11
Carboxymethyl cellulose, sodium gel	7.00	5	HCl	↓	0.09 ± 0.12
Hydroxyethyl cellulose gel	5.11	9	NaOH	↑	1.15 ± 0.21

3.2.3 *Buffer System in the Skin*

The skin has an acidic milieu on the surface and is also associated with an inherent ability to resist exogenous acidic/alkaline aggression [4]. When the formulation's pH differs considerably from the skin's, the skin interacts with the formulation by exchanging endogenous materials to modulate the pH of the formulation to a physiologically favorable range. This physiological buffering phenomenon of the skin is attributable to fatty acids, amino acids, keratin, acidic components, and other soluble peptides present in the skin layers or expressed due to exposure to products of extreme pH [5, 6]. A wealth of literature evidence exists about the acid milieu present in the skin layers and the buffering ability of the skin [4, 7]. Based on in-house experimental results and comprehension of literature reports, the complex interaction of skin with the formulation can be described stepwise as follows (Fig. 3.6).

(a) Solvent evaporation could change the formulation's pH due to the change in its composition.
(b) The formulation components will diffuse into the stratum corneum layers, leading to a tentative drastic change in the surface pH.
(c) Some of the pH-modulating exogenous and endogenous components will diffuse from the stratum corneum into the viable layers.
(d) This will trigger biochemical processes in the viable layers to generate endogenous buffering components, eventually transported into the stratum corneum.

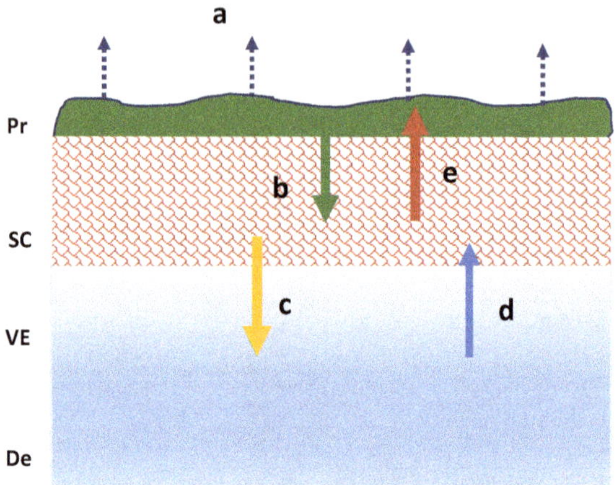

Fig. 3.6 Sequential steps of skin interaction with the topical formulation. *Pr* Topical product, *SC* Stratum corneum, *VE* Viable epidermis, and *De* Dermis

(e) The preexisting endogenous buffering components and the components supplied by the viable layers would diffuse into the formulation, gradually changing the formulation pH.

The authors performed an in-house human volunteer study to evaluate skin interaction with the formulation. First, the application site on the forearm was washed with deionized water three hours before dosing with a topical formulation. Different pH gels and cream formulations were prepared and applied to the forearm area. The surface pH of the skin was measured using a surface probe before application. The formulation was recovered after 30 min, and the pH was measured (Tables 3.3 and 3.4). Invariably, all the formulations with a pH >5.5 tend to decrease, while those with a pH <4 showed an increase in the formulation pH (Table 3.3). When a formulation of about pH five was applied to the skin, the pH of the formulation did not change notably. It is evident from this data that the skin has an inherent ability to modulate the pH of products applied to the skin to render it compatible with its pH. This natural physiological response of the skin is viewed

Table 3.3 pH changes in formulations at 30 min after application to human volunteers, in which pH modulators were not incorporated

Formulation	Initial pH of the formulation	Increase↑/decrease↓	ΔpH of the formulation 30 min after application on the skin ($n = 6$, SD)
Cream 1	6.98	↓	0.81 ± 0.05
Cream 2	4.36	↑	1.22 ± 0.06
Carboxymethyl cellulose, sodium gel	7.00	↓	0.60 ± 0.06
Hydroxyethyl cellulose gel	5.11	–	No significant change

Table 3.4 pH changes in formulation measured 30 min after application to human volunteers, in which pH modulators were incorporated

Formulation	Initial pH	Adjusted pH	pH modulator used	Increase (↑)/ decrease (↓)	Delta pH due to 50% solvent evaporation
Cream 1	6.98	4	HCl	↑	0.61 ± 0.01
		5	HCl	↓	0.1 ± 0.034
		9	TEA	↓	1.32 ± 0.35
Cream 2	4.36	4	HCL	↑	0.98 ± 0.36
		9	NaOH	↓	1.23 ± 0.34
Carboxymethyl cellulose, sodium gel	7.00	9	NaOH	↓	1.22 ± 0.11
Carboxymethyl cellulose, sodium gel	7.00	5	NaOH	–	No significant change
Hydroxyethyl cellulose gel	5.11	9	NaOH	↓	1.55 ± 0.13

as a positive characteristic meant to protect the skin [8]. However, in the case of formulations, the change in pH might impact the permeation of APIs.

Typically, buffers contain a weak acid or base and its respective salt, and their buffer capacity depends on their concentrations. They work effectively on the pH scale of pKa ± 1. Phosphate, citrate, or borate buffers are commonly seen in dermal, ophthalmic, and vaginal formulations. The buffer systems can significantly improve the formulations' IBF_a and IBF_b, thus enabling them to sustain the adjusted pH.

A carboxymethyl cellulose (CMC) gel prepared in different ionic strength PBS (1X, 0.5X, 0.1X, and 0.01X) was applied on the skin of human subjects ($n = 6$), and the change in the formulation pH after 30 min was measured. The pH changes at 30 min in the case of gels prepared with 1X and 0.1X buffers were 0.28 ± 0.04 and 0.75 ± 0.05 units, respectively ($n = 6 \pm$ SD), indicating that the use of higher buffer capacity buffers will enable the formulation to resist pH change more strongly compared to gels prepared with lower strength buffers.

3.2.4 *Viscosity*

The viscosity of the formulation determines the rate at which the drug is released from the formulation. The structural components of the formulation (such as polymers, gums, lipids, fatty acids or esters, fatty alcohol, glycols, and other viscosifying agents) contribute to viscosity building in a formulation. The product's microstructure is also responsible for its consistency and rheological properties. When the formulation is applied to the skin, the solvents leave the system, thus increasing the concentration of the viscosity builders. The cellulose polymers would

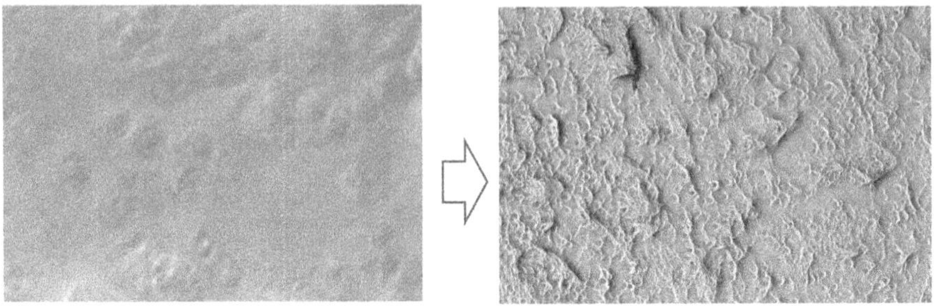

Fig. 3.7 The microscopic appearance of Carbopol gel immediately after application on a gelatin-coated slide and after losing about 50% of its water content, respectively

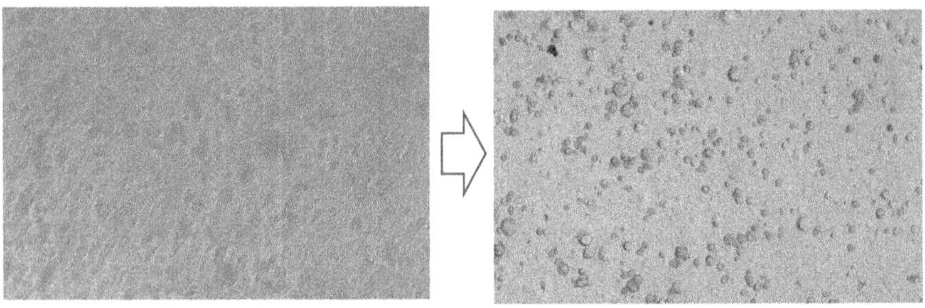

Fig. 3.8 The microscopic appearance of gels prepared with Sepineo P 600 immediately after application on a gelatin-coated slide and after losing about 50% water content, respectively

get entangled, forming a thicker gel network [9]. The gels shrink in volume, and some polymers form a continuous film, whereas some form a porous layer on the skin surface (Figs. 3.7 and 3.8) [10–12]. In a cream or lotion, the globule density increases, followed by phase separation. The dispersed phase changes in a cream or lotion are discussed in detail in the next section. In products containing surfactants, the increase in concentration could lead to the formation of micelles of different shapes, sizes, and aggregate numbers. If the formulation contains dispersed solids, the solvent evaporation might change their size distribution.

A gradual increase in the viscosity can impact the drug's permeation by decreasing the diffusivity of API molecules within the formulation. Incorporating non-volatile cosolvents would enable the system to maintain consistent API diffusivity even after a significant water loss [13]. This can be used as an approach to overcome the effect of viscosity changes on the drug release. Hygroscopic substances such as glycerol, polyethylene glycols, hydrophilic polymers, and propylene glycol are some of the excipients that can be used. Owing to their hydrophilic and hygroscopic nature, they can also retain a significant amount of water in the gel for a long time.

Incorporating salts in small amounts could also help hydrogels to retain moisture longer. However, salts could loosen the gel network when present in adequate amounts due to the osmotic and salting-out effect. When the salt concentration increases to a threshold point during solvent evaporation, it dehydrates/salts out the polymer, making remnant solvent available for the API to diffuse.

In one of the studies reported by Cross and coworkers, the formulations with higher viscosity favored API permeation when applied in finite doses due to slower drying of the thicker formulation. On the other hand, when studies were performed with infinite doses, the higher viscosity formulations retarded the drug permeation compared to low viscosity formulations [14].

3.2.5 Dispersed Phase Globules

The creams, emulgels, and lotions are emulsion systems containing at least one dispersed phase within a continuous phase. O/W creams have oil, and W/O creams have water in their dispersed phase. The formulation needs to be stabilized using suitable exogenous or in situ-forming emulsifying agents to achieve a reasonable thermodynamically stable system. The globules in a cream or a lotion are polydisperse. The globule size distribution greatly determines a topical product's sensory, textural, and aesthetic properties [15].

Rubbing products as a thin layer on the skin surface can lead to friction-induced deformation, coalescence, and rupture of some globules. However, if applied gently, the product's microstructure is less affected (Fig. 3.9).

Solvent evaporation starts instantaneously after application, and the drying process might take a few minutes (in formulations with alcoholic solvents) to a few hours. The formulation's drying rate is rapid initially due to the direct evaporation of the solvent. It decreases gradually due to the formation of a thin layer of the dry region at the air-formulation interface. The solvent evaporating from the core layers must diffuse across the peripheral dry region.

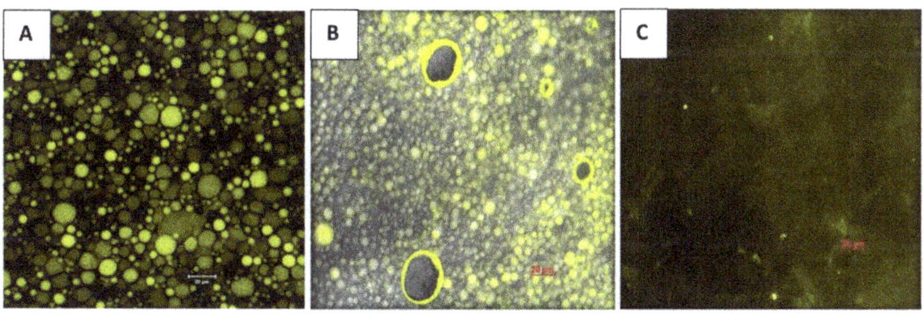

Fig. 3.9 Confocal fluorescent microscopic pictures of the cream loaded with Nile red in the dispersed phase (**a**) were rubbed with different intensities of shear on the cadaver skin. The globule size decreased when the cream was rubbed with moderate intensity. (**b**) The globule size was reduced to a much smaller diameter. (**c**) Some globules even ruptured when rubbed aggressively

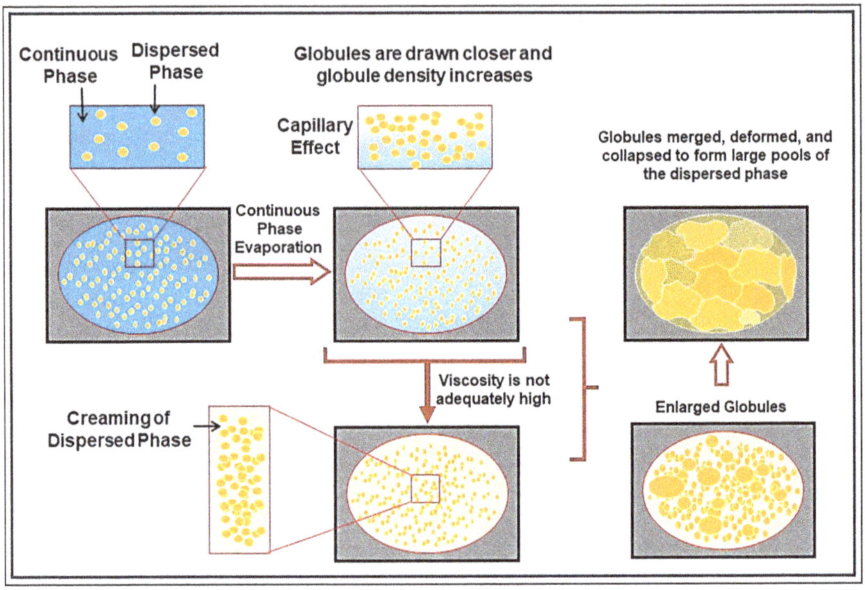

Fig. 3.10 Schematic presentation of the different stages of transformation of dispersed phase globules due to evaporation of continuous phase solvents

Fig. 3.11 The real-time pictures of dispersed phase dynamics on a glass slide coated with gelatin. The first slide shows distinct globules at time zero. The second slide shows the globules with diminished borders, coalesced and collapsed globules, and the last picture shows the patches formed due to the calescence of globules forming islands separated by a precipitated emulsifying agent

Most topical emulsions are of the o/w type. Therefore, the metamorphosis process discussed in this section is relevant to o/w dispersed systems. The changes in the dispersed phase when an o/w product is applied to the skin, occur in two phases. The fluid dynamics of the continuous phase predominantly drive the changes in phase 1. The second phase comprises changes after most of the continuous phase solvent system is evaporated (Fig. 3.10) [16]. The collapsing behavior of a custom-made o/w cream was investigated microscopically on a Gelatin coated glass slide (Fig. 3.11) and on a cadaver skin surface (Fig. 3.12).

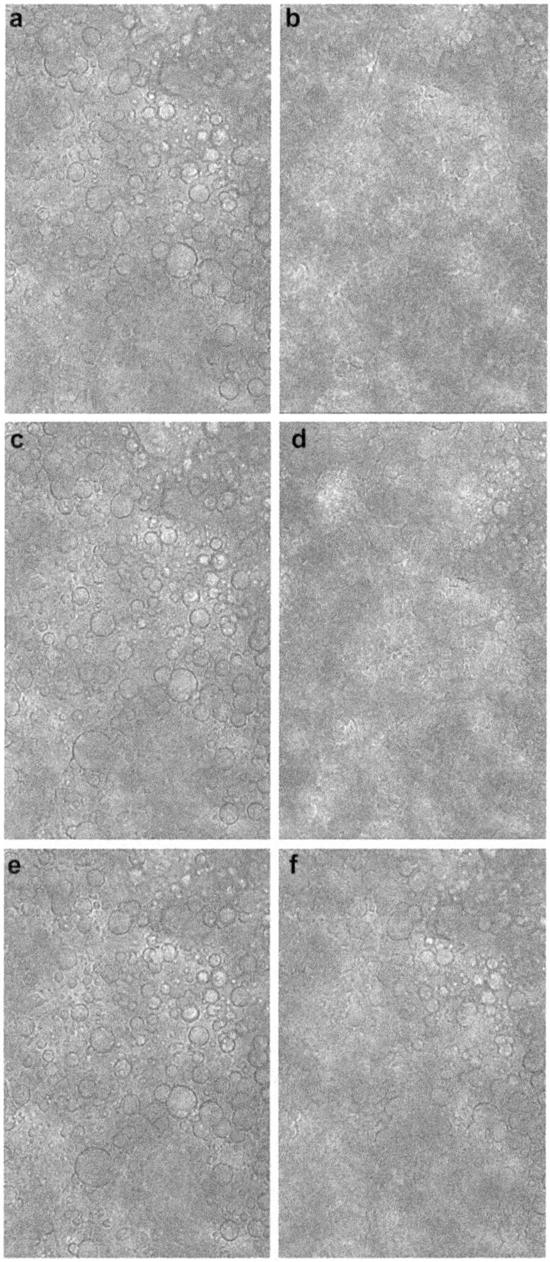

Fig. 3.12 The coalescence of globules on the skin captured real-time at different stages: (**a**) The globules are intact and discrete when applied; (**b**) the globules are drawn closer due to capillary effect caused by evaporation of continuous phase; (**c**) the globules start to expand in size; (**d**) the borders appear diminished either due to coalescence or collapse of globules; (**e**) the coalescence of globules forms larger pools of dispersed phase; (**f**) the complete loss of continuous phase leaves behind the patches of dispersed phase separated by the precipitated emulsifying agent

As the continuous phase evaporates, to some extent, the redistribution of the continuous phase liquid takes place due to the capillary effect (unless the liquid film is too thin and the evaporation rate is so high that the capillary effect fails to equilibrate). The convection of solvent would often draw the globules closer to each other, leading to an increased globule density (number of globules per unit area) appearing in patches. If the viscosity of the cream or lotion is not adequate to mitigate the creaming of the dispersed phase, the globule density also increases due to the creaming. When the viscosity is high, the contribution of creaming to the increase in the globule density would be negligible. As the evaporation continues, the oil globules would start deforming and growing in size due to coalescence.

In the second phase of metamorphosis, the globules will change in size and shape. When a solvent evaporates, some globules will rupture to form large pools of oil, separated by the emulsifier that precipitates due to attaining maximum solubility. Towards the final drying stage, the glass slide or skin is covered with islands of non-coalesced large emulsion droplets or oil pools. On a hydrophobic substrate (uncoated slides), unlike the hydrophilic substrate, the globules form a continuous patch-like layer. The sequence of disruption of globules on the skin surface was similar to that observed on the gelatin-coated glass slide. The islands formed during the oil phase depend on the globule density, globule size, and concentration of surfactant. The entire process of drying of cream formulations is schematically shown in Fig. 3.10.

Other ingredients, such as surfactants, salts, and cosolvents/emollients, significantly influence the rate and pattern of disruption of globules. The presence of a higher weight fraction (>30% w/w) of glycols would slow down the globule deformation, whereas volatile solvents like ethanol would hasten the process. When glycols are present in considerable amounts in the creams, after evaporation of water, the glycol forms a layer underneath the oil phase. If the API is present in the oily dispersed phase, the performance will likely be impaired significantly as there would be a disconnect between the skin and the drug-rich dispersed phase layer.

3.2.6 *Particle Size Distribution of API*

When the drug's solubility is poor, an excess amount of API is often dispersed in the formulation. The dispersed drug serves as a reservoir to maintain the API saturation in the formulation. When applied to the skin, the dissolved API precipitates in most cases due to solvent evaporation (Fig. 3.13a).

In some cases, the dissolved drug deposits on the crystals that have already formed or preexisted in the formulation, enlarging their size further. The submicron-size crystals might dissolve and deposit the API on the larger crystals, analogous to the Ostwald ripening phenomenon in suspensions (Fig. 3.13b).

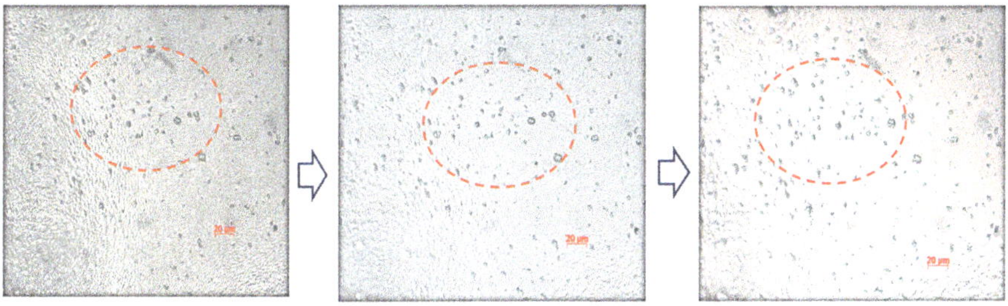

Fig. 3.13a The particle density increases when the solvent evaporates due to precipitation of the dissolved API

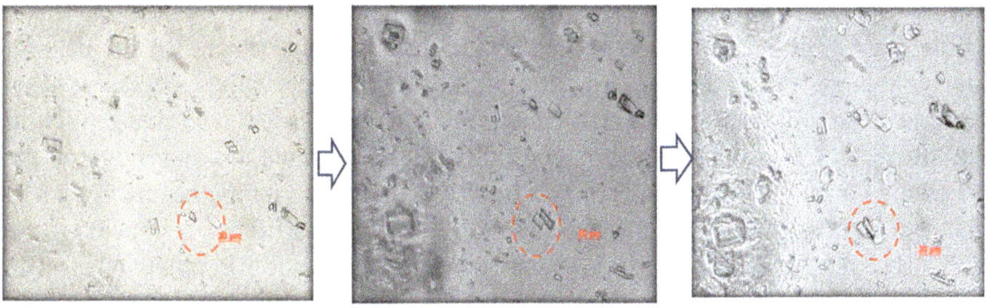

Fig. 3.13b The smaller particles increases due to solvent evaporation owing to the deposition of the dissolved API on the existing crystals

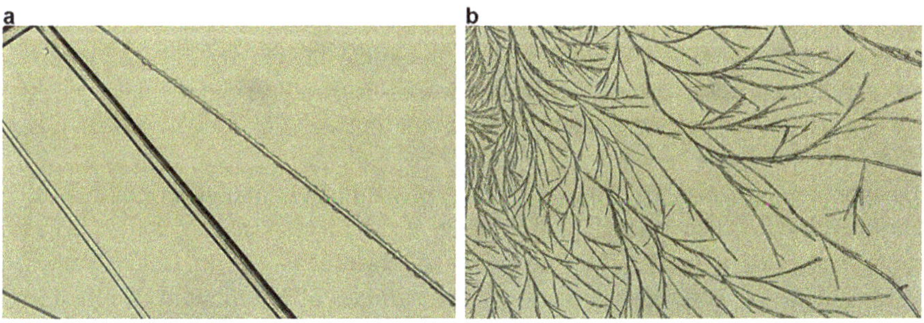

Fig. 3.14 Metronidazole crystals precipitated from aqueous solution appear like elongated linear crystals (a), whereas when the Carbopol gel with 0.75% of metronidazole was dried, the API manifests a branched pattern (b)

The API precipitated due to solvent evaporation at the site of application may or may not manifest in the same polymorphic form. The residue may not even retain the API's inherent crystal shape. For example, APIs of the azole class compounds tend to form spiral structures when the gel is dried [17]. Metronidazole forms a unique branched pattern when Carbopol gel with 0.75% API is dried, as shown in Fig. 3.14.

The transformation of the dissolved drug to particles leads to an increased number of particles or an alteration of the size distribution of the particles. These changes significantly impact the formulation's performance. Due to the diverse solid-state nature of the APIs, the formulator must understand the potential particle-related changes due to solvent loss on the skin.

3.2.7 Drug Distribution Between the Dispersed and Continuous Phase

The APIs partition to distribute between the oil and aqueous phases in an emulsion formulation. The partition coefficient, or the distribution constant (in the case of ionizable drugs), is an inherent property of the API that determines this distribution between two immiscible liquids. The terms are often used interchangeably in the field. However, the partition coefficient in a formulation depends on the product's composition and microstructure. To differentiate the partition coefficient that we observe in the formulations from the inherent partition coefficient of the API, the term "apparent partition coefficient" (Papp and log Papp) is used. The Papp of an API could be different in different formulations due to compositional differences. The Papp can also differ in formulations with the same composition but different microstructure and quality attributes. The "partition coefficient" and "apparent partition coefficient" are attributes of the API when the entire system is at equilibrium thermodynamically.

In a system with two immiscible phases, the solute attaining equilibrium involves two simultaneous and opposite diffusion processes across the interfacial boundary [18–20]. The transport of a solute from a continuous (Cc) to a dispersed phase (Cd) can be called a forward process, and the rate constant associated is k_{cd}. The transport rate constant from the dispersed phase to the continuous phase is k_{dc}. The partition coefficient (P) is the quotient of the forward and the backward rate constants (k_{cd}/k_{dc}). The log P would be log (k_{cd}/k_{dc}) [21].

In a study by Baxevanis and coworkers, the rate of partitioning of nifedipine into a fat phase was found to fit first-order kinetics [22, 23]. The first-order rate of the partition constant of nifedipine was doubled when fat concentration increased from 5 to 25% w/v, which can be attributed to the larger available area for diffusion when higher fat volumes are used, as described by Fick's first law of diffusion. When the solvent evaporates from the formulation on the skin, the composition and properties of a topical product may change, leading to the redistribution of the API between the two phases. The system will need a longer duration to equilibrate due to the constantly changing microenvironment in the formulation (dispersed and continuous phase) to reach equilibrium.

Table 3.5 The time course of the amount of drug partitioned into the aqueous and oil phases in an o/w cream

Phase	6h	12h	24h	One week (Control)
Oil phase (ug/g)	356.12 ± 13.68	318.4 ± 22.31	256.13 ± 9.75	257.13 ± 3.86
Aqueous phase (ug/g)	131 ± 2.79	116.87 ± 6.18	234.98 ± 9.81	233.17 ± 11.10

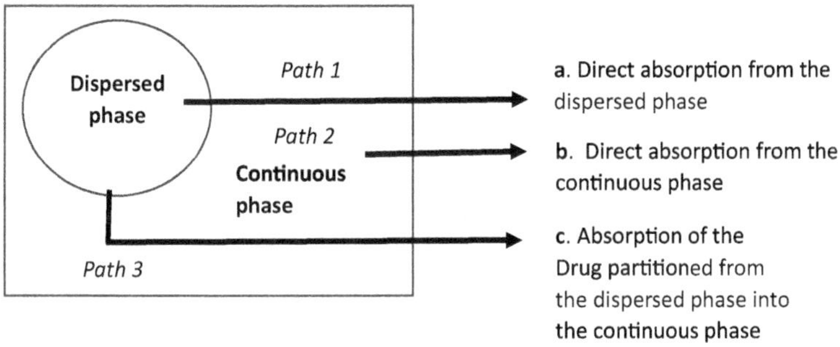

Fig. 3.15 If the drug is highly lipophilic and is localized in the oil phase (in o/w emulsion), the drug can be absorbed directly (path 1). Conversely, if the API is highly hydrophilic and mainly exists in the aqueous continuous phase, the absorption occurs via path 2. Moderately lipophilic molecules, distributed in significant amounts in both phases, can undergo dermal absorption via paths 1, 2, and 3

In one of the in-house studies, the partitioning of clobetasol propionate in the oil phase increased with time and equilibrated after 24 h, indicating the kinetic nature of the thermodynamic equilibration. The experimental control was the same cream that was aged longer (Table 3.5).

The ever-changing microenvironment caused by solvent evaporation on the skin impacts the drug distribution. However, in an oil-in-water emulsion system, if the molecule is highly lipophilic (meaning that the amount of drug in the dispersed phase is much greater than in the aqueous phase) or extremely hydrophilic (meaning that the amount of drug in the dispersed phase is much lesser than in the aqueous phase), the evaporation-driven change on the redistribution of the API could be negligible.

As shown in Fig. 3.15, drug release and permeation could occur via different pathways in an emulsion system.

3.2.8 *Water Activity*

Water activity (a_w) is the thermodynamic energy status of water, represented as a ratio of the partial vapor pressure of water (ρ) to the vapor pressure of pure water (ρ_0) at a constant temperature. Some of the fundamental aspects of water activity are already discussed in Chap. 1. Water activity influences the rate of drying of

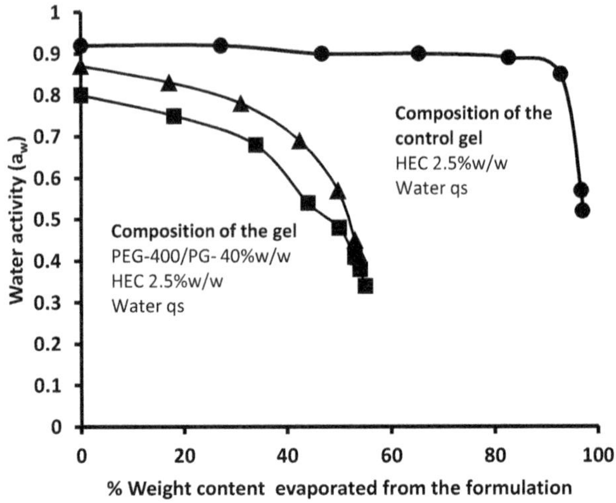

Fig. 3.16 The water activity of a formulation changes as the formulation loses varying amounts of solvent, mainly water, through evaporation. The water activity of gels containing propylene glycol (square) and PEG 400 (triangle) decreased gradually as the solvent evaporated from the formulation. In the case of the control gel, the water activity remained constant until all the free water content was completely depleted

a formulation on the skin. An inverse relationship has been found between water activity and the rate of drying gels and creams [24].

In the skin, the water activity in the inner layers is almost at unity. The water activity of the atmosphere is significantly lower and more variable. Due to the concentration (or thermodynamic activity) gradient, there is constant water loss across the skin. When a formulation with adequate water content is applied to the skin, the concentration gradient across the skin decreases or completely vanishes [25]. However, when the formulation dries out, the water activity gradient recovers gradually.

Topical formulations usually include cosolvents. Glycols are less volatile than water, so water evaporates more quickly when applied to the skin than glycols. This leads to an increase in the weight fraction of glycol in the remaining formulation over time. Eventually, this process leaves behind a thin layer of glycol and other excipients. Figure 3.16 illustrates the water activity profile from topical gel formulations due to the loss of solvents, mainly water.

The formulation's water activity determines the extent of the skin's hydration. Water equilibration between the formulation and the skin occurs quite rapidly. Therefore, any change in the formulation's water activity due to metamorphosis would impact the extent of hydration of the stratum corneum [26, 27]. This could significantly affect the skin's permeability to API [28]. A very low water activity ($aw<0.5$) is also known to cause detachment of skin layers, causing poor drug permeation in absence of glycols [24].

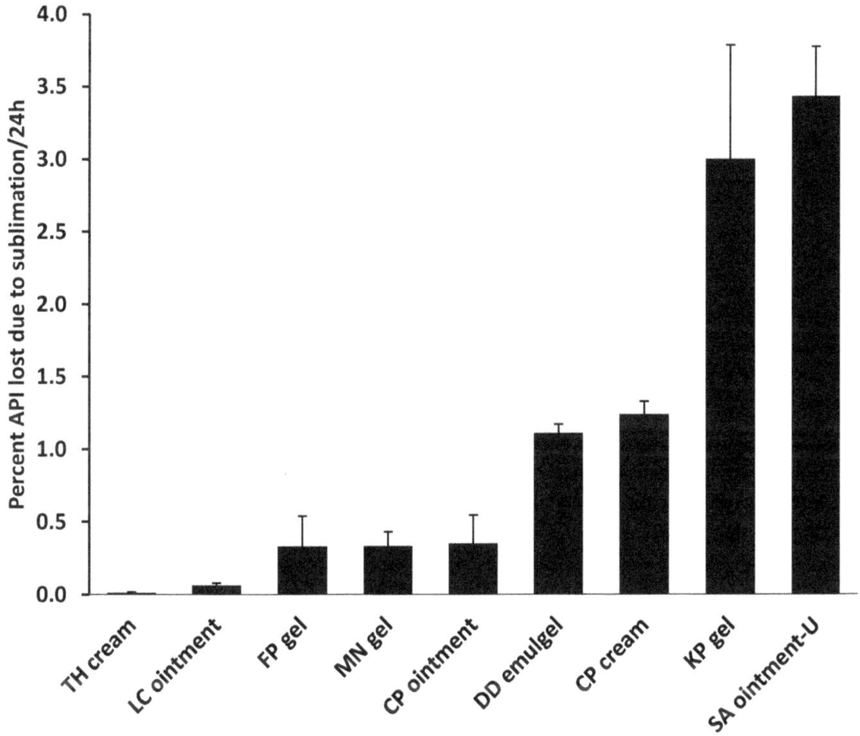

Fig. 3.17 Sublimation of APIs from different marketed topical products ($n = 3 \pm$ SD). *TH* terbinafine hydrochloride, *LC* Lidocaine hydrochloride, *FP* Flurbiprofen, *MN* Metronidazole, *CP* Clobetasol propionate, -Diclofenac dimethylammonium, *KP* ketoprofen, *SA-U* salicylic acid ointment with urea (salicylic acid)

3.3 Sublimation of API

Sublimation is the transformation of solid substances into vapor phase directly without forming a liquid. Some APIs can undergo sublimation after the solids precipitate due to significant solvent evaporation. The extent of loss due to sublimation of API from the application site depends on the thermochemical properties, such as melting point, heat of fusion, and molecular weight. Sublimation of some of the topically administered APIs was investigated by Pragathi and coworkers [29, 30]. The extent of sublimation depends on the formulation composition, as shown in Fig. 3.17. The type of formulation can also influence the extent of API sublimation. A study found that the sublimation of salicylic acid was lesser in creams than in ointment and gel formulations (Fig. 3.18).

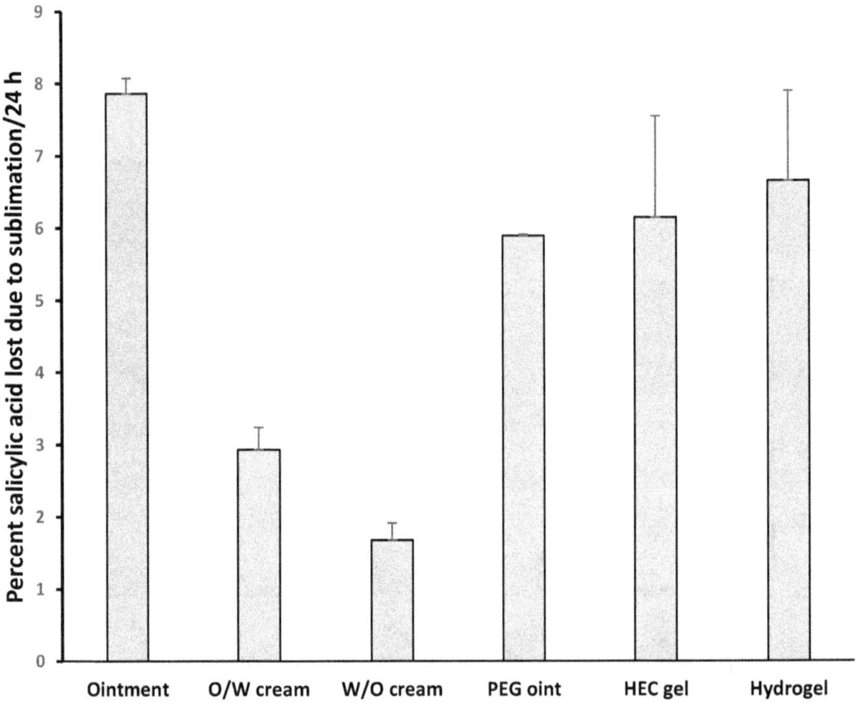

Fig. 3.18 Effect of formulation type/composition on the extent of sublimation of 3% w/w salicylic acid ($n = 3 \pm$ SD)

3.4 Conclusion

As soon as the formulation is applied to the skin, the solvent evaporates, resulting in various changes in the formulation's physical and chemical makeup. The microstructure of the formulation changes and its quality attributes gradually change too. The performance of topical products is determined by the time course of changes in these attributes. Therefore, when developing topical formulations, it is crucial to predict how the properties of the topical products will change in order to create a high-quality product that performs optimally.

Annexure 1: Composition of topical formulations

	Quantity % W/W			
Ingredients	Cream 1	Cream 2	HEC gel	Sodium CMC gel
CSA 70	7	–	–	–
Glyceryl monostearate	2.5	–	–	–
Tween 80	4.2	2	–	–
MCT 70	4.5	10	–	–
Liquid paraffin light	4.5	–	–	–
Kollicream OA	3	–	–	–
Glycerin	3.3	–	–	–
PEG 8000	–	2	–	–
PEG 6000	–	2	–	–
Sepineo P 600	–	1	–	–
Propylene glycol	–	10	–	–
PEG 400	–	5	–	–
Sodium CMC	–	–	–	3.5
HEC 250 HHX	–	–	2.5	–
Carbomer 974P	–	–	–	–
TEA	–	–	–	–
0.8 N NaOH	–	–	–	–
Water Qs	100	100	100	100

CSA Ceto stearyl alcohol, *MCT* Medium-chain triglyceride, *PEG* Polyethylene glycol, *HEC* Hydroxyethylcellulose, *CMC* Carboxymethyl cellulose, *TEA* Triethanolamine, *NaOH* Sodium hydroxide, *OA* Oleyl alcohol

References

1. Jin X, Imran M, Mohammed Y. Topical Semisolid Products-Understanding the Impact of Metamorphosis on Skin Penetration and Physiochemical Properties. Pharmaceutics. 2022, 14(11), 2487.
2. Rusdi M, Moroi Y, Nakahara H, Shibata O. Evaporation from Water–Ethylene Glycol Liquid Mixture. Langmuir. 2005. 21(16), 7308–7310.
3. Dubrovskii A V, Kim A L, Musin E V, Ramazanov B R, Tikhonenko S A. The Discovery of the Buffer Capacity of Various Types of Polyelectrolyte Microcapsules. Polymers. 2021. 13(22), 4026.
4. Ayer J, Maibach H I. Human Skin Buffering Capacity against a reference base Sodium hydroxide: in vitro Model. Cutan Ocul Toxicol. 2008. 27(4), 271–281.
5. Pascher G, von Steinruck G, Spier H. The water-soluble ingredients of the peripheral horny layer of the skin; quantitative analyses. V. Inhomogeneous distribution of alpha-amino-nitrogen, lactic acid, chloride, potassium (and sodium) in the stratum disjunctum. Arch Klin Exp Dermatol. 1957.204: 140.
6. Hachem J P, Crumrine D, Fluhr J, Brown B E, Feingold K R, Elias P M. pH directly regulates epidermal permeability barrier homeostasis, and stratum corneum integrity/cohesion. J Invest Dermatol. 2003. 121: 345–353.

7. Zheng Y, Sotoodian B, Lai W, Maibach H I. Buffering capacity of human skin layers: in vitro. Skin Res Tech. 2012.18(1), 114–119.
8. Burckhardt W. Protective function of the skin against the external environment, with special reference to its buffer capacity against alkalis and acids. Schweiz Med Wochenschr. 1957.87: 1525–1529.
9. Adams D. Does Drying Affect Gel Networks? Gels. 2018. 4(2), 32.
10. Waje S S, Meshram M W, Chaudhary V, Pandey R, Mahanawar P A, Thorat B N. Drying and shrinkage of polymer gels. Braz. J. Chem. Eng. 2005. 22(2), 209–216.
11. Hawlader M N A, Ho J C, Qing Z. Temperature and moisture distribution within materials undergoing shrinkage during drying, Proceedings of the 10th International Drying-Symposium. 1996. Vol. A, p. 120.
12. Jomma W, Aregba W, Puiggali J R, Quintard M. Drying of highly shrinking products: application to gels. Drying. 1991. 91, p. 110.
13. Di Colo G, Carelli V, Giannaccini B, Serafini M F, Bottari F. Vehicle effects in percutaneous absorption: In-vitro study of influence of solvent power and microscopic viscosity of vehicle on benzocaine release from suspension hydrogels. J Pharm Sci. 1980. 69:387–391.
14. Cross S E, Roberts M S, Jiang R, Benson H A E. Can Increasing the Viscosity of Formulations be used to Reduce the Human Skin Penetration of the Sunscreen Oxybenzone?. J Invest Dermatol. 2001.117(1), 147–150.
15. Murthy R, Rangappa S, Repka M A, Vanaja K, Shivakumar H N, Murthy S N. Infrared thermal measurement method to evaluate the skin cooling effect of topical products and the impact of microstructure of creams. J Drug Del Sci Tech. 2017. 39, 296–299.
16. Kapilashrami A, Eskilsson K, Bergström L, Malmsten M. Drying of oil-in-water emulsions on hydrophobic and hydrophilic substrates. Colloids Surf. A: Physicochem. Eng. Asp. 2004.233(1–3), 155–161.
17. Ajjarapua S, DeBoyaceb K, Sarabua S, Shankara V K, Rangappaa S, Wildfongb P L D, Murthy S N. Metronidazole crystal patterns formed during the metamorphosis of topical carbopol gels. J. Excipients and Food Chem. 2003.14 (1). 2150–2668.
18. de Haan F H N, Jansen A C. The influences of solvent properties on interfacial transfer kinetics m stirred two-phase systems of organic solvent-water. Int J Pharm. 1986. 29, 177–90.
19. de Haan F H N, de Vringer T, Van de Waterbeemd J T M, Jansen A C A. Transport rate constants and transport rate parameters in various organic sol vent-water systems. Int J Pharm. 1983. 13, 75–87.
20. Kubinyi H. Drug partitioning relationships between forward and reverse rate constants and partition coefficients. J pharm Sci. 1978.67, 262–3.
21. Dong Y, Hengst L, Patel D, Hunt R, Qu H, Choi S, Ashraf M, Cruz C N, Xu X. A Kinetic Approach to Determining Drug Distribution in Complex Biphasic Systems. J. Pharm. Sci. 2019.1–10.
22. Baxevanis F, Zarmpi P, Kuiper J, Fotaki N. Investigation of drug partition kinetics to fat in simulated fed state gastric conditions based on drug properties. Eur J Pharm Sci, 2020.146. 105263.
23. Fahelelbom K M S, Quigley J M, Timoney R F, Corrigan O I. Thermodynamics of the Distribution and Interfacial Transfer Kinetics of Antimicrobial Iminophenazines in a Two-Phase System of N-Octanol/Aqueous Buffer. Proceedings of the Royal Irish Academy. Section B: Biological, Geological, and Chemical Science. 1989. 89B, 287–298.
24. Angamuthu M, Shankar V K, Murthy S N. Water Activity and Its Significance in Topical Dosage Forms. J Pharm Sci. 2018. 107(6), 1656–1666.
25. Crowther J M, Sieg A, Blenkiron P, Marcott C, Matts P J, Kaczvinsky J R, Rawlings A V. Measuring the effects of topical moisturizers on changes in stratum corneum thickness, water gradients and hydration in vivo. Br J Dermatol. 2008. 159(3): 567–577.
26. Albèr C, Buraczewska-Norin I, Kocherbitov V, Saleem S, Lodén M, Engblom J. Effects of water activity and low molecular weight humectants on skin permeability and hydration dynamics—a double-blind, randomized and controlled study. Int. J. Cosmet. Sci. 2014. 36(5), 412–418.

27. Kerdudo A, Fontaine-Vive F, Dingas A, Faure C, Fernandez X. Optimization of cosmetic preservation: water activity reduction. Int. J. Cosmet. Sci. 2015. 37(1), 31–40.
28. Nadarzynski A, Scholz J, Schröder M S. Skin Barrier Enhancing Alternative Preservation Strategy of O/W Emulsions by Water Activity Reduction with Natural Multifunctional Ingredients. Cosmetics. 2022.9(3), 53.
29. Kasting G B, Miller M A. Kinetics of finite dose absorption through skin 2: Volatile compounds. J. Pharm. Sci. 2006. 95(2), 268–280.
30. Pragathi S G, Echanur A V, Matadh A V, Rangappa S, Shivakumar H N, Murthy R N, Ranganath V S, Ureña-Benavides E E, Maibach H, Murthy S N. Sublimation of Drugs from the Site of Application of Topical Products. Mol. Pharmaceutics. 2023.20, 6, 2814–2821.

Chapter 4
Optical Microscopic Techniques for Evaluation of Dermal Drug Absorption from Topical Products

Dandan Tu, Ting Chean Khoo, Yuxiao Wei, Saara Luna, Benjamin A. Kuzma, and Conor L. Evans

4.1 Introduction

4.1.1 Dermal Drug Evaluation

The topical application of therapeutics is a desired method to avoid potential adverse systemic side effects while achieving therapeutic active pharmaceutical ingredient (API) concentrations at the local site of action. Achieving adequate target site concentrations is paramount for therapeutic success. While API pharmacokinetic (PK) considerations are more thoroughly discussed elsewhere in this book, in brief, they are commonly referred to as the API's absorption, distribution, metabolism, and excretion/elimination (ADME). The permeation of the API from the formulation and into the skin is the primary rate-limiting step for API exposure [1]. This process is similar in the complexity of an orally delivered API, yet more must be investigated. Once the API has permeated the stratum corneum (SC), the metabolism and elimination into the various layers can be monitored independently from topical application to inform about the absorption process. While the metabolism and elimination are thought unlikely to be modified by the excipients in the formulation, these processes have not been characterized for dermal drug delivery.

Concentration-time profiles are important for estimating PK parameters, which are critical for bioavailability (BA) and bioequivalence (BE) assessment. However, quantifying concentrations locally within the skin has been challenging to date. HPLC and mass spectrometry-based techniques, which have achieved success in quantifying therapeutics' concentrations in blood, need to be used with other

D. Tu · T. C. Khoo · Y. Wei · S. Luna · B. A. Kuzma · C. L. Evans (✉)
Wellman Center for Photomedicine, Charlestown, MA, USA
e-mail: Evans.Conor@mgh.harvard.edu

S. N. Murthy (ed.), *Topical Products and Dermal Drug Delivery*,
AAPS Introductions in the Pharmaceutical Sciences 2,
https://doi.org/10.1007/978-3-031-80525-7_4

methods to retrieve the local site information on the skin. As discussed in the other chapters, there are a multitude of methods under investigation for obtaining local API information after topical application, including in vitro permeation testing, dermal open-flow microperfusion [2, 3], dermal microdialysis [4, 5], and tape stripping [6]. These techniques have shown a lot of usefulness and advantages and are more informative than using the systematic concentrations in blood for a rough estimation of the local concentrations in the skin; however, they quantify local API concentrations at the macroscale level.

Because the target site of topical dermatological therapeutics is at the cellular level within the skin's microstructure, microscale techniques that quantify and visualize topical API permeation and biodistribution within the skin at subcellular resolution would be highly beneficial. In addition to the lateral resolution, precisely evaluating the API's axial distribution at different skin layers without damaging the upper SC layer is important for a comprehensive evaluation of the API.

In recent decades, optical microscopic techniques, such as fluorescence imaging, infrared imaging, Raman imaging, and coherent Raman imaging, have been explored in the evaluation of dermal drugs. Compared with these non-microscopic techniques, optical microscopic techniques show capability in visualizing microscale information. Microscopic techniques can provide valuable insight into the ADME processes at a cellular level that can aid in new chemical entities and formulation optimization and BE implications. This chapter introduces different optical microscopic techniques used in dermal drug absorption studies. The theory of each technique and the instrumentation for building the imaging system are described. Examples of using each technique for evaluating dermal drugs, the unique features of each technique, and its corresponding advantages are discussed. At the end of the chapter, the challenges of current optical microscopic techniques, as well as opportunities aside from these challenges, are discussed.

4.1.2 Skin Preparation for Optical Imaging

Human skin is a challenging sample for optical imaging due to light absorbers in the skin, such as hemoglobin and melanin, and highly scattering substances, such as keratin in the epidermis and collagen in the dermis [7]. Notably, the absorption and scattering effect is wavelength-dependent. Longer wavelength light can penetrate deeper into the skin, and light penetration of $\sim$5 mm can be achieved by near-infrared light [8].

Imaging intact skin samples in vivo is ideal for dermal drug studies because it most closely mimics daily product usage. This type of imaging is often achieved by positioning an optical source and a detector on the skin's surface. In this case, the light source penetrates the skin to a certain depth and excites the target. The excited light needs to propagate back through the skin and reach the detector. Because of low optical backscattering and high signal loss in the tissue before reaching the detector, optical imaging can detect only a limited signal in intact skin in vivo.

Imaging methods using backscattered infrared light can excite signals at a depth up to 2 mm [9]. Among different imaging techniques, an inescapable trade-off exists such that the higher the achievable image resolution, the smaller the imaging depth. To achieve cellular resolution, the imaging depth is typically at the papillary dermis layer ($\leq$200 μm) [9].

Because of challenges related to in vivo imaging, ex vivo imaging of excised human skin is more widely used to study topical products. Ex vivo human skin samples can be collected using different methods. Biopsies and skin removal surgeries are the most common sources of excised human skin. Dermatomes can excise thin skin samples with controlled thickness from the human body [10]. Tape stripping is another method for collecting very thin layers (each tape excised 0.1–1.9 μm) of epidermis skin from body regions [11, 12].

More generally, for the skin from biopsies and surgeries, post-preparation procedures are often needed to reduce the thickness of the skin, in particular for transmission mode detection geometries. The microtome is a powerful tool for slicing thin skin samples with controlled thickness (as thin as 2 μm) [10]. However, microtome protocols often require fixatives or optimal cutting temperatures (O.C.T.), which are exogenous substances that may interfere with imaging and change the API's PK features. A dermatome is a tool made for slicing skin samples with a controlled thickness (100–400 μm). To slice excised skin, the skin needs to be frozen and cut with a cold dermatome [10]. Another skin processing method using forceps, scalpels, and microsurgical scissors can decrease the thickness of fresh skin down to hundreds of μm [13]. In addition, heat, chemicals (e.g., sodium thiocyanate), or enzymes (e.g., trypsin and dispase) can be used to separate the epidermis from the dermis [14]. Epidermis separation by chemicals and enzymes introduces exogenous substances. In contrast, mild heating of the full-thickness skin for around a minute in a water bath (around 55 °C) can produce a thin sample of epidermis [10]; the relatively low heat could prevent the thermal denaturation of dermal collagen and ensure minor distortion of the skin's compositions and structures.

4.1.3 Optical Microscopic Techniques

4.1.3.1 Fluorescence Imaging

Fluorescent dyes have often been used in biomedical studies for the visualization of contrast between biotargets, cells, or tissue elements. Following the same concept, fluorescence imaging can be used to image fluorescent molecules to evaluate cutaneous PK and estimate PK parameters. Although any APIs can be labeled with exogenous fluorophores, the labeling process may induce toxicity or alter the API's PK features. Only dermal drugs with naturally fluorescent APIs can achieve label-free fluorescence imaging. Several common fluorescence methods used to study APIs in topical products are: conventional fluorescence [15, 16],

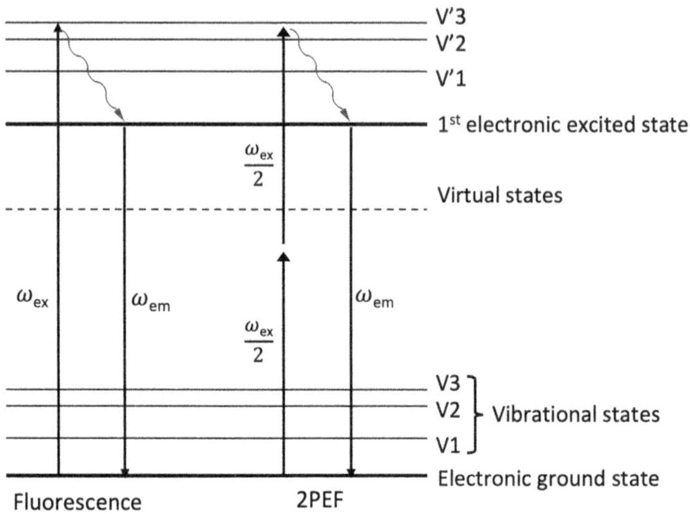

Fig. 4.1 Energy diagram of fluorescence and two-photon excited fluorescence (2PEF) interactions

confocal fluorescence [17–19], two-photon excitation fluorescence (2PEF) [20–25], and fluorescence lifetime imaging microscopy (FLIM) [26–30]. Conventional fluorescence, confocal fluorescence, and 2PEF are intensity-based methods that provide spatial information on fluorophores with unique spectral properties. On the other hand, FLIM generates images with contrast from the lifetime information of fluorophores, enabling the ability to distinguish autofluorescence from desired fluorophore signals.

Theory The theory of fluorescence is based on the principle of light absorption and emission. When a molecule absorbs light energy, the molecule will be excited from its ground state to a higher energy state. Since the higher energy state is unstable, the molecule will quickly decay and emit light as it relaxes to its ground state. This is what is known as fluorescence emission and can be represented in an energy diagram (Fig. 4.1). The energy of the released photon is always lower than the absorbed photon as the energy may be converted into molecular vibrations and/or heat during the decay process. As a result, the fluorescence emission photon will always have a longer wavelength compared to the absorption photon. Such a shift in wavelength is known as the Stokes shift, named after Sir George Gabriel Stokes, an Irish-born physicist and mathematician. Fluorescence was first described by Stokes in 1852 when he observed the mineral fluorspar fluoresce when illuminated with ultraviolet light. Stokes then published his findings in a paper titled "On the Change of Refrangibility of Light" in the journal Philosophical Transactions of the Royal Society of London [31]. Conventional fluorescence and confocal fluorescence microscopy are both based on the basic theory of fluorescence.

Maria Goeppert-Mayer announced the theory of two-photon emission in her PhD dissertation in 1930; the theory predicted the ability of a molecule to reach an excited state by simultaneous absorption of two photons with sufficient combined energy (Fig. 4.1). This remained theoretical until the invention of lasers in 1960. In the 1990s, W. Denk, J. Strickler, and W. Webb constructed the first two-photon microscope and demonstrated 2PEF [32]. 2PEF requires ultrashort pulsed near-infrared or infrared laser sources to excite a molecule. The 2PEF signal intensity is proportional to the square of the excitation laser power, as two photons are now used for excitation. The probability of two-photon absorption is the probability of a molecule absorbing the first photon multiplied by the probability of absorbing a second photon, which results in an extremely small probability compared to single-photon excitation. To compensate for the low probability of absorption, high excitation power density (MW/cm^2 to GW/cm^2) [33–37] is used in two-photon excitation. This high excitation power density allows localized excitation only at the focal spot where high peak powers drive two-photon excitation. To prevent damage to the specimen from high powers, mode-locked femtosecond lasers are typically employed to achieve low average power. As an example, a typical laser for 2PEF microscopy produces 100 fs pulses every 10 ns (a repetition rate of 100 MHz); the duty cycle is: 100 fs/10 ns = 0.001%, leading to a peak power of 100 kW, while the average power is only 1 W. Overall, 2PEF microscopy does not suffer from off-focus emission and provides higher penetration depth and lower photo damage compared to single-photon fluorescence. The penetration depth is further improved by the development of three-photon excitation fluorescence (3PEF) [38]. As an emerging imaging technique, 3PEF is currently used in neuroscience and oncology and has not yet been established for dermal drug studies yet.

As mentioned before, when fluorophores absorb light energy, the molecule is excited to a higher energy state, before decaying to the lowest vibrational energy level of its excited state and then relaxing back to the ground state. FLIM takes advantage of this mechanism for contrast by measuring the fluorescence lifetime (τ), the average time the molecule spends in an excited state before it relaxes back to its ground state. Fluorescence lifetime can be characterized by the decay rate constants k_r (radiative) and k_{nr} (non-radiative) with equation:

$$\tau = \frac{1}{kr + knr}.$$

The fluorescence lifetime is also defined as the time where the fluorescence intensity of an excited fluorophore population has decreased to 1/e ($\sim$37%) from its initial value, which is mathematically equivalent to the average time excited [39].

Instrumentation In conventional fluorescence microscopy, a parallel beam passes through an excitation filter to illuminate the specimen with a specific excitation wavelength. The fluorescence signal is collected in the backward (epi) direction; dichroic mirrors and optical filters are used to separate the excitation and fluorescence light. Conventional fluorescence imaging microscopy is best for imaging thin specimens because it lacks the capability to block extra fluorescence signals from out of focus planes.

Unlike conventional fluorescence microscopy, confocal fluorescence microscopy is designed to block the fluorescence signal from out-of-focus planes, which eliminates the background signal observed with conventional fluorescence imaging. In a confocal fluorescence microscope, the sample is scanned with a laser beam, which is focused on the sample using an objective lens, enabling imaging at specific X, Y, and Z coordinates within the sample. This prevents the entire sample from being illuminated at once and confines fluorescence signal generation to the focal region(s), reducing much of the photobleaching, phototoxicity, and photodamage observed with conventional fluorescence, which are important considerations for skin imaging. Spatial filters or confocal pinholes are used to allow only the fluorescence light emitted from the focal spot to pass through, whereas the signal from off-focal planes is blocked. The focal plane can be adjusted by changing the distance between the sample and the objective lens [40, 41].

While the fluorescence signal is spatially filtered in confocal fluorescence imaging, the signal may still travel through the pinhole if generated outside of the plane of interest or scattered outside the pinhole, resulting in a reduced signal of interest and an increased background signal. Confocal fluorescence also suffers from reduced performance in turbid samples, especially deep within specimens, as light scattering reduces the amount of light transmitted through pinholes. 2PEF has significant advantages for deep tissue imaging: The longer near-infrared wavelengths used to excite fluorophores experience considerably reduced light scattering and absorption and can thus penetrate deeper in tissue [42]. Moreover, as fluorescence excitation is only generated at the focus, there is no need for pinholes or spatial filters. Most 2PEF microscopes use ultrashort pulsed near-infrared or infrared lasers as sources. The most common 2PEF laser source has traditionally been the Titanium:Sapphire laser, though other sources such as optical parametric oscillators, Nd:YLF, Yb:KGW, Er-doped fiber Cr-, and Kr-ion lasers have been employed for 2PEF imaging [43–45].

FLIM is often carried out using two-photon excitation to take advantage of both mechanisms. As FLIM generates contrast images with lifetime information of fluorophores, it can be used to differentiate fluorophores with similar emission spectra. FLIM modalities have been incorporated into many commercial imaging systems [46]. Time-correlated single-photon counting (TCSPC) is a common technique to measure fluorescence decays. TCSPC uses fast electronics and/or fast detectors to measure the arrival time of individual photons. An ultrashort pulse of light is used to excite the fluorophores, and the time required for the emitted fluorescence photon to arrive at the detector is measured. This process is repeated to form a histogram with photon count vs. time as all the photons with similar arrival time are stacked together. The histogram will show a decay curve that corresponds to fluorescence decay. Photon collection is highly affected by the dead time of the detector and electronics, which is a period of time that no photon collection is registered after the initial trigger or a reset time for the detector after one measurement. If the photon emission rate is too high, the detector may miss the photons during its dead time, and the decay may be calculated incorrectly [47, 48]. Although TCSPC is a well-established technique in FLIM, analog mean delay (AMD) is another recently developed technique that does not suffer from dead time effects and offers considerably greater imaging speed [49, 50].

4.1.3.2 Infrared Imaging

Infrared (IR) spectroscopy is based on the detection of IR absorption when IR light is irradiated onto the specimen. It is a nondestructive method whose contrast arises from the intrinsic vibrational mode of molecules in a sample, so labeling using exogenous dyes is unnecessary. This advantage makes IR imaging useful for dermal drug study without the challenges associated with labeling.

Theory IR spectroscopy is a technique based on the interaction between IR radiation and the vibrational modes of molecules. When molecules are irradiated with infrared light, the molecular vibrations with chemical moieties can undergo vibrational absorption. Different chemical bonds or functional groups have different absorption frequencies associated with "positions" in the infrared spectrum [51]. An energy diagram (Fig. 4.2) can also be used to describe the transition of IR signal.

Early-stage dispersive IR microscopy used diffraction gratings in monochromators to select individual wavelengths of light to irradiate the specimen. Dispersive IR microscopy became less popular after the invention of Fourier transform IR (FTIR) microscopy, which is based on the Michelson interferometer [52]. Compared to dispersive IR, FTIR can provide faster imaging, better spectral quality, and better signal-to-noise ratio (SNR).

Instrumentation To generate an FTIR spectrum, an infrared beam is split into two equal beams by a beam splitter. One beam is reflected by a fixed mirror, and the other beam is reflected by a movable mirror. The two beams of light reflect off their respective mirrors and recombine to create interference waves, which are directed onto the specimen. When the interfering beam passes through the specimen, radiation of a specific frequency is absorbed by the specimen due to the excited vibrations of the molecular bonds. The transmitted signal is then recorded by a detector, digitized by an analog-to-digital converter, and stored in a computer. An FTIR interferogram is generated by plotting the intensity of the resulting beam versus the movable mirror position or optical path difference (OPD). Adjusting the distance of the movable mirror causes a phase shift of the reflected beam and leads to OPD changes to the resulting interference beam. As the collected interferogram is a time-domain spectrum, it is difficult to interpret and analyze. Transforming the spectrum into frequency domain with Fourier transform provided more readily interpreted data that can used for qualitative and quantitative analysis [53].

The two most commonly used detectors for recording IR signals are the deuterated triglycine sulfate (DTGS) detector and the mercury cadmium telluride (MCT) detector. DTGS detectors detect the change of IR intensity as a change of capacitance in voltage. When IR radiation strikes the DTGS detector, the temperature at the detector changes, triggering a change in the dielectric constant, which alters the capacitance in the DTGS. In contrast, the MCT detector is a semiconductor device that induces a current when IR light strikes it, causing the electrons to move from the valence to the conduction band. The induced electrical current is proportional to the IR intensity. DTGS detectors are cost-effective but have relatively slow response and low sensitivity. MCT detectors provide more sensitivity, faster response time, and higher SNR compared to DTGS [54, 55].

Another important aspect of IR instrumentation is selecting suitable objective lenses. Modern objectives combine numerous internal glass lenses to generate high-focus and magnification images. Using modern objectives in an IR instrument should be avoided, as glass will absorb IR radiation. Specially designed objectives made of materials that are transparent within the IR range, such as Germanium, Silicon, Zinc selenide, Zinc sulfide, and Gallium arsenide, are needed to preserve the IR signal [56, 57]. Alternatively, all reflective objective designs can be used.

In skin imaging, IR signals can be acquired by transmission, reflection, or attenuated total reflection (ATR). Transmission FTIR usually requires preparing a thin-tissue cross-section and depositing it on infrared transparent windows. In reflection mode, a good reflective IR signal requires a highly reflective sample. Skin samples are usually not highly reflective, and polishing the skin sample will damage the skin barrier. In contrast, ATR-FTIR spectroscopy is a versatile technique that requires minimal or no sample preparation. The imaging depth of ATR-FTIR is usually $< 20\ \mu m$ [58]. ATR-FTIR leverages the concept of total internal reflection to perform reflection measurements on the specimen. The specimen is placed in contact with a high refractive index crystal, such as diamond, germanium, or zinc selenide. A beam of infrared light aimed at the crystal at an angle greater than the critical angle to achieve total internal reflection. The IR light travels inside the crystal and interacts with the specimen at the reflection point. Evanescent waves form on the reflection point, and a small amount of IR light is absorbed by the specimen, causing a slight attenuated total reflection. The IR spectrum of the specimen is obtained by measuring the reflected light intensity and its wavelength. The penetration depth of the evanescent wave in the specimen depends on the wavelength of light, while the intensity of the IR spectrum is fundamentally dependent on the number of reflections occurring in the specimen, the penetration depth of the evanescent wave, the concentration or amount of the molecule of interest, and the respective absorptivity of the specimen [59].

Another IR technique is atomic force microscopy (AFM)-based IR spectroscopy. AFM provides high-resolution surface sensing on the nanometer scale, which is more than 1000 times better than the optical diffraction limit. AFM is operated by measuring the bending of a cantilever with a very sharp tip at one end. A close-range attractive force acts between the cantilever tip and a sensing surface, causing the tip to stick to the surface. Nonuniform surfaces cause the cantilever to bend, and these small changes are detected by reflecting an incident laser beam off the flat top of the cantilever. As cantilever bending results in slight changes in the reflecting beam, a position-sensitive photodiode (PSPD) can be used to track these changes [60].

IR spectroscopy can collect an image at one vibrational band, or it can be operated to collect a full spectrum at each pixel. Spectra data are often preprocessed to remove noise and background and implement normalization, and different preprocessing methods can be accessed in a previous review [61]. In addition, data analysis methods can be used for analyzing the spectra to distinguish signals from the skin, APIs, and excipients in a topical product. Specifically, to classify different molecules, discriminative analysis methods can be used: principle component analysis (PCA), hierarchical cluster analysis (HCA), and linear discriminant analysis (LDA). In addition, there are more complicated methods: support vector machines, genetic algorithms, neural networks, decision trees, and generalized linear models. To quantify molecules, multivariate regression analysis methods can be used: partial least squares (PLS), direct classical least squares (DCLS), principal component regression (PCR), and multiple linear regression (MLR). Details about these multivariate analysis methods can be accessed in several reviews [62–64].

4.1.3.3 Spontaneous Raman Imaging

Since the first experimental observation of the Raman phenomenon in 1928, the Raman technique has undergone rapid development. More than 25 different types of Raman techniques have been developed for various applications [65]. Among these Raman techniques, spontaneous Raman and coherent Raman techniques are the two main techniques that have been used in studying dermal drugs. This section will focus on spontaneous Raman. Spontaneous Raman is noninvasive, label-free, chemically selective, and hyperspectral. Spontaneous Raman imaging is capable of studying chemical structures, molecular compositions, orientations, and concentrations. For dermal drug studies, spontaneous Raman imaging has been used in characterizing skin compositions, topical product compositions, and the BA and BE of topical products.

Theory Spontaneous Raman spectroscopy is a vibrational spectroscopic technique that measures inelastic light scattering after light interacts with a molecule. Inelastic scattering occurs when incident light is scattered at a different wavelength or frequency. When the scattered light has a longer wavelength than the incident light, it is called Stokes scattering. When the scattered light has a shorter wavelength than the incident light, it is called anti-Stokes scattering. The Stokes Raman signal is stronger than the anti-Stokes signal and is more often used in spontaneous Raman instruments [65].

Based on the classical theory, the Raman signal is related to the modulation of polarizability. The charge distribution of a molecule starts at an equilibrium state. The incident laser light, which is an oscillating electromagnetic wave with a frequency of ω_p, perturbs the molecule's charge distribution. The molecule becomes polarized, meaning the molecular charge distribution is changed from the equilibrium state. A molecule's polarizability, which refers to the ability to be polarized by external perturbation, can change during certain vibration modes [65]. A vibrational mode is considered Raman active when the polarizability can change during vibration. The periodical polarizability change oscillates at the same frequency as the vibrating frequency (ω_{vib}). This process leads to the scattering

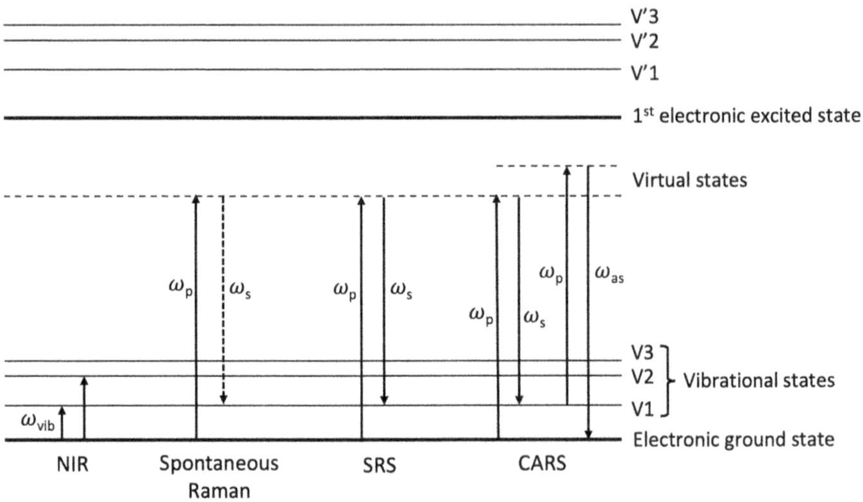

Fig. 4.2 Energy diagram of near-infrared (NIR), spontaneous Stokes Raman scattering, stimulated Raman scattering (SRS), and coherent anti-Stokes Raman scattering (CARS) interactions

of photons with different frequencies: ω_p(elastic scattering), $\omega_p - \omega_{vib}$(Stokes scattering), and $\omega_p + \omega_{vib}$(anti-Stokes scattering).

An energy diagram (Fig. 4.2) can be used to describe the mechanism of Stokes Raman scattering. For Stokes Raman scattering, the incident photon interacts with a molecule at the ground electronic state to reach a virtual energy state; a photon with lower energy is emitted when the molecule relaxes from the virtual state to an excited vibrational state [66]. Different molecular vibrational modes have different vibrational energies.

The difference in energy between the emitted and incident light, expressed in the unit known as the wavenumber, is used as the x-axis in a Raman spectrum, and this is called the Raman shift. The characteristic Raman spectrum of a molecule is determined by its vibrational modes, making it widely used in characterizing chemical compositions and molecular structures. In addition to its chemical specificity, the Raman signal does not photobleach. Moreover, the Raman spectrum has narrow peaks, which are usually 10–100 times narrower than the bands in the fluorescence spectrum [67–69]. The non-photobleachable signal enables long-term tracking of molecules, which is useful in studying the absorption of dermal drugs through time. The narrow vibrational bands help to better resolve multiple molecules, which enables the simultaneous study of APIs and/or excipients. The Raman scattering intensity I_{Raman} is proportional to the number of molecules of interest [70]. This linear relation between the intensity and the molecules' abundance enables spontaneous Raman scattering to quantify target molecules. However, it should be noted that the spontaneous Raman scattering effect is very weak, emitting only 10^{-6}–10^{-9} of the intensity of incident light [69].

Instrumentation Similar to IR spectroscopy, dispersive Raman spectrometers and Fourier transform Raman spectrometers are the two types of Raman spectrometers typically used for collecting Raman spectra [66]. The dispersive Raman spectrometer uses diffraction gratings to separate different wavelengths from signal light. In contrast, the Fourier transform Raman spectrometer uses an interferometer and creates an interference pattern that can be Fourier-transformed to recover the spectrum. Dispersive Raman spectrometer is more widely used than Fourier transform Raman spectrometer because of its better performance in biological samples [71]. A general dispersive Raman spectrometer comprises three primary components: laser source, sample light delivery and collection, and dispersive element with detector [72].

Different types and wavelengths of laser can be used for exciting Raman signals. Conventional solid-state and gas lasers have been used in Raman systems, such as the continuous mode Titanium:sapphire (Ti:Sapph) laser and helium–neon (He-Ne) laser [66, 72]. These conventional lasers have high output powers and can enable near-diffraction-limit imaging; however, they are bulky and require complicated electronic and cooling components. In contrast, diode lasers are smaller, less expensive, and more efficient [71]. As a trade-off, diode lasers require highly stabilized thermal and current control, and they often have more output power fluctuations and generate elliptical beams with astigmatism [72]. Because the Raman signal is measured from skin tissue when studying dermal drugs, near-infrared and red laser sources are more often used. These longer wavenumber lasers will yield spectra with reduced autofluorescent background from skin than shorter wavelength lasers [72], though this comes at the cost of reduced Raman scattering efficiency as the Raman signal is proportional to $1/\lambda^4$. Most dermal studies use Raman system with 785 nm lasers [73–82]. As short-wavelength lasers generate stronger Raman signals than long wavelengths, many studies still use 532 nm lasers [83–88].

Light delivery and collection can be achieved using free-space optical components or optical fibers. Fiber-based Raman systems usually have smaller footprints, are more flexible, are better for accessing anatomical sites, and are more suitable for clinical Raman systems [72]. It is worth mentioning that Raman systems that use only fibers for light collection are often not sensitive to spatial information (i.e., lateral position and axial depth). Because of this, they are not ideal for high-spatial-resolution mapping of drug distribution. Instead of only using fiber, a system combining fiber and free-space components can retain some spatial information while having a smaller footprint. For example, a recently developed Raman system with a handheld probe used fibers together with free-space components [89]. The objective can move in the axial direction using a miniaturized linear motor, and the system can collect Raman spectra at different skin depths in vivo. However, there is no motor for controlling the lateral movement of the probe. Additional mechanical arms or motors are needed in this system for high-precision lateral imaging.

The scattered light collected from the sample comprises both elastic and inelastic scattering. Notch and edge filters can be used to eliminate the strong elastic scattering (also known as Rayleigh scattering) [71]. Notch filters can specifically filter out light with the same wavelength as the laser, allowing both anti-Stokes and Stokes scattering light to pass. However, notch filters are usually more expensive than edge filters. For a Raman spectrometer that only measures Stokes scattering, a less expensive long-pass filter is more often used to allow Stokes scattering light to pass.

Diffraction gratings are typically used as dispersive elements to separate wavelength components from the input light. There are different types of diffraction gratings, including reflective gratings and transmission gratings. Reflective gratings can cover a wide range of spectrum with high spectral resolution; however, they have low optical throughput and require longer integration times. Transmission gratings have a high throughput and acquire a wide range spectrum in a short time period [72]. An appropriate grating is often determined by the required spectral resolution, spectral range, diffraction efficiency, excitation wavelength, and the wavelength of the maximum diffraction efficiency [90]. The specific grating used in a Raman instrument depends highly on the specific instrument design and the optical components in the instrument.

After passing through the diffraction grating, the incident signal light is dispersed into separated wavelength constituents. A detector then records the spatially separated constituents. Silicon charge-coupled device (CCD) detectors and complementary metal oxide semiconductor (CMOS) detectors can be used. The CCD detector makes use of chips with arrays of pixels that convert photons into electrons [72]. The horizontal axis of the chip is used to measure photons from different wavelengths/wavenumbers. The signals from pixels at the vertical axis are accumulated for the specific wavelength/wavenumber. The electrons from all pixels transport across the chip and are finally read at the corner of the array. Although this mechanism helps decrease signal noise and increase light sensitivity, it makes signal acquisition slow. The electronic circuitry outside the pixel array consumes large amounts of power and generates heat [91], and as such, a thermoelectric (TE) multistage Peltier system is usually used to cool the camera to decrease dark noise. CCDs have high photon sensitivity, high quantum efficiencies for visible and near-infrared light, are compact, and require low maintenance [71]. In contrast, a CMOS detector comprises an array of photo sensors. Each sensor has circuitry around that can immediately convert photons into electrons and amplify the signal. Because of this, the CMOS detector has much faster acquisition speeds. In addition, CMOS detectors consume less energy and have lower costs than CCD detectors. However, CMOS detectors have several limitations, including higher noise, less uniform pixel response, lower quantum efficiencies, larger pixel size, and smaller dynamic range [91]. In current dermal drug studies, the CCD detector is more frequently used in Raman instruments to achieve high sensitivity, high resolution, and consistent performance.

The dispersive Raman spectrometer can be incorporated with different designs for measuring Raman signals from the skin. Different types of Raman instruments are available, such as fiber-optic Raman spectrometers, spatially offset Raman spectrometers (SORS), and confocal Raman microscopes (CRM) [72]. Most of these Raman instruments have only been used to study the skin's chemical composition and morphological features. Descriptions of these different instruments and their applications are well summarized in several reviews [72, 92, 93]. Among these various instruments, the CRM is most often used to study dermal drug absorption and evaluate the BA and BE of dermal drugs [94].

A CRM combines a dispersive Raman spectrometer with a confocal microscope. It usually comprises a laser, dichroic beam splitter, lens, pinholes, beam expander, filters, and a dispersive spectrometer with a CCD detector [65]. The monochromatic laser first passes through beam expanders and is then reflected by a dichroic beam splitter and focused by an objective lens on a sample. The backscattered light from the sample then passes the lens and propagates through the dichroic beam splitter. Pinholes are used in the optical path to only collect light from the focal plane. Light from the sample that is from out-of-focus planes is spatially filtered out. The light that passes through the pinhole is then reflected by a mirror and passes a notch/edge filter to remove light from elastic scattering. The filtered beam is then dispersed into wavelength/wavenumber constituents, measured by the CCD detector, and recorded as a spectrum. Notably, some Raman instruments use an optical fiber, which acts as a confocal pinhole, to collect the scattered light from the objective and reject signal from out-of-focus planes [95].

CRMs use point-scanning systems, which collect a full spectrum at each pixel and scan all pixels within a field of view. The pixel scanning can be achieved via raster scanning the laser using mirrors or moving the sample under the fixed objective using motorized stages [71, 96]. The focal volume determines the part of the sample excited and measured at each pixel. In addition to scanning in the lateral direction, scanning in the axial direction helps to measure the signal at a specific sample depth. The axial movement can be achieved by moving the objective or the sample using a high-precision motorized translation stage. The narrow focal plane of the CRM enables the excitation and collection of signals from a very thin layer of the skin. The depth profiles of the studied molecule can be depicted by axial scanning from the uppermost layer to a deeper layer. Because of the low efficiency of the spontaneous Raman scattering effect, signal acquisition in each pixel usually takes tens of milliseconds to seconds, and thus, a hundreds-pixel by hundreds-pixel image measurement takes hours. This makes CRM a slow method for high-resolution imaging. This should be carefully considered during a study design to have a good trade-off between image spatial resolution and temporal resolution.

Spontaneous Raman imaging is a hyperspectral method that measures the spectra at each pixel of a sample. In addition to using only one peak, multiple peaks or the full spectrum can be used for analyzing the target molecule(s). Similar to IR spectra, multivariate analysis approaches based on multiple variables/features can be used in Raman spectra analysis. PCA, HCA, LDA, support vector machines, genetic algorithms, neural networks, decision trees, and generalized linear models can be used to classify different molecules. PLS and PCR can be used to quantify molecules. Details about these multivariate analysis methods can be accessed in several reviews [62, 64, 70, 72, 97].

4.1.3.4 Stimulated Raman Imaging

While CRM enables highly specific chemical analysis of molecular vibration modes, the spontaneous Raman scattering effect is too weak to allow image acquisition speeds sufficient for real-time monitoring of dynamic events such as dermal drug absorption. The invention of pulsed lasers with high peak power facilitated the discovery of coherent Raman scattering (CRS) phenomena to enhance sensitivity and offer much faster image acquisition capabilities [98, 99]. CRS imaging is typically implemented as coherent anti-Stokes Raman scattering (CARS) or stimulated Raman scattering (SRS) microscopy, where both techniques utilize synchronized pulsed lasers to coherently excite vibrational modes within a sample. When the energy difference between the two pulsed lasers matches the vibrational modes' energy of the sample molecule, CARS and SRS can be detected within the sample. CARS is detected via the emission of photons at the anti-Stokes frequency, while SRS detects the energy transfer from the higher frequency beam (Pump) to the lower frequency beam (Stokes) either through stimulated Raman loss (SRL) or stimulated Raman gain (SRG). CARS is technically easier to implement and has seen widespread use in imaging cells and tissues [100, 101]. However, in contrast to SRS, CARS is a parametric process and thus suffers from a significant nonresonant background and spectral distortion [102]. Additionally, SRS signal intensity scales linearly with the concentration of Raman scatterers, which enables quantitative analysis [103]. As such, SRS is preferred over CARS in a variety of applications, including PK studies of dermal drugs [13, 104–107].

Theory As mentioned previously, spontaneous Raman Stokes scattering is very weak; however, in SRS, highly efficient scattering occurs such that 10% or more of the energy of the incident pump beam is converted into the Stokes frequency and produces a signal that is several orders of magnitude greater than spontaneous Raman scattering [108]. The physical phenomena can be understood intuitively through an energy level diagram (Fig. 4.2). In SRS, two beams, pump (ω_p) and Stokes (ω_s), incident on the sample generate stimulated emission when the energy difference $\Delta\omega = \omega_p - \omega_s$ matches a molecular vibration in the sample. Stimulated emission can be thought of as the excitation of a molecule from the ground state to a virtual state by the pump photon with the emission of an additional Stokes photon as the molecule spontaneously falls to a vibrational state. The incident Stokes beam ensures a higher initial number of photons in ω_s, which enhances the stimulated emission process and increases the rate of ω_s photon emission [109]. The SRS interaction can be measured through the loss of photons in ω_p (stimulated Raman loss, SRL) or the gain of photons in ω_s (stimulated Raman gain, SRG). For a more rigorous description of the theory and principles of CRS, refer to [110, 111]. A clear and practical difference between spontaneous Raman and CRS is that while spontaneous Raman uses one laser to probe all vibrational bands, CRS typically uses two lasers, whose frequencies are tuned to match a specific vibrational mode to achieve much more efficient signal generation.

For the application of SRS microscopy to dermal drug delivery, it is important to understand how SRS can be used to quantify relative or absolute drug concentration. One of the distinct advantages of SRS is that signal strength is directly proportional to target molecule concentration:

$$S_{SRS} \propto C * V_{focal} * \sigma_{molecule} * I_{pump} * I_{stokes},$$

where C is the concentration of the target molecule, V_{focal} is the focal volume, $\sigma_{molecule}$ is the differential Raman cross section of the molecule, and I_{pump} and I_{Stokes} are the intensities of the pump and Stokes beams, respectively [103]. In theory, a linear calibration curve can be obtained by varying C and keeping all other factors constant. In practice, however, factors such as light scattering, non-Raman background, and cross talk must be considered. Light scattering by the sample reduces the amount of light available to generate SRS signal at the focus, and this effect cannot be directly measured, so it must be addressed by obtaining calibration on similar samples or by using a known internal standard for correction. Non-Raman background can occur due to two-photon absorption, thermal lensing, the resonance-induced Kerr effect, and cross-phase modulation [112], all of which are not chemically specific and reduce SRS contrast. The impacts of these effects vary by sample and application and make absolute quantification difficult. As a result, relative concentration measurements account for most SRS quantitative studies, including those investigating dermal drug delivery. In the presence of low levels of non-Raman background, such as in tissue samples with low levels of melanin, hyperspectral SRS in combination with computational unmixing can account for background, and this has been implemented for PK analysis in skin tissue [13, 105, 107, 113].

Instrumentation Initial SRS microscopy was reported using a femtosecond amplified laser, split into a narrowband femtosecond pulse and a broadband white light pulse for excitation, and multiplex detection with a photodiode array [114]. However, the amplified laser system is unsuitable for imaging biological specimens due to high peak powers causing sample damage and low repetition rates limiting image acquisition speed. The first practical SRS microscope for biomedical imaging was developed using a high repetition rate (76 MHz) laser with four orders of magnitude lower peak power, modulation transfer, and lock-in detection [115]. In a modulation transfer scheme, either one of the Stokes or pump beams is modulated such that the generated SRS signal (SRG at the Stokes wavelength or SRL on the pump wavelength, respectively) is also modulated at the chosen frequency. Lock-in amplification at the frequency of Stokes (or pump) modulation is critical to detect the small SRL (or SRG) signals over laser noise, enabling signal detection at four orders of magnitude more sensitive than the previous multiplex detection scheme.

Most current SRS microscopy systems employ dual-color picosecond or femtosecond laser sources in the near-infrared region, a laser scanning microscope, and a photodetector with lock-in amplification. The laser source can be a free-space optical parametric oscillator (OPO) system, which offers automatic pulse synchronization and high spectral resolution at the expense of high cost and necessary cooling systems, or an all-fiber laser, which can be less expensive and easier to maintain but often provides less power and has higher laser noise. Single-frequency SRS microscopes are often operated with two picosecond sources or two spectrally focused femtosecond beams. Spectral or hyperspectral SRS can be implemented with two tunable picosecond lasers [107, 116] or with two femtosecond sources with linear chirping or pulse shaping [117–121], along with other techniques [122, 123].

Either the pump or Stokes beam is modulated at megahertz frequency using an acousto-optic modulator (AOM) or electro-optical modulator (EOM). Pump and Stokes beams are then combined, fed into a home-built or commercial laser scanning microscope, and focused onto the sample, typically using an objective with a high numerical aperture (NA). The pump and Stokes beams must be spatially aligned (propagating in the same direction and overlapped) and temporally aligned (path-length matched). In the transmission-detection scheme, the SRS signal propagated in the forward direction is collected through a condenser, optically filtered to remove the modulated excitation beam, and detected by a photodiode. In an epi-detection scheme, backscattered SRS signal is collected through the focusing objective. A polarization beam splitter placed just before the objective is typically used to pass the linearly polarized excitation light and reflect the polarization-scrambled backscattered light to the photodiode [115, 118, 124–126]. In either case, the electrical signal from the photodiode is sent to a lock-in amplifier that extracts the modulated signal from the noisy raw signal. The extracted electrical signal, corresponding to the SRS signal, is then read and output for image formation.

As mentioned previously, SRS typically probes one vibrational mode at a time. For dermal drug studies, it is often more desired to probe multiple vibrational modes to detect multiple APIs or API and excipients and account for skin background signals. Imaging by probing multiple vibrational modes is called hyperspectral imaging and has been executed in several ways. Sequential frequency tuning is the simplest implementation of hyperspectral SRS and is achieved by obtaining a series of SRS images while changing the frequency of either the pump or the Stokes beam [105, 115, 116, 127]. The speed of this method is directly limited by the speed required to tune the beam and the number of frequencies measured. Another method is spectral focusing [119, 128, 129], in which time delay is tuned between two chirped broadband pulses. Yet another implementation is through multiplexing wavelengths [117, 130, 131], where multiple SRS wavelengths are simultaneously excited by narrowband Stokes pulses and broadband pump pulses and spectral decomposition is used to generate individual chemical maps. Spectral decomposition is performed by algorithms such as multivariate curve resolution (MCR), alternating least squares (ALS), nonnegative least squares (NNLS), and vertex component analysis (VCA), among others [116, 119, 132]. Other variations of hyperspectral SRS include the use of more than two colors by time gating, adding single-color fields, or with multicolor broadband sources [120–123, 133].

4.2 Dermal Drug Evaluation Using Optical Techniques

As described in Sect. 4.1.3, these optical microscopic techniques probe different information from the chemical molecule of interest and/or utilize different instrumental setups. Differences between techniques and instruments largely determine the advantages/limitations of each technique in dermal drug studies. In this section, examples of using each technique for evaluating dermal drugs are discussed. Findings from these examples, covering the features and limitations of each technique, specifically in skin imaging and dermal drug PK study, are summarized in Table 4.1.

4.2.1 Fluorescence Imaging

Several fluorescence imaging modalities are available for assessing dermal drugs, including conventional fluorescence, confocal fluorescence, and more advanced methods, such as 2PEF and FLIM. Among these modalities, conventional fluorescence is frequently used for ex vivo skin imaging, while the rest can be used to perform in vivo imaging.

Autofluorescence A major challenge of fluorescence imaging is the presence of autofluorescence. Skin naturally contains many fluorescent molecules, such as collagen and nicotinamide adenine dinucleotide (phosphate) (NADH and NADPH) [42]. A series of studies performed by Masters et al. using multiphoton excitation microscopy in in vivo human skin from skin surface to a depth of 200 μm suggested that NADPH is the main source of skin autofluorescence at 730 nm, with additional contributions from flavoproteins and other chemical species at other wavelengths [134–136]. Changes in fluorescence intensity due to these species may depend on cellular and tissue metabolic oxidative processes [135]. While the presence of these molecules can provide important information about the skin composition and state, they often have overlapping excitation/emission profiles with fluorescent molecules of interest, which may confound fluorescence imaging of drugs in skin [42]. Autofluorescence levels also vary within skin, as certain structures, such as the pilosebaceous unit, have varying or high levels of autofluorescence, and are distributed nonuniformly throughout the skin [42, 137]. There is also significant variation in autofluorescence levels between individuals [42, 138], as autofluorescence may be impacted by age, diet, health conditions, etc. Hermsmeier et al. noted large variations in endogenous fluorescence in a group of more than 35 individuals, calculating at least a six-fold difference in autofluorescence levels from facial skin samples [138]. The authors used an integrating sphere, which converts captured photons into a quantifiable photocurrent, to quantify autofluorescence levels between skin samples. A large variation in endogenous fluorescence between individuals was observed in this study [138]. Therefore, sample or donor variance should be taken into consideration in dermal drug study designs, and screening

Table 4.1 Summary of imaging techniques and their application in dermal drug studies

Imaging modality	Instrument type	Spatial resolution (lateral)	In vivo imaging	Imaging speed	In vivo imaging skin layers	BA study available?	BE study available?	Active pharmaceutical ingredients and excipients measured in BA and BE studies
Fluorescence	Conventional fluorescence imaging	<250 nm	N (usually)	Fast		Y	N	Tazarotene, minocycline, doxorubicin, fluorescein-isothiocyanate-labeled dextran
	Confocal fluorescence imaging	200–400 nm	Y	Fast	SC, VE (viable epidermis)	Y	N	Lipid nanocapsules containing curcumin, sebamed lotion containing sodium fluorescein, nile red labeled nanoemulsions with asiaticoside loading
	Two-photon fluorescence imaging (2PEF)	400 nm	Y	Fast	SC, VE (viable epidermis), superficial dermis	Y	N	Minocycline, silver nanoparticles; CD-P3OH nanocarrier, RBITC loaded MIL-100, SLS and oleic acid visualized using fluorophores (SRB and RBHE)
	Fluorescence lifetime imaging microscopy (FLIM)	300 nm	Y	Fast	SC, VE (viable epidermis), superficial dermis	Y	N	Anti-inflammatory (GSK2894512), silver nanoparticles, zinc pyrithione (ZnPT), minocycline, tazarotene
Infrared	Reflected IR	>1 μm	Y	Slow	SC	Y	N	Econazole nitrate, estradiol, urea;
	Transmission FTIR	>1 μm	N	Slow		Y	N	Tacrolimus; deuterated oleic acid, deuterated DMSO
	AFM-FTIR	<200 nm	N	Slow		Y	N	Ibuprofen; Janus nanoparticle (JNP)

	Synchrotron FTIR	<1 μm	N	Slow		Y	N	–
	ATR-FTIR	>1 μm	Y	Slow	SC	Y	N	Ibuprofen, 4-cyanophenol, diclofenac sodium, anthocyanin, terbinafine, genistein; PG, DPPG, IPM, ISIS, vesicles, DMSO, SDS, SLES, STZ, Lipoid® S75, jojoba oil, medium chain triglycerides, sunflower oil, liposomes
Spontaneous Raman	Confocal Raman microscopy (CRM)	≤300 nm	Y	Slow	SC, VE (viable epidermis)	Y	Y	Ibuprofen, trans-cinnamaldehyde, retinyl acetate, alpha-tocopheryl acetate, trans-retinol, lidocaine, hyaluronic acid (HA), flufenamic acid (FluA), quinidine, resorcinol, caffeine, 5-fluorouracil, acyclovir, 4-cyanophenol, procaine-HCl, dexamethasone (Dx), tocopheryl acetate; Petrolatum, jojoba oil, almond oil, soybean oil, avocado oil, olive oil, white mineral oil, paraffin oil, PG, ethanol, acetone, mycosporine-like amino acids (MAAs) gel, gadusol (Gad) gel, hydrophilic active glycerol, lipophilic octyl methoxycinnamate (OMC), caprylic/capric acid triglyceride (MYRITOL® 318)
Stimulated Raman	Transmission SRS	≤300 nm	N	Fast		Y	Y	Ibuprofen, ketoprofen, zinc pyrithione, climbazole, ruxolitinib, deuterated betamethasone dipropionate, ectoine, isotretinoin, retinoic acid, tazarotene, 4-cyanophenol, loxoprofen, lidocaine; DMSO, PEG 200, deuterated PG, deuterated oleic acid
	Backscattered SRS	≤300 nm	Y	Fast	SC	Y	N	Trans-retinol, retinoic acid, vitamin E; DMSO

samples or donors with low autofluorescence levels may be used to reduce this confounding effect.

Conventional Fluorescence Imaging While conventional fluorescence imaging is a simpler modality, requiring less expertise, training, and expensive equipment than more advanced fluorescence imaging tools, such as FLIM, it can still be instrumental in dermal drug uptake and PK studies.

Thin tissue slices are optimal for conventional fluorescence imaging, since any fluorescence signal generated within the sample may be detected. Thicker samples result in more background fluorescence, causing blurring [42]. This low signal-to-noise ratio typically limits the use of conventional fluorescence imaging to high dosage experiments. Despite this limitation, conventional fluorescence can still be instrumental in dermal drug uptake studies. Blagus et al. performed conventional fluorescence imaging in in vivo mouse skin, developing a novel in vivo fluorescence-based monitoring system using a fluorescence stereomicroscope to evaluate the topical and transdermal delivery of fluorescent molecules (fluorescein-isothiocyanate-labeled dextran and doxorubicin) following electroporation of mouse skin [16]. By imaging the region of drug application and a distant segment of the mouse tail, the authors were able to assess the relative quantities and distribution of the applied compounds delivered topically and transdermally, respectively. This monitoring system allowed them to determine that electroporation, especially with increased applied electric pulse amplitude, leads to increased penetration of the compounds through the skin [16].

Deep-learning-based approaches are highly useful in processing imaging data. It can be used to reduce the impact of autofluorescence on overwhelming API fluorescence signatures, thus enabling dermal drug experiments for not only very high "infinite" dosages but also daily clinical doses. Evans et al. utilized a conventional fluorescence microscope alongside a convolutional neural network image processing pipeline to study the uptake and distribution of two naturally fluorescent APIs, tazarotene and minocycline, in human ex vivo facial skin [139]. The tazarotene and minocycline emission images with conventional epi-fluorescence imaging were collected after topically applying doses of two APIs. Minocycline emission images after a UV light bleaching step were also collected to remove the tazarotene signal since tazarotene and minocycline have different bleaching rates. Based on all these images, human experts identified the region of two APIs and the skin's collagen and generated annotated images. Originally collected images and the annotated images were used to train the neural network to identify each API's signals. Results showed that the trained neural network was able to identify and measure regions of API uptake in the skin, even detecting the presence of the API administered at a clinically relevant dose. This work displayed that conventional fluorescence microscopy, together with computational image processing, could successfully be used to detect and quantify the uptake of two different APIs in ex vivo skin after topically applied clinically relevant doses. This presents a simple and accessible means of assessing dermal drug PK or formulation penetration [139].

Confocal Fluorescence Imaging Confocal fluorescence imaging enables three-dimensional imaging of a block of tissue, expanding the pool of compatible imaging samples from two-dimensional thin tissue slices for conventional fluorescence to thicker tissue samples, such as biopsies, and in vivo organisms. Confocal fluorescence imaging can image up to 50–70 μm in human skin, depending on fluorophore properties, skin barrier characteristics, and fluorophore concentration. Compared to conventional fluorescence, confocal fluorescence has also improved resolution [42]. In addition, confocal fluorescence imaging confines fluorescence signal generation to the focal plane, eliminating much of the autofluorescent background signal and reducing much of the photobleaching, phototoxicity, and photodamage observed with conventional fluorescence, which are important considerations for skin imaging.

Confocal fluorescence has been used in many investigations to study dermal drug uptake and PK. One such study by Nguyen et al. utilized fluorescence confocal spectral scanning microscopy to evaluate the potential of lipid nanocapsules as drug carriers in the uptake of hydrophobic molecules into the skin [17]. Lipid nanocapsules are thought to improve the topical delivery of certain APIs, especially for hydrophobic molecules, enabling them to be delivered through hydrophilic vehicles, which may promote patient comfort. The authors investigated the effectiveness of lipid nanocarriers in transporting curcumin, a fluorescent dye and antioxidant, in human ex vivo abdominal skin, noting that curcumin preferentially accumulated in the SC and viable epidermis [17].

A separate study by Sattler et al. highlights the real-time near-histopathological imaging resolution of skin in ex vivo and in vivo for fluorescence confocal laser scanning microscopy (CLSM) [18]. In this study, CLSM was used to track the absorption kinetics of a topically applied sebamed lotion (oil-in-water lotion) containing sodium fluorescein in human volunteers over 24 h. The authors could assess fluorescence levels by semiquantitative scoring of images and intensity measurements, showing that most participants displayed fluorescence signal even after 24 h [18]. The use of confocal fluorescence imaging enabled the authors to visualize the kinetics of the lotion in the skin of human volunteers, highlighting the potential of this method for both ex vivo and in vivo PK studies.

Another study by Choi et al. used a self-constructed imaging system capable of 2PEF and fluorescence CLSM imaging by sharing optical scanning components and objective lenses to investigate the transdermal delivery of two nanoemulsions in ex vivo porcine skin [19]. Nanoemulsions, or mixtures of two liquids where one has been emulsified into nanometer-scaled droplets within the other, have recently been explored for their potential to improve transdermal drug delivery. Nanoemulsions also show promise for enabling targeted drug delivery in the skin through the incorporation of specific ligands in the nanoemulsion. In this study, the authors used 2PEF to obtain skin structural information and fluorescence CLSM to visualize the uptake of two types of nanoemulsions containing the fluorescent dye Nile Red, one containing the ligand asiaticoside to target specific cells and one nontargeted emulsion. By imaging Nile Red distribution, the targeted emulsions were observed to penetrate deeper into the skin, suggesting that ligand binding to keratinocytes leads to improved penetration [19].

Two-Photon Fluorescence Imaging Two-photon excitation is highly localized within the imaging sample, only occurring at the focus of the excitation, which minimizes photobleaching and phototoxicity. This also helps reduce background signal and further constricts fluorescence generation to a single focal plane, providing greater imaging clarity than obtained with confocal fluorescence imaging. These features make 2PEF an excellent tool for imaging live cells and tissues in vivo. 2PEF may detect fluorescent compounds and skin structures up to approximately 200 μm [42, 140].

Several studies have been performed investigating the effectiveness of drug delivery vehicles in skin using 2PEF or other forms of multiphoton microscopy. One such study by Kirejev et al. utilized 2PEF to evaluate the use of nanocarriers to deliver topical compounds in human ex vivo skin [20]. The authors studied two nanocarrier drug delivery systems, CD-P3OH (a BETA-cyclodextrin 3-hydroxy-phenyl porphyrin conjugate) and MIL100 (iron (III) carboxylate metal-organic framework). MIL100 was loaded with a fluorophore, rhodamine B isothiocyanate (RBITC), for fluorescent imaging. By imaging fluorescence signal from the nanocarriers, CD-P3OH was shown to deliver to the upper skin layers, despite its large size, showing that this nanocarrier could be used to deliver large hydrophobic molecules by increasing their water solubility, whereas the ability of MIL100 to deliver to the skin remained inconclusive [20]. While 2PEF enabled imaging of the distribution of these nanocarriers in ex vivo skin, it only provided qualitative information. Using additional methods with increased sensitivity, such as TPFCS (two-photon-based fluorescence correlation spectroscopy), which combines 2PEF with fluorescence correlation spectroscopy to measure diffusion and the number of fluorophores in the focal volume [20, 22], can provide additional information about drug biodistribution, localization, and diffusion, which may enable a quantitative analysis of drug distribution dynamics [20].

2PEF has also been used to visualize and investigate transdermal transport pathways in the skin. Kushner et al. used dual-channel two-photon imaging to study transport pathways in the skin and their enhancement using low-frequency ultrasound or treatment with a permeation enhancer (sodium lauryl sulfate, SLS) by tracking the transdermal penetration of sulforhodamine B (SRB) and rhodamine B hexyl ester (RBHE), which are hydrophilic and hydrophobic fluorophores, respectively [21]. Dual-channel two-photon fluorescence imaging was performed in order to visualize SRB, RBHE, and skin autofluorescence to determine where compounds were located relative to skin structure, such as corneocytes. The authors confirmed that treatment of ex vivo human skin with ultrasound or ultrasound and SLS creates discrete regions in the skin that are especially permeable, termed localized transport regions (LTRs), and found that SRB and RBHE had significantly increased SC penetration compared to untreated skin and penetrated into the viable epidermis. Imaging both SRB and RBHE fluorescence and skin autofluorescence also suggested that the transcellular pathway was used by both compounds in ultrasound and ultrasound/SLS treated cases [21].

A series of studies [23–25] have been performed to use 2PEF to visualize and investigate changes in oleic acid-induced transport pathways and transport dynamics. Oleic acid is a permeation enhancer that influences topical compound transport through the skin barrier by altering lipid phase transitions in the lipid bilayer. In 2001, Yu et al. used two-photon scanning fluorescence microscopy to study changes in concentration gradients and vehicle-skin partition coefficients for SRB and RBHE following oleic acid treatment in ex vivo human skin [23]. Comparisons with mathematical models describing transdermal transport allowed the authors to quantify changes in the skin diffusion coefficient and barrier diffusion length due to oleic acid application. Increased permeability was observed for both fluorophores, likely attributable to increased vehicle-skin partition coefficients for both and increased diffusion coefficient in the skin for SRB [23]. In 2003, Yu et al. expanded upon this study by using a dual-channel high-speed 2PEF microscope to simultaneously measure fluorophores and skin autofluorescence signal, enabling them to study fluorophores' localization and transport through the skin [24]. They found that oleic acid caused increased transcellular diffusion through corneocytes for SRB and increased RBHE localization to the lipid-rich intercellular space [24]. Tseng et al. expanded upon this work by investigating transport dynamics and skin polarization changes resulting from oleic acid application using TPEF timelapse imaging [25]. The authors tracked SRB permeation through ex vivo mouse skin and assessed skin polarization changes using Laurdan, a fluorescent polarity-sensitive dye, and found that oleic acid treatment increased both skin polarity in both intra- and intercellular regions in the SC and SRB permeation, especially in the intercellular lipid regions, which represents one of the most common transdermal transport pathways [25]. These studies described how 2PEF can enable not only the investigation of the effect of permeation enhancers but also the quantification of changes occurring in transdermal transport mechanisms.

Fluorescence Lifetime Imaging Microscopy FLIM is a popular and highly useful tool in fluorescence imaging. This technique generates contrast by measuring fluorescence lifetimes of fluorophores, distinguishing fluorophores with similar emission spectra but different fluorescence lifetimes. FLIM can be used for imaging samples with multiple sources of fluorescence, and it is highly useful in distinguishing fluorophores of interest from endogenous fluorescence, such as skin autofluorescence [42]. FLIM is often used to image cellular and subcellular structures and may also be used to evaluate cell metabolic and drug activity, since fluorophore lifetimes can change depending on the environment and binding state [42].

Because of its ability to distinguish between multiple fluorophores and skin autofluorescence, FLIM is a valuable tool for studying dermal drug uptake and distribution in skin, and several dermal drug studies have been performed with FLIM. One such study performed by Alex et al. used FLIM to investigate the spatial distribution and residency time of a topically delivered anti-inflammatory designed to treat atopic dermatitis and psoriasis (GSK2894512) [26]. GSK2894512 has natural fluorescent properties and a long fluorescent lifetime, allowing it to be distinguished from skin autofluorescence. In this study, GSK2894512 was applied daily to the skin of human volunteers over seven days, and FLIM images were taken prior to, throughout, and following the drug application sequence. FLIM was used to image structures in the epidermis and dermis up to 200 μm deep, providing three-dimensional maps of API localization over time by fitting fluorescence signal decay with a standard multiparameter decay model. The authors observed significant variation in fluorescence levels between applications and a rapid decrease in signal over the first few days after GSK2894512 application ceased, which was found to be consistent with punch biopsies taken at the same time. Most GSK2894512 fluorescence was observed in the furrows of the skin and not in deeper layers, an observation unattainable with more traditional methods such as biopsy, although this may have been confounded by photobleaching or metabolic breakdown and localization may well have occurred in the furrows, SC, and hair follicles [26]. This supports using FLIM as a noninvasive tool to assess API distribution and retention, especially as FLIM may also be used to provide biochemical information about the skin target region and assess cellular and skin pharmacodynamic (PD) responses [26].

When combined with other fluorescence imaging modalities, such as 2PEF, FLIM can provide an additional layer of imaging contrast. Zhu et al. used two-photon tomography with FLIM (TPT-FLIM) to assess the penetration of silver nanoparticles through porcine skin [27]. Silver nanoparticles are frequently used in food, medical device, and cosmetic product sanitation and sterilization procedures, but may have toxic effects, so it is important to assess their topical uptake [27, 141, 142]. The authors used a DermaInspect [143–146] optical imaging device to perform TPT-FLIM, which combines the imaging depth achieved by 2PEF with the ability to distinguish between fluorophore species based on their fluorescence decay rates with FLIM. Fluorescence lifetime analysis enabled the distinction between silver nanoparticles in the skin and skin autofluorescence. Using TPT-FLIM, the maximum penetration depth of silver nanoparticles into porcine skin was 14 μm, with significant signal attenuation from the surface of the skin to 4 μm [27].

Mangion et al. used multiphoton excitation microscopy, in addition to FLIM, to study the delivery of zinc pyrithione (ZnPT), an antifungal agent, into hair follicles and to assess different ZnPT formulations in ex vivo human abdominal skin [28]. Dissolution of ZnPT was determined by using Zinpyr-1, a fluorescent probe used to detect free zinc. The authors contrasted massaging and not massaging the formulation into the skin and corresponding FLIM images showed an increase in penetration depth for massage conditions, although this was not significant. However, the formulation type and ZnPT particle size were shown to influence the penetration and delivery of ZnPT [28]. While the authors used both multiphoton excitation imaging and FLIM, they noted that FLIM would be an optimal imaging method due to the additional contrast afforded by distinguishing between fluorescence lifetime spectra.

Jeong et al. used 2PEF and FLIM with phasor analysis to visualize the distribution of a topical minocycline gel in ex vivo human facial skin [29]. Minocycline is a naturally fluorescent anti-acne drug, and it is important to evaluate the delivery and PK of minocycline in the pilosebaceous unit, which is the target site for acne treatment. The authors noted high minocycline uptake levels in the epidermal layers using 2PEF; however, this was sometimes confounded by skin autofluorescence, especially for the pilosebaceous unit, especially since minocycline has a much weaker fluorescence signal [29]. To combat this, FLIM was used to provide additional contrast by separating minocycline and skin autofluorescence signals based on differences in fluorescence lifetime. Non-Euclidean phasor analysis was used to help further distinguish signal from API and autofluorescence signal. With phasor analysis, decay patterns are transformed into phasor coordinates, which, when plotted in a phasor space, appear as clusters depending on the fluorescence lifetime pattern exhibited. This enabled the authors to extract the minocycline signal from skin autofluorescence and identify the localization of minocycline signal in the skin, as minocycline has a unique lifetime pattern. This study revealed that minocycline was delivered to the pilosebaceous unit and that higher minocycline uptake occurred with higher minocycline dosage [29]. Notably, the non-Euclidean phasor only allows obtaining the relative concentration of API and mapping of the API's spatial distribution, rather than providing a quantitative metric of concentration. Following this work, Jeong et al. developed a novel multicomponent phasor analysis algorithm to visualize the distribution of two APIs, minocycline and tazarotene [30]. Multicomponent phasor analysis enabled a better visualization of multiple APIs compared to the previous phasor analysis. Tazarotene and minocycline are thought to have a synergistic effect in the treatment of acne, but their joint uptake and diffusion had not been understood at the time due to a lack of appropriate methods. By developing a multicomponent phasor analysis algorithm, Jeong et al. assessed the local distribution of both APIs in the skin using FLIM following joint application in a gel. This showed a high concentration of API localizing in the pilosebaceous unit, the intended targets of both APIs [30]. While this method holds great promise in studying dermal drug PK, especially in the joint delivery of APIs, it still remains a qualitative method and cannot be used to calculate drug concentrations or PK parameters, but progress in this method may eventually enable API quantification [30].

4.2.2 Infrared Imaging

IR imaging is a nondestructive method to image skin, API, and excipients. IR imaging attracted significant attention due to its abundant spectral information related to the compositional and structural information of imaged molecules. Generally, IR spectral imaging has three different measurement modes: transmission, reflection, and attenuated total reflection (ATR) [147]. Transmission mode requires the skin sample to be thin, so skin preparation is required. ATR-FTIR spectroscopy is the most often used in dermal drug studies, because it is versatile and requires minimal or no sample preparation. ATR-FTIR is suitable for full-thickness skin ex vivo, and in vivo [148, 149]. However, the signal from this mode is mainly from the outer surface of the sample (depth of a few μm). To image skin in deeper layers, the tape stripping method can be used to remove the outer surface and reach deeper skin layers [150–155]. In most dermal drug studies, IR imaging is used to either collect a full spectrum from each pixel and/or map multiple pixels of a sample using one peak of a vibrational mode. The full spectrum is better for evaluating changes in chemical bonds or functional groups in skin, API or excipients, and the mapping image is better for visualizing skin microstructures or the distribution of API and/or excipients.

Skin structures and compositions Skin IR spectra contain abundant information about the skin's endogenous components [147]. For example, nucleic acids, carbohydrates, and phosphates can be characterized using spectra within 900–1300 cm^{-1}. Lipids can be characterized using spectra within 1450–1470 cm^{-1} and 2800–3000 cm^{-1}, which is attributed to CH_2 and CH_3 stretching, and the band near 1740 cm^{-1}, which is attributed to the C=O stretching vibrations of the ester functional groups in phospholipids. The spectra within 1500–1750 cm^{-1} include the amide I band, which is due to amide C=O stretching, and the amide II band, which is due to amide N-H stretching. Amide I and Amide II bands, and the band near 1400 cm^{-1} attributed to symmetric vibrations of C=O, can be used to characterize skin proteins [147]. Amide III (1175–1350 cm^{-1}) can also be used to show fibrous protein distribution [156]. IR images showing lipids and proteins are often used to delineate the skin's corneocytes and the surrounding lipid matrix composed of ceramides, fatty acids, and cholesterol.

Many dermal drug studies used IR spectra change to indicate the skin's interaction with various components in formulations. For thin skin samples, conventional FTIR can be used to map the skin's lipid change after the treatment with water, acetone, and sodium laureth sulfate (a surfactant) [157]. In one study, porcine skin was first dermatomed to around 500 µm. After applying the formulations to the skin, the skin was frozen and cut into a transverse 6-µm-thick sample for imaging. The spectra bands within 2800–3000 cm^{-1} corresponding to the CH_2 stretching were analyzed. Peak shifts were observed on the skin after acetone treatment, indicating that the remaining intercellular SC lipids presented a more ordered organization. This observation differed significantly from the water or sodium laureth sulfate-treated skin. The organized lipid distribution was related to increased barrier function and decreased skin permeability, which explained the low permeation of caffeine and ibuprofen in the acetone-treated skin observed in this study [157].

Compared to the conventional FTIR, a modified FTIR using synchrotron radiation helps to generate stronger signals and achieve better spatial resolution (sub-micrometer) [158, 159]. Synchrotron-based FTIR microspectroscopy has been used to study the molecular changes on different skin layers caused by the various penetration enhancers [160, 161]. One study investigated the effect of propylene glycol (PG) on the skin [161]. 6-µm-thick porcine skin samples were prepared for FTIR imaging. The collected IR spectra were analyzed by classical least squares (CLS) method to identify the sample's SC, epidermis, and dermis layers. Wavenumber shifts were observed for all three layers of the skin after PG treatment. Specifically, the peaks at 2850 cm^{-1} (CH_2 symmetrical stretching) and 2920 cm^{-1} (CH_2 asymmetrical stretching) shifted to higher wavenumbers. This change indicated that PG treatment caused more disordered lipid structures.

An AFM with a tunable IR laser source has been used to achieve high-spatial-resolution IR imaging (~200 nm) [162]. AFM mapped the morphology based on the photothermal effect after the adsorption of the IR light. In this study, IR images at specific wavenumbers were collected to map the endogenous constituents on the skin: 2930 cm^{-1} for lipids and 3290 cm^{-1} for keratin. In this study, IR images showed that a delipidization process reduced the total lipid content and damaged the human SC morphological structure.

Different from conventional FTIR, synchrotron FTIR, and AFM-FTIR, ATR-FTIR is able to measure both thin and thick skin samples. ATR-FTIR is also the technique that is used the most in studying dermal drugs among these IR measurement methods. There are many ATR-FTIR-based studies [163–165]. For example, Niu et al. showed that ATR-FTIR could be used to study the interaction between ethosomes, a drug carrier, and the skin, on 1.2-mm-thick skin sample. Kopečná et al. used ATR-FTIR to evaluate the effect of penetration enhancers on the skin barrier [166]. Zhang et al. used ATR-FTIR to evaluate the effect of APIs (i.e., (S)-flurbiprofen than (R)-flurbiprofen) on the skin [167]. Among these studies, ATR-FTIR was mostly used to collect the full spectra of a spot for characterizing compositional changes in the skin, instead of mapping for images.

Bioavailability in Thin Skin Samples In addition to studying skin structures and compositions, IR imaging techniques have been used to study the absorption of dermal drugs into the skin. When measuring transmitted light, thin skin samples allow more light to pass through, making signal acquisition easier compared to thick skin samples.

Transmission FTIR imaging has been used to map the lateral diffusion of different vehicles (oleic acid (OA) and dimethylsulfoxide (DMSO)) in skin. One study used the thin SC layer tissue isolated from ex vivo human skin [168]. Deuterated OA and DMSO were applied to the skin sample, and IR spectra in glyph-rich and glyph-poor areas were collected, respectively. The BA of deuterated OA or DMSO was quantified using the C-D band (2000–2300 cm^{-1}). Comparing the IR spectra collected in glyph-rich and glyph-poor areas, the OA was found to spread within the grooved glyphs predominantly. This study showed the capability of IR spectral imaging for studying the glyph pathways for the lateral spreading of topically applied formulations.

In addition to vehicle studies, transmission FTIR was also used to visualize API (tacrolimus) distribution at different depths [169]. Tacrolimus-loaded thermosensitive solid lipid nanoparticles and a reference Protopic® ointment were applied to ex vivo and in vivo rat dorsal skins. After 24-h permeation, the skin was excised and cut into 50-μm thin slices for imaging. The tacrolimus's distribution was imaged using 3450 cm^{-1}, attributed to O-H stretching. This study showed that the solid lipid nanoparticle helped the API to penetrate a deeper skin layer than the reference ointment. Tacrolimus was observed to penetrate more than 300 μm deep in the skin in the ex vivo and in vivo penetration tests.

AFM-IR can image skin morphology with a higher spatial resolution than transmission FTIR. In one study, AFM-IR was used to study skin penetration of Janus nanoparticle (JNP), a type of nanocarrier [170]. Cryostat was used in the study to freeze and section ex vivo skin samples to sections of around 1–3 μm thickness. The study first used AFM-IR to show high-resolution skin morphology before applying JNP. Different structures, such as SC, viable epidermis, dermo-epidermal junction, and dermis structures, were clearly depicted. The more chemical-specific IR image using amide band vibration (1640 cm^{-1}) delineated the protein distribution. Chemical images at 1740 cm^{-1} (C=O band of the oil phase of JNP) and 1640 cm^{-1} (O-H band of the aqueous phase of JNP) were collected from the skin after applying JNP. Results showed that, after 24 h of cutaneous application, the JNP permeated the skin and had a gradient distribution from the uppermost SC layer to the deepest SC layer.

ATR-FTIR spectroscopy has also been used to study API and vehicles in dermal drugs using thin skin samples. ATR-FTIR spectroscopy was used to study the permeation of API and vehicles on ex vivo human skin [171]. The isolated SC layer after skin preparation was used in this study. The API, cyanophenol, in different vehicles, was applied to skin samples. The specific vehicles were propylene glycol diperlargonate (DPPG), isopropyl myristate (IPM), isostearyl isostearate (ISIS), and water. Amide I and II bands (1500–1700 cm^{-1}) were used to characterize the SC. The CN stretch (2227 cm^{-1}) was used to quantify cyanophenol's BA. Using a single peak was not suitable for measuring the vehicles (IPM, ISIS, or DPPG), because the position and intensity of their characteristic peak (carbonyl band near 1742 cm^{-1}) changed when interacting with cynaophenol. A more stable approach obtaining information from a range of wavenumbers was used for measuring the vehicle. This study used ATR-FTIR to collect spectrum instead of mapping for images. Specifically, multivariate target factor analysis of the spectra from 1478 to 1350 cm^{-1} was used to deconvolve the vehicle's signal from the skin and cyanophenol. Results showed that DPPG, IPM, and ISIS diffuse faster across the skin than the API. In addition, DPPG led to a significant increase in API distribution in the SC, indicating its strong penetration-enhancing effect. IPM resulted in a moderate level of cyanophenol penetration, similar to water delivery. Using ISIS as the vehicle demonstrated the lowest level of cyanophenol penetration, indicating its penetration retarding effect.

The studies mentioned above required complicated skin preparation methods to obtain a thin skin sample. Another option is to use tape stripping, a simpler method for obtaining a thin skin sample for imaging. ATR-FTIR combined with the tape stripping method has been widely used in studying the BA of constituents in topical formulations including APIs such as 4-cyanophenol [12] and ibuprofen [172] as well as excipients such as vesicles [173]. It is noticed that, in these studies, ATR-FTIR was mainly used to collect the full spectra of a spot on the stripped tapes, instead of mapping the drug on tapes.

The tape stripping combined method was also used to study the metamorphosis of APIs in topical formulations. Crystallization of diclofenac sodium on the skin was evaluated using this combined method [174]. Spectra in the range 1500–1700 cm^{-1} covered the peaks for carboxylate asymmetric stretching vibrations of diclofenac sodium. This study showed that peaks within this range changed when diclofenac sodium crystals dissolved in solvents. Interestingly, the peak changes were different when dissolved in PG and DMSO. Tape strips from different depths of SC were collected from ex vivo porcine skin after applying diclofenac sodium formulations. The measured spectra were analyzed using hierarchical cluster analysis (HCA) and principal component analysis (PCA) to classify the tape strips with API crystals and those without crystals. This study observed that crystalline diclofenac sodium was located at 0.7–1.3 μm depth in a PG vehicle. This study used ATR-FTIR to collect spectra instead of mapping for images.

AFM-IR combined with tape stripping has also been used to investigate the crystallization of ibuprofen after topical application on porcine skin [175]. Ibuprofen formulations with different ratios of water and PG were topically applied. The tape-stripped skin samples were then imaged. Results confirmed the presence of ibuprofen crystals by the presence of C=O stretching vibrations at around 1720 cm^{-1}.

In addition to ex vivo skin samples, tape stripping has enabled the study of dermal drug BA in vivo. Studying API penetration on human skin in vivo using ATR-FTIR combined with tape stripping has been shown in multiple studies [176, 177]. Formulations with 4-cyanophenol were applied to the ventral forearm of human subjects. Tape strips from the skin at different depths in SC were collected and measured. The area under the curve between 2250 and 2200 cm^{-1}, which is attributed to C ≡N stretching, was used to quantify 4-cyanophenol. These studies confirmed ATR-FTIR's capability in assessing API BA after topical application.

Computational methods can be used to improve spectral analysis for ATR-FTIR. Compared with HPLC, which measures the absolute drug concentration in the skin, the IR signal reflects the relative drug concentration. To enable absolute quantification of drug concentrations, regression analysis, which uses a reference sample to associate the IR signal intensity to absolute drug concentrations, can be used. Specifically, partial least squares regression (PLS-R) analysis has been used to analyze ATR-FTIR spectra and quantify API's distribution in the skin [178]. Ex vivo porcine ear skin and in vivo skin region on volar forearms of human subjects were applied with lipstick formulations containing anthocyanins. Ex vivo and in vivo tape strips from these regions were measured using ATR-FTIR. The depth distribution, diffusivity partition parameters, permeability coefficients, and steady-state fluxes were calculated. These results from PLS-R-assisted ATR-FTIR correlated well with those from HPLC-MS, demonstrating its ability to quantify API in the skin.

Bioavailability in thick skin samples Full-thickness ex vivo skin can be measured using reflected IR and ATR-FTIR. These studies mainly used IR spectroscopy to acquire spectra at different depths of the skin, which characterized the spatial distribution in axial direction but not the lateral direction.

Medendorp et al. used reflected IR methods to study econazole nitrate and estradiol in ex vivo human skin [179]. Estradiol in ethanol, econazole nitrate in PG, and an econazole nitrate cream were applied to the skin, respectively. IR spectra from the surface of skin on the epidermal side and the dermal side were collected and analyzed. The penetrated API quantified using IR demonstrated a strong correlation to the HPLC results, indicating the IR method's capability in quantifying API located on the skin's surface.

Because reflected IR and ATR-FTIR only measured the skin's outer surface, tape stripping can be combined to remove the outermost layers and reach deeper layers in the SC. In one study, ATR-FTIR was used to study different excipients penetration after applying to ex vivo porcine skin [150]. The tape stripping method was used to help reach deeper skin depths in SC. FTIR spectra were collected on the skin tissue surface after each tape strip. The results showed that the excipient's amount decreased steadily with increasing skin depth. The penetration depth of these tested excipients, ranked in ascending order from the shallowest to the deepest: DMSO $>$ SDS(sodium dodecyl sulfate) $>$ SLES(sodium laureth sulfate) $>$ STZ(sulfathiazole sodium). Moreover, this study demonstrated a good agreement between ATR-FTIR and confocal Raman spectroscopy results. ATR-FTIR combined with tape stripping enabled studying depth distribution of emulsifiers (Lipoid® S75 and sodium dodecyl sulfate) and oil components (jojoba oil, medium chain triglycerides, and sunflower oil) on the ex vivo pig ear skin [151]. Another study used a similar tape stripping combined approach to study the penetration of liposomes on full-thickness porcine ear [152].

In addition to ex vivo full-thickness skin samples, reflected IR and ATR-FTIR can track topical formulations in vivo in human subjects and live animals. In one study, reflected IR spectra were collected on human skin in vivo using an FTIR spectrometer with a fiber-optic probe [180]. Human heels were measured after treatment with urea-containing cream. Water was measured using 1900 nm, and urea was measured using the maximum relative peak height of the spectra from 1949 to 1985 nm. The study showed urea in SC significantly increased 1 h after the treatment and then gradually decreased. In addition, water mobility increased 1–4 h after topical application of the urea-containing cream. This study related the high water mobility and water content in the SC to treatment with urea-containing cream.

In another study, ATR-FTIR achieved in vivo imaging of API on hairless mice [154]. The tape stripping method was used to achieve API imaging at different skin depths. Lyotropic liquid crystal genistein-based formulation was first applied on the mouse's skin. Different voltages of electroporation pulses and duration of electroporation were then used on the skin. The application site was then stripped with adhesive tape, and the stripped areas on the mouse were measured. The peaks specific to genistein were used to quantify the API: C1=C2 stretch at 1652 cm^{-1}, and the C=O stretch, O-H bend, and ring C-C stretch at 1518 cm^{-1}. In this study, ATR-FTIR and the tape stripping method enabled mapping depth distribution of genistein and evaluating the synergistic effect of electroporation on transdermal drug delivery.

Several studies have used ATR-FTIR to study the in vivo BA of APIs and the impact from different penetration enhancers [155, 181]. In one study, terbinafine formulations with different penetration enhancers (oleic acid, 2-pyrrolidone, and urea) were applied to the ventral forearms of human subjects [155]. ATR-FTIR spectra were collected in vivo; deeper layers were exposed and measured after tape stripping. The peak at 774 cm^{-1} attributed to the stretching vibration of aromatic C-H was used to indicate terbinafine. Results from ATR-FTIR showed that urea and 2-pyrrolidone did not significantly enhance the penetration of terbinafine across the SC. In contrast, the results showed that oleic acid enhanced the penetration of terbinafine across the SC. The penetration-enhancing effect from oleic acid was also observed in another study using ATR-FTIR [181]. Moreover, this study mapped the in vivo distribution of the API (4-cyanophenol) and the vehicle (PG) with and without the addition of oleic acid [181]. The formulations were applied to the midventral forearm of human subjects. Distribution of 4-cyanophenol was measured using $C \equiv N$ stretching vibrations at 2230 cm^{-1}, while PG was measured using C-O stretching at 1040 cm^{-1}. The results showed that both the 4-cyanophenol and PG decreased more rapidly when the formulation contained oleic acid.

4.2.3 Spontaneous Raman

Spontaneous Raman spectroscopy provides noninvasive and, importantly, nondestructive measurements for in vivo and ex vivo skin. Raman spectroscopy is capable of extracting important information such as skin structure, skin composition, BA and BE measurements of dermal drug (APIs and excipients as listed in Table 4.1) applied on skin [78–82, 85–88, 182–192]. CRM is one of the most common and widely used techniques applying Raman spectroscopy to studying drug formulations. Compared to conventional Raman microscopy, CRM provides higher spatial resolution and the ability to acquire depth profiles in three-dimensional thin structures [77, 84, 193]. Because of the low efficiency of the spontaneous Raman scattering effect, acquiring a high-resolution mapping image would largely impair the time resolution in dermal drug studies. Thus, in dermal drug studies, CRM is more often used to collect a full spectrum of a pixel/a sample spot instead of imaging a region. While it is feasible to use CRM to map and visualize skin microstructures or distribution of API and/or excipients, this comes along with a trade-off between image spatial resolution and temporal resolution.

Skin structures and compositions Mapping skin samples using CRM enables visualizing skin's microstructures. CRM can be used to obtain precise correlations between biochemical characteristics and histological structure of ex vivo human skin tissue. Investigating Raman spectral features and performing univariate and multivariate analyses allow precise correlations between biochemical characteristics and histological structure of ex vivo human skin tissue. Raman images of collagen, lipids, melanin, and phenylalanine in ex vivo human skin were mapped by integrating their characteristic Raman peaks at 1248, 1300, 1580, and 1608 cm^{-1}, respectively. Raman images showed high collagen content in the dermis region, while the epidermis contained significant aggregation of melanin, and large amounts of lipid and phenylalanine signals were found in the SC. The data also showed a relatively uniform distribution of collagen in the dermis below the hair follicle, while melanin was found to be deposited in the epidermis and hair follicles. Applying three-group K-mean clustering to the spectral region of interest enabled the identification of the mean spectra of each skin layer. Raman bands, including 853, 940, 1248, 1453, and 1670 cm^{-1} peaks, corresponded to the typical spectral signatures of the dermis layer. Meanwhile, the spectral characteristics of the SC and epidermis corresponded to the peaks representing lipids at 1064, 1106, and 1300 cm^{-1} [194].

Recent studies have shown the application of CRM to visualize the microstructural changes that occur when transdermal drugs interact with the ex vivo porcine skin. CRM coupled with Empty Modeling (EM) component analysis was used to reveal components in the nicotine patch. Raman image repainted with principle component analysis (PCA) enabled the segmentation of lipid and protein regions. This study also demonstrated the application of Raman spectroscopy in studying how drug levels and solubility vary in different parts of the film, by mapping Raman spectra and depth slice images of the interface between skin and film-forming systems. Tested film including Klucel, Eudragit with different types of plasticizer: triethyl citrate (TEC) and medium chain triglyceride (MCT). Raman-based depth slice images showed that the plasticizer, TEC, and applied drug, Betamethasone-17-valerate (BMV), were evenly distributed in the Klucel polymer. Similar TEC and BMV distribution was observed in Eudragit-based polymer. The nonuniform distribution between the plasticizer and BMW in Eudragit polymer occurred when the plasticizer was replaced with MCT. An MCT cluster accumulated in the film, causing the nonuniform distribution of BMV in the polymer. Inhomogeneous distribution of the plasticizer in the polymer, which led to the nonuniform distribution of applied drugs, might disturb the drug release/delivery process. This study provides an illustrative overview of Raman spectroscopy as an imaging tool that enables the detection and characterization of topical or transdermal drug delivery in skin [76].

Bioavailability in skin samples CRM has been used in several BA studies of dermal drug [12, 73, 75, 83]. These studies mostly used CRM to collect a full spectrum of a sample spot instead of mapping a region, ensuring a short acquisition time on one sample.

CRM can be used with the tape stripping method to evaluate the BA of API. SC in ex vivo porcine skin, applied with 170 mg mL^{-1} of 4-Cyanophenol (CP) in different water v.s. propylene glycol (PG) ratios for 0.5, 1, and 2 h, was extracted with tape stripping. FTIR and Raman spectra were collected from extracted SC tape strips, and the relative CP concentration profile was extracted by high-performance liquid chromatography with ultraviolet detection (HPLC-UV) from all tape strips. The results demonstrated a high degree of correlation between the spectroscopy data and chemical quantification of the tape strip. Both spectroscopy and chemical quantification of the tape strip data showed an increasing uptake of CP with time of CP application, followed by decreasing CP concentration profiles with increasing depth [12].

Another study used CRM to test API diffusion across the skin barrier and API accumulation in the tissue models. The study also included the measurement of the chemical composition of rat skin, human skin, and human-reconstructed epidermis by measurement of ceramides, cholesterol, lactate, urea, protein, and water at different depths of the tissues. The results showed that the transdermal diffusion of quinidine was low, and the skin accumulation was comparable between the APIs (A hydrophilic drug caffeine and the P-glycoprotein substrate quinidine). The reconstructed skin contained less water and more urea and protein, while human skin was abundant in ceramides and cholesterol. The study demonstrates the application of CRM in testing API penetration and skin study [75].

CRM can also be applied in the quantification of API in human SC. Isolated human SC was immersed in resorcinol solutions in PBS with concentrations of 0.5, 1, 2.5, 5, 10, 25, and 50 $g\,L^{-1}$ individually for 24 h. Raman spectra were collected on each SC with different resorcinol concentrations, and the spectra were analyzed with PCA, area under the curve (AUC), and PLSR. PCA is a multivariate analysis used to evaluate the similarities/differences of the data and to simplify a complex data set of multiple dimensions into a scatter plot. PCA scatter plot showed that the collected spectra were organized according to the concentration of resorcinol in SC as each spectrum collected from the same resorcinol concentration clustered together in the scatter plot. The AUC results showed that the peak intensities of Raman spectra were linearly correlated to the concentration of resorcinol in the SC and PLSR analysis, demonstrating the reliability in accurate quantification of resorcinol in human SC [83].

Another study demonstrated a quantification of concentration profiles of skin components and topically applied oils by tailored multivariate curve resolution-alternating least squares (tMCR-ALS) method couples with confocal Raman spectroscopy [73]. This study processed the Raman spectra with the tMCR-ALS method and used the same data but applied a different quantification method to show its ability to quantify concentrations of drug penetrating into the skin (in vivo). Experiment results showed that the water concentration (mass %) increased from $\sim$34% to 63% from the surface of SC to the bottom of the SC, protein concentration decreased from $\sim$63% (surface) to 35% (bottom), and lipid concentration decrease from $\sim$7% (surface) to 2% (bottom) in SC. Also, the almond oil, jojoba oil, paraffin, and petrolatum signal terminated at 40% depth of SC thickness. The topically applied oils showed the highest concentration on the surface of SC, 4.3 ± 3.1 for almond oil, 5.9 ± 3.8 for jojoba oil, 8.2 ± 2.4 for paraffin, and 12.6 ± 5.6 (mass %) for petrolatum.

Bioequivalence in skin samples CRM has also been used to assess BE of topical products. Two commercially available topical products, Nurofen Max Strength 10% Gel and Ibuleve Speed Relief Max Strength 10% Gel, were used to study the permeation of ibuprofen (IBU). A finite dose, 10 $\mu L/cm^2$, of 10% (w/w) IBU formulations was applied over selected areas of five volunteers. No significant difference in IBU signal was detected for both formulations after evaluating the area under the depth profiles curves (AUC). This study concluded that both commercial formulations delivered IBU across the skin in vitro over 24 h [74].

CRM can be used to characterize topical semisolid formulations and their penetration into human skin ex vivo. Raman spectroscopy can provide composition details of formulation structure, arrangement, and structural characteristics of substances in dosage forms, polymorphic forms of substances that may be present in topical products, and other attributes that can be used for direct comparison of physicochemical and composition. In addition to characterizing and comparing product quality attributes, CRM has also been shown to be useful for evacuation and calculating product performance by characterizing the kinetics of entry of topical semisolid products into the upper layers of human skin. In one study, CRM was used to investigate the polymorphic form of acyclovir in the formulation as there were distinct differences between the Raman spectrum of an acyclovir cream, its polymorphic forms I, II, and V, and acyclovir crystal. The differences were tracked using Raman peaks associated with the drug penetration enhancer propylene glycol (PG) at $840\,\mathrm{cm}^{-1}$ and peaks significant for the distinction of the different polymorphs of acyclovir at $650\,\mathrm{cm}^{-1}$. CRM also monitored the penetration of the excipient PG from the acyclovir cream product into excised human skin by recording Raman spectra through line scans with a step size of $2\,\mu\mathrm{m}$ from the skin surface downward. Results showed no significant differences between the PG depth penetration for six acyclovir products at 4 h after applying the topical products on the skin. Most of the products showed higher PG peak intensity at a greater depth by 24 h after dose application, indicating that a larger amount of PG had penetrated deeper into the skin. These results demonstrated the application of CRM to monitor changes in the penetration kinetics at a selected skin depth as a function of time, which can be used in the study of the BE between topical products [195].

4.2.4 Stimulated Raman Imaging

SRS enables label-free imaging of APIs within the skin. SRS has high specificity to target vibrational bands alongside temporal resolution for video-rate imaging at high spatial resolutions to delineate subcellular-level features. SRS is a quantitative imaging technique; however, it is currently mostly used in semiquantitative studies of the API in skin due to the challenge of determining absolute concentration at different skin depths. A custom-built inverted microscope is the commonly used instrument for studying formulations on ex vivo skin samples. Most current studies measure the forward direction SRS transmitted through the skin, which is hundreds of micrometers thick. SRS microscopes can also be set to measure backscattered SRS signals from the samples. Backscattered SRS signal can be collected from intact in vivo skin and ex vivo thick skin tissues. The following subsections discuss some examples of using SRS for studying skin structures and compositions, BA in ex vivo models, and BA in in vivo models.

Skin structures and compositions SRS microscope can be used to map lipid and protein distribution [127]. This information is useful for characterizing skin integrity, which plays an important role when developing novel polymeric nanocarriers for dermal drug delivery. A study used SRS imaging to locate SC and viable epidermis on the skin and calculate the protein-to-lipid ratio [196]. SC had high lipid signal at around 2850 cm^{-1}, and the viable epidermis had low lipid signal but high protein signal at 2930 cm^{-1}. An ex vivo human abdominal skin sample was frozen in a media and sliced into 7 μm thickness using a freeze microtome. The study compared the impact of DMSO and newly developed sulfoxide-functionalized nanogels on ex vivo human skin. The skin applied with DMSO showed an obvious change in the protein and lipid organization in the SC, as well as a significant decrease in the protein-to-lipid ratio. The sulfoxide-functionalized nanogels performed similarly to water, which did not cause apparent changes on the skin. The study used the SRS microscope to prove the developed nanogel improved the topical delivery of the target drug without causing any skin damage.

SRS microscopy can help to display the chemical composition distribution of skin cluster and canyon microstructure [197]. It provides not only the lateral distribution of the chemical compositions at a depth but also the axial distribution at different depths. The ex vivo porcine skin sample thickness of around 500 μm was prepared using an electric dermatome. Three substances were mapped using different Raman wavenumbers: 3340 cm^{-1} for water, 2950 cm^{-1} for protein and lipid, and 2850 cm^{-1} for lipid. The cluster regions were found to have higher water content. The water-to-protein ratio increased with the depth of the skin. The ratio increased drastically between the SC layer and the upper layer of the viable cells of the epidermis and reached a plateau at deeper layers. In contrast, the canyon region was water-deficient and sometimes had an accumulation of lipids near the top of the canyon. In this study, the penetration of deuterated oleic acid was also analyzed to study canyons' role in the delivery of formulation into viable skin cells. The deuterated oleic acid was imaged at 2105 cm^{-1} (the C-D stretching). The distribution of oleic acid 25 min after application was overlaid on the skin's cluster and canyon microstructure, which showed an accumulation of the oleic acid in the canyon. These results displayed oleic acid could rapidly reach deep layers via canyons.

Another study used SRS to show the skin structure and used another method, scanning transmission X-ray microscopy (STXM), to quantify APIs in a formulation [198]. This study investigated ethanol's effect on API absorption on different skin matrices: in vitro reconstructed human skin, ex vivo human skin, and ex vivo mouse skin. The ex vivo skin was sectioned into 5–10 μm-thick slices for imaging in the experiment. STXM was used to quantify API (dexamethasone) concentrations. SRS images were mainly used to show the protein (2934 cm^{-1}) and lipid (2850 cm^{-1}) structures of the skin, as well as to calculate the protein-to-lipid ratios. Ex vivo human skin had the lowest protein-to-lipid ratio among the three metrics, and it also showed the least change when exposed to ethanol for 300 min. Mouse skin showed an increased protein-to-lipid ratio when exposed to 35% and 70% ethanol. The difference in changes between human and mouse skin is likely caused by the different lipid structures of mouse skin and its higher dissolving rate by ethanol. In vitro reconstructed human skin showed a large increase in protein-to-lipid ratios when exposed to 12% ethanol, indicating high lipid dissolution. Interestingly, its ratios showed a very small change when exposed to 35% and 70% ethanol, indicating the lipid and protein are affected by ethanol to a similar degree. Because of ethanol's impact on the skin structures, ethanol exposure increased the penetration of dexamethasone. The penetration-enhancing effects of ethanol happened in all three metrics, and the effect in mouse skin and in vitro reconstructed human skin was slightly higher than in ex vivo human skin.

Bioavailability in thin skin samples SRS imaging can image the APIs' distribution in the skin in addition to its morphological changes. One study used SRS to assess the permeation of ibuprofen and ketoprofen into porcine skin [199]. Ibuprofen and ketoprofen are nonsteroidal anti-inflammatory drugs. After applying a solution formulation with ibuprofen or ketoprofen in propylene glycol on the skin, the ibuprofen and ketoprofen crystallized out on the skin's surface during the 3.5-h study. In addition, the movement or swelling of the skin during formulation uptake was tracked by monitoring the CH_2 stretching signal from skin lipids. The SRS imaging also helped to monitor the morphological change of the ingredients in the formulations.

Another study used SRS to map the distribution of two APIs in an antifungal formulation on ex vivo porcine skin [200]. The two APIs, zinc pyrithione and climbazole, were imaged and distinguished using two wavenumbers: 1540 cm^{-1} for the C=N pyridine stretch of zinc pyrithione and 1580 cm^{-1} for the imidazole ring vibration of climbazole. In addition, imaging the CH_2 stretch mode of lipids helped to delineate the skin's microstructures, such as hair follicles, SC, stratum lucidum, stratum granulosum, stratum spinosum, and the stratum basale. Imaging different vibrational modes enabled depicting the distribution of API in different skin morphological features. The zinc pyrithione penetrated up to 10–20 μm into the skin and formed 1–4 μm crystals on the skin's surface. In contrast, climbazole diffused into the infundibular spaces surrounding the hair shaft within the skin (up to 60 μm), and climbazole predominately resided in lipid-rich crevasses.

SRS imaging assisted with computational image analysis method has been used to track APIs on ex vivo mouse skin [105]. Specifically, deep learning based on convolutional neural networks was used to identify cells, cell junctions, and cell types within the skin. Segmenting different cell structure regions enables analysis of API distributions inside and outside these regions. Moreover, it enables studying different mechanisms of API permeation (e.g., across the skin barrier, through hair follicles). Instead of the laborious and time-consuming manual labeling process, the deep-learning-based image analysis method is accurate, automated, and well-suited for large image data sets. After topically applying the formulations, the intensity-time profile was collected using the SRS microscope. Ruxolitinib and deuterated betamethasone dipropionate (BMDP-D) were imaged using 2250 cm^{-1} (the nitrile stretching vibration and the C-D band). AUC, t_{max}, and C_{max} were calculated using the obtained intensity-time profile. The study showed that ruxolitinib in gel formulation permeated faster in the skin than in transcutol formulation. Using lipid imaging, SC barrier disruption caused by the gel was observed. This disruption led to an enhanced API permeation, thus a smaller t_{max} value from the gel formulation. The intensity-time profile recorded the rise and fall of the API concentration in different cellular structures at different skin depths. For example, the SRS intensity reached a high initial level in the SG because the API rapidly traversed the follicle; the intensity then decreased rapidly as the API diffused out of the SG into the surrounding dermis. SRS image segmentation allowed the comparison of API in both lipid-rich and lipid-poor cellular structures. Notably, ruxolitinib in the gel formulation was found to deposit more at the lipid-rich junctions between the corneocytes and the lipid-rich structures like SG, AD, and white adipose cells of the subcutaneous fat. Moreover, microscale partition factors of the API could be estimated using measured distribution in lipid-rich and lipid-poor regions. Partition factors of ruxolitinib and BMDP-D were compared, and no significant difference was observed in this study.

Higher contrast and sensitivity API detection can be achieved using a time-resolved phase-modulated SRS microscope (PM-SRS) [201]. Compared to the commonly used amplitude-modulated SRS, the nonspecific background caused by cross-phase modulation, photothermal lensing effects, and transient absorption are negligible for PM-SRS [201]. In addition, the time-resolved signal measurements address the difficulties in detecting a small amount of API in the fingerprint region (500–1800 cm^{-1}) where a strong background signal from skin tissues is present. The time-resolved signal detection method can suppress signals with a short coherence lifetime (nonresonant background and unwanted SRS signals from the skin); thus, the signal with a long coherence lifetime (SRS signal from APIs) becomes prominent [202]. The PM-SRS microscope was used to map the skin moisturizing agent (ectoine) in a reconstructed human epidermis tissue, showing improved signal contrast ($\geq$40 times in the spectral fingerprint region) than that of the conventional amplitude-modulated SRS [203]. The results showed that ectoine molecules penetrated the skin within 180 min. Attributed to the decrease in background noise, PM-SRS enabled imaging of low-concentration drugs on the skin in the millimolar range. In addition, PM-SRS was also used to study a local anesthetic drug (lidocaine hydrochloride) and a nonsteroidal anti-inflammatory drug (loxoprofen sodium). The two APIs could be imaged using 1092 cm^{-1} (methyl rock, CH_2 twist) for lidocaine hydrochloride and 1184 cm^{-1} (C-C stretching) for loxoprofen sodium salt. After 240 min, the loxoprofen sodium concentration reached about 47.3 mM on the skin's surface. In contrast, negligible penetration was observed for the lidocaine. Notably, this study used in vitro reconstructed human skin samples; more studies using ex vivo skin samples are needed to confirm these observations.

In addition to APIs, SRS imaging can also image other components in the formulations, such as the vehicles. In one study, two APIs, ibuprofen and ketoprofen, and a commonly used vehicle, propylene glycol, were imaged on ex vivo mouse ear skin [204]. Ketoprofen could be imaged using CH bond stretching at 1599 cm^{-1}. Ibuprofen and propylene glycol do not have a primary peak, so deuterated molecules were used to image them at 2120 cm^{-1}(C-D stretching band). CH_2 stretching at 2845 cm^{-1} could map the lipid distribution, which helped to represent the skin's structures, such as the intercellular lipids around corneocytes and hair follicle regions around hair shafts. This study showed the different penetration rates for ketoprofen and PG through the intercellular pathway across the SC. PG showed more rapid penetration via SC than ketoprofen. In addition, different penetration pathways (follicular pathway and intercellular pathway) for PG were compared. PG showed rapid penetration via the hair shaft and reached a steady state within 26 min post-application. In contrast, PG slowly penetrated the skin via the intercellular pathway across the SC during the 2-h experiment. SRS imaging also enabled visualizing crystal formation on the skin 25 min after applying the ibuprofen formulation on the skin. This crystal formation was due to the evaporation or the fast penetration of PG caused by oversaturation of the ibuprofen.

Most current studies have been focused on imaging APIs that have unique vibrational bands that are either intense or are nonoverlapping to the skin's background signal. However, a few APIs have such unique vibrational bands, necessitating a means to collect SRS spectral information. Two approaches capable of collecting SRS spectra have been tested in cutaneous BA studies.

One SRS system was built using a fast-tuning all-fiber OPO laser to tune to different wavenumbers and sparsely collect the SRS spectra [116]. The SRS system uses a turn-key fiber laser system, is highly automated and rapid in wavenumber switching (<5 ms switching time), has a high spectral resolution (<4 cm^{-1}), and a wide tunable range (750–3200 cm^{-1}). Assisted with sparse spectral sampling SRS (S^4RS) approach, a sparse subset of Raman wavenumbers capable of sample discrimination can be determined. Using a subset of Raman wavenumbers helps to save imaging time and thus achieve the high-temporal-resolution mapping of multiple APIs or skin components with highly overlapped peaks. This approach successfully distinguished three retinoids from each other, even though their peaks were close and overlapping. Specifically, three APIs were mapped using their unique Raman bands (1568 cm^{-1} for isotretinoin, 1582 cm^{-1} for retinoic acid, and 1594 cm^{-1} for tazarotene). It was observed that the SG could be distinguished from the subcutaneous fat and the AD in S^4RS spectra, due to increased saturated fatty acid and wax ester content of the sebum in SG. The approach was also tested for mapping a formulation with Tazarotene and PEG 200 on a fresh mouse ear skin. This approach enabled rapid mapping of multiple components: the API tazarotene, the inactive ingredients PEG200 (Polyethylene glycol), and the lipid from the skin structure. From this study, the API and PEG200 were located in the same lipid-rich skin regions and were not found to permeate deeper than the SGs within 2 h after topical application. This approach was found useful for studying the permeation mechanism of API and the effect of inactive ingredients on its localization.

Spectral processing methods can be introduced to help better distinguish API's SRS spectra from skin's spectra. Specifically, a multivariate curve resolution-alternating least squares (MCR-ALS) method was used to help decompose skin's and API's signal [113]. In this study, human abdominal skin was applied with 4-cyanophenol in propylene glycol/water (1:1) solution. The skin was then frozen and sliced to 30 μm using a cryo-microtome. Because 4-cyanophenol also has SRS signal in the silent region at 2235 cm^{-1} (nitrile group), and the skin has no SRS signal in this region, SRS image at 2235 cm^{-1} was collected as an accurate distribution reference. Fingerprint region SRS spectra of the skin with 4-cyanophenol were then collected and analyzed by MCR-ALS. The reconstructed 4-cyanophenol distribution showed a good match with the reference distribution. This result presented an example that spectral unmixing methods can help to better interpret the SRS spectra for studying topical formulations on the skin.

Another approach to collecting the SRS spectra is using femtosecond sources with pulse shaping. A study used this approach to collect SRS spectra from live mouse skin in vivo [117]. DMSO solution was applied to a mouse ear. The mouse ear had a thin thickness for measuring the transmitted signal; thus, no skin preparation was needed. MCR was also used to help decompose the spectrally overlapped signal from the DMSO and the skin's lipid in the range of 2800–3100 cm^{-1}. Because of the fast spectra collection speed, the study successfully showed the penetration of DMSO into the extracellular area in the adipose tissue within 58 s after application.

Bioavailability in thick skin samples In addition to transmitted SRS signals, backscattered SRS signals can also be measured to study the BA of dermal drugs. Skin preparation and thickness control are not required when imaging backscattered SRS signals. SRS images can be collected from intact in vivo skin and ex vivo thick skin tissues. One approach to measuring the backscattered SRS signals is using a large-area photodetector with a small hole in the center [205, 206]. Lipids (2845 cm^{-1}) and water (3250 cm^{-1}) distribution on thc skin in living mice were achieved using this approach. The imaging also showed the penetration of trans-retinol via the hair shaft into the SG. This study also showed the accumulation of deuterated DMSO (2120 cm^{-1}) in the hair shaft of the skin of living humans [205].

Another approach to collect backscattered SRS signals is through the focusing objective and a polarization beam splitter. This approach was used to monitor a permeation enhancer, DMSO (670 cm^{-1}), and an API, retinoic acid (1570 cm^{-1}), on a fresh full-thickness ex vivo mouse skin [115].

Remote-focusing configurations can be added to the SRS microscope with the polarization beam splitter for volumetric imaging with a rapid axial scan rate [207]. The total acquisition time for a 150 $\times$ 150 pixels image at four depths and two wavenumbers was 1.9 s. The system was used for in vivo imaging of chemical penetration into human sweat pores. DMSO was applied on an upper palm. Two wavenumbers were used for imaging: DMSO (2914 cm^{-1}) and skin (2884 cm^{-1}). The system successfully captured DMSO's quick penetration (within 15 s) from the surface to the deep layers (up to 60 μm).

Using the spatial frequency multiplexing technique, SRS spectra can be acquired on the sample rapidly (each pixel takes tens of microseconds) [131]. The multiplex SRS microscope was used to image vitamin E distribution on a live mouse's skin in vivo. Using the SRS spectra and multivariate curve resolution (MCR) analysis, vitamin E, lipids of the skin, and proteins of the skin were decomposed from the collected SRS images. This study showed the modified SRS technique can map the API in a spectrally overlapped tissue in vivo. A fiber-based handheld SRS system that measures backscattered SRS has been developed [208]. The handheld imaging probe design helped reach most areas of the human body, including those that are difficult to access by a benchtop unit. The system was briefly tested to show its performance on in vivo imaging of the skin on a human's inner arm. A commercial cosmetic of vitamin E mixed alpha-tocopherol was applied to the skin. Even though the API had a spectrum highly overlapping with the skin's spectrum, its signal was decomposed using MCR analysis. The SRS system mapped the cosmetic formulation's distribution on the SC layer of the live human skin.

Bioequivalence in skin samples Transmission SRS has been used in a study to assess BE of topical products in ex vivo human skin [209]. In this study, commercially available topical products, including a reference-listed drug (Tazorac® cream 0.1%) and an approved generic drug (Tazarotene cream 0.1%), were measured. In addition, a lab-made formulation (0.1% tazarotene in PEG 200) with a different dosage form from these creams was also measured in this study. The permeation of tazarotene was measured using SRS imaging. The high spatial and temporal resolution of SRS imaging enabled visualizing API's distribution in different skin microstructures during the permeation process. Intensity-time profiles related to API's concentration in different skin microstructures can be collected. BE assessment of tazarotene-containing formulations was achieved using PK parameters obtained within different skin microstructures. Analysis of AUC and Cmax values confirmed the bioequivalence between the approved generic drug and the corresponding reference-listed drug; additionally, the result showed that the lab-made tazarotene formulation, with a distinct vehicle from the cream, is not bioequivalent to the reference-listed drug. This study demonstrates the successful application of transmission SRS in monitoring dermal drug delivery in different spatial regions.

4.3 Outlook

4.3.1 Challenges and Opportunities

For each optical method, one major common challenge is the depth at which skin samples can be imaged. When utilized in vivo, most methods are limited to several hundred microns. This is especially true for ATR-FTIR imaging, which is essentially limited to measuring the skin's outermost layer (a few μm deep). However, there are workarounds when conducting PK skin BA studies. One such method is to utilize optical clearing methods to reduce the potential of light-scattering materials interfering with API quantification [210]. Adding skin-clearing materials, glycerol, or a PEG formulation could allow for deeper API quantification in vitro, ex vivo, or in vivo, yet the implications on the API concentration-time profiles and, thus, PK parameters are not well yet understood. Another method is to improve the imaging depth using modified instrument design, such as using adaptive optics to shape the light beam to image deeper. One additional method that has been used in some dermal drug studies mentioned above is using flash-freezing skin samples treated with the topical formulation at predetermined time points and then sectioning the samples using a microtome to obtain a cross-sectional view of the skin. Mapping these cross-sectioned samples enables visualizing the depth profile of the dermal drug.

Another general challenge of optical methods for BA assessments in skin is absolute concentration quantification. Current studies mostly use optical techniques to measure relative concentration and evaluate dermal drugs semiquantitatively. One potential way to bridge the gap between relative and absolute concentrations is to correlate optical intensity with concentrations measured by more established methods, such as dermal microdialysis or dermal open-flow microperfusion, as concentrations are quantified using LC/MS. The measured optical intensities can be correlated with macroscale concentration values measured by these established quantitative methods. However, light scatterers and absorbers in the skin alter the optical intensity, and thus, the heterogeneity of skin makes the prediction or correction of these factors challenging. In addition, light emitted at different skin depths, affected by different depths of the light propagates through, is more challenging to use to quantify absolute concentration. This becomes even more challenging if a light-scattering formulation is applied to the skin. One potential way to overcome these challenges and bridge the gap between relative and absolute concentrations is to utilize sectioned thin samples. It would be possible to have the API of interest in various solution concentrations to build a calibration curve of light intensity vs. solution concentration and compare cross-sectioned imaging samples. Accounting for the skin heterogeneity, imaging mass spectrometry can be used to spatially correlate the optical intensity with absolute concentrations. Cross-sectioned skin samples can be used to develop calibration of the light change as a function of depth. These ways may potentially help push the optical imaging methods toward quantifying absolute concentration.

Among the optical techniques mentioned in the chapter, IR and spontaneous Raman suffer from slow acquisition time to obtain high-resolution mapping images. They can still image samples collected at a predetermined time point, such as tape strips from the skin at a certain time after topical application. However, they encounter challenges in mapping dynamic drug distribution on skin ex vivo and in vivo. For compounds and formulations that permeate slowly into the skin [2], they are highly informative, but for compounds/formulations that permeate on the scale of seconds to minutes [116], a rapid imaging method is required.

While speed is one facet that may impact the researcher's decision to use a specific method, the specificity of the method for the molecule of interest is also important. Not all APIs are suitable for each method. For example, fluorescence imaging requires the measured molecule to be fluorescent. Limited APIs and excipients are natively fluorescent, and labeling the API might profoundly impact the PK, which remains to be investigated with complementary approaches for labeled and native molecules. In addition, autofluorescence from endogenous components of the skin poses a challenge to fluorescence imaging and Raman imaging. The autofluorescence background leads to detector saturation in the signal collection and reduces signal specificity. For IR, spontaneous Raman, and SRS imaging that probe vibrational bands, the bands of the APIs may overlap with the skin or excipients. Overlapped signals also reduce signal specificity. As shown in some examples in this chapter, computational analysis methods can help unmix components' signals and recover the specificity to some extent. The components that can be unmixed, however, are limited. To enable studying the synergic effect between multiple APIs and excipients in a highly multiplexed study, further work is still needed, and there are opportunities to develop more data analysis methods and/or instruments to overcome this challenge better.

In vivo imaging is ideal for studying dermal drugs. It allows comprehensive ADME effect from the live subject and is closest to the scenario of the topical product's daily application. The four imaging techniques in this chapter have instrument types capable of in vivo imaging. For SRS imaging, there are limited studies using in vivo imaging. Among these reported studies, they are currently limited to investigating high-concentration API/excipients with strong peaks. Improving the sensitivity of instruments for backscattered SRS signal may help enable in vivo imaging for a broad range of dermal drugs using this imaging modality.

While many of these optical methods have been around for quite a while, many have only recently had applications in cutaneous PK for BA and BE estimated. Therefore, as with other cutaneous PK approaches discussed in other chapters, there are numerous opportunities for technology to mature and, therefore, these modalities also to mature. The limitations mentioned above for optical methods can be addressed with targeted research; however, the implementation and widespread use of these optical methods will depend on the "plug and play" ability of the methods to have limited user interaction of the technical aspects.

References

1. Santos LL, Swofford NJ, Santiago BG. In vitro permeation test (IVPT) for pharmacokinetic assessment of topical dermatological formulations. *Current Protocols in Pharmacology*, 91(1):e79, 2020.
2. Bodenlenz M, Tiffner KI, Raml R, Augustin T, Dragatin C, Birngruber T, Schimek D, Schwagerle G, Pieber TR, Raney SG, *et al.* Open flow microperfusion as a dermal pharmacokinetic approach to evaluate topical bioequivalence. *Clinical pharmacokinetics*, 56(1):91–98, 2017.

3. Bodenlenz M, Augustin T, Birngruber T, Tiffner KI, Boulgaropoulos B, Schwingenschuh S, Raney SG, Rantou E, Sinner F. Variability of skin pharmacokinetic data: insights from a topical bioequivalence study using dermal open flow microperfusion. *Pharmaceutical Research*, 37(10):1–11, 2020.
4. Kuzma BA, Senemar S, Ramezanli T, Ghosh P, Raney SG, Stagni G. Evaluation of local bioavailability of metronidazole from topical formulations using dermal microdialysis: Preliminary study in a Yucatan mini-pig model. *European Journal of Pharmaceutical Sciences*, 159:105741, 2021.
5. Kuzma BA, Senemar S, Ramezanli T, Ghosh P, Raney SG, Stagni G. The dose-duration effect on cutaneous pharmacokinetics of metronidazole from topical dermatological formulations in Yucatan mini-pigs. *European Journal of Pharmaceutics and Biopharmaceutics*, 2022.
6. Pensado A, Chiu WS, Cordery S, Rantou E, Bunge A, Delgado-Charro M, Guy R. Stratum corneum sampling to assess bioequivalence between topical acyclovir products. *Pharmaceutical Research*, 36:1–16, 2019.
7. Lister T, Wright PA, Chappell PH. Optical properties of human skin. *Journal of Biomedical Optics*, 17(9):090901–090901, 2012.
8. Ash C, Dubec M, Donne K, Bashford T. Effect of wavelength and beam width on penetration in light-tissue interaction using computational methods. *Lasers in Medical Science*, 32:1909–1918, 2017.
9. Wan B, Ganier C, Du-Harpur X, Harun N, Watt F, Patalay R, Lynch M. Applications and future directions for optical coherence tomography in dermatology. *British Journal of Dermatology*, 184(6):1014–1022, 2021.
10. Jacques SL, Alter C, Prahl SA. Angular dependence of HeNe laser light scattering by human dermis. *Lasers in the Life Sciences*, 1(4):309–333, 1987.
11. Olesen CM, Fuchs CSK, Philipsen PA, Hædersdal M, Agner T, Clausen ML. Advancement through epidermis using tape stripping technique and reflectance confocal microscopy. *Scientific Reports*, 9(1):12217, 2019.
12. Tabosa MAM, Vitry P, Zarmpi P, Bunge AL, Belsey NA, Tsikritsis D, Woodman TJ, White KJ, Delgado-Charro MB, Guy RH. Quantification of chemical uptake into the skin by vibrational spectroscopies and stratum corneum sampling. *Molecular Pharmaceutics*, 20(5):2527–2535, 2022.
13. Kuzma BA, Pence IJ, Ho A, Evans CL. Visualizing and quantifying pharmaceutical compounds within skin using coherent Raman scattering imaging. *Journal of Visualized Experiments: JoVE*, (177), 2021.
14. Jian L, Cao Y, Zou Y. Dermal-epidermal separation by heat. *Epidermal Cells: Methods and Protocols*, pages 23–25, 2020.
15. Evans CL, Hermsmeier M, Yamamoto A, Chan KF. Visualizing topical drug uptake with conventional fluorescence microscopy and deep learning. *Biomed Opt Express*, pages 6864–6880, 2020.
16. Blagus T, Markelc B, Cemazar M, Kosjek T, Preat V, Miklavcic D, Sersa G. In vivo real-time monitoring system of electroporation mediated control of transdermal and topical drug delivery. *Journal of Controlled Release*, 172(3):862–871, 2013. ISSN 0168-3659.
17. Nguyen HTP, Munnier E, Perse X, Vial F, Yvergnaux F, Perrier T, Soucé M, Chourpa I. Qualitative and quantitative study of the potential of lipid nanocapsules of one hundred twenty nanometers for the topical administration of hydrophobic molecules. *Journal of Pharmaceutical Sciences*, 105(10):3191–3198, 2016. ISSN 0022-3549.
18. Sattler E, Kästle R, Arens-Corell M, Welzel J. How long does protection last? – in vivo fluorescence confocal laser scanning imaging for the evaluation of the kinetics of a topically applied lotion in an everyday setting. *Skin Research and Technology*, 18(3):370–377, 2012. ISSN 1600-0846.
19. Choi S, Kim JW, Lee YJ, Delmas T, Kim C, Park S, Lee H. Evaluation of transdermal delivery of nanoemulsions in ex vivo porcine skin using two-photon microscopy and confocal laser-scanning microscopy. *Journal of Biomedical Optics*, 19(10):106006, 2014. ISSN 1083-3668.

20. Kirejev V, Guldbrand S, Bauer B, Smedh M, Ericson MB. Novel nanocarriers for topical drug delivery: investigating delivery efficiency and distribution in skin using two-photon microscopy. *SPIE Proceedings*, 7903(1):455–463, 2011. ISSN 16057422.
21. IV JK, Kim D, So PT, Blankschtein D, Langer RS. Dual-channel two-photon microscopy study of transdermal transport in skin treated with low-frequency ultrasound and a chemical enhancer. *Journal of Investigative Dermatology*, 127(12):2832–2846, 2007. ISSN 0022-202X.
22. Yu L, Lei Y, Ma Y, Liu M, Zheng J, Dan D, Gao P. A comprehensive review of fluorescence correlation spectroscopy. *Frontiers in Physics*, 9:644450, 2021. ISSN 2296424X.
23. Yu B, Blankschtein D, Langer R, Dong CY, So PT. In vitro visualization and quantification of oleic acid induced changes in transdermal transport using two-photon fluorescence microscopy. *Journal of Investigative Dermatology*, 117(1):16–25, 2001. ISSN 0022-202X.
24. Yu B, Langer R, Blankschtein D, Kim KH, So PT. Visualization of oleic acid-induced transdermal diffusion pathways using two-photon fluorescence microscopy. *Journal of Investigative Dermatology*, 120(3):448–455, 2003. ISSN 0022-202X.
25. Tseng TY, Hsu HC, Yang CS, Tsai TH, Dong CY. Investigation of transport dynamics in oleic acid–induced transdermal drug delivery by two-photon fluorescence microscopy: an ex-vivo study of mouse skin. *Journal of Biomedical Optics*, 18(9):096016, 2013. ISSN 1083-3668.
26. Alex A, Frey S, Angelene H, Neitzel CD, Li J, Bower AJ, Spillman DR, Marjanovic M, Chaney EJ, Medler JL, Lee W, Johnson LSV, Boppart SA, Arp Z. In situ biodistribution and residency of a topical anti-inflammatory using fluorescence lifetime imaging microscopy. *British Journal of Dermatology*, 179(6):1342–1350, 2018. ISSN 0007-0963.
27. Zhu Y, Choe CS, Ahlberg S, Meinke MC, Ulrike A, Lademann JM, Darvin ME. Penetration of silver nanoparticles into porcine skin ex vivo using fluorescence lifetime imaging microscopy, Raman microscopy, and surface-enhanced Raman scattering microscopy. *Journal of Biomedical Optics*, 20(5):051006, 2015. ISSN 1083-3668.
28. Mangion SE, Sandiford L, Mohammed Y, Roberts MS, Holmes AM. Multi-modal imaging to assess the follicular delivery of zinc pyrithione. *Pharmaceutics*, 14(5):1076, 2022. ISSN 1999-4923.
29. Jeong S, Hermsmeier M, Osseiran S, Yamamoto A, Nagavarapu U, Chan KF, Evans CL. Visualization of drug distribution of a topical minocycline gel in human facial skin. *Biomedical Optics Express*, 9(7):3434, 2018. ISSN 2156-7085.
30. Jeong S, Greenfield DA, Hermsmeier M, Yamamoto A, Chen X, Chan KF, Evans CL. Time-resolved fluorescence microscopy with phasor analysis for visualizing multicomponent topical drug distribution within human skin. *Scientific Reports*, 10(1):1–12, 2020. ISSN 2045-2322.
31. Stokes GG. On the change of refrangibility of light. *Proceedings of the Royal Society of London*, pages 195–200, 1854.
32. Grzybowski A, Pietrzak K. Maria goeppert-mayer (1906–1972): Two-photon effect on dermatology. *Clinics in Dermatology*, 31:221–225, 2013.
33. Giordmaine JA, Rentzepis PM, Shapiro SL, Wecht KW. Two-photon excitation of fluorescence by picosecond light pulses. *Appl Phys Lett*, 11(7):216–218, 1967.
34. He GS, Bhawalkar JD, Zhao CF, Prasad PN. Optical limiting effect in a two-photon absorption dye doped solid matrix. *Appl Phys Lett*, 67(17):2433–2435, 1995.
35. König K, Göhlert A, Liehr T, Loncarevic I, Riemann I. Two-photon multicolor fish: A versatile technique to detect specific sequences with in single DNA molecules in cells and tissues. *Single Molecules*, 1(1):5–86, 2000.
36. He GS, Lin TC, Dai J, Prasad PN, Kannan R, Dombroskie AG, Vaia RA, Tan LS. Degenerate two-photon-absorption spectral studies of highly two-photon active organic chromophores. *J Chem Phys*, 120(11):5275–5284, 2004.
37. Wee TL, Tzeng YK, Han CC, Chang HC, Fann W, Hsu JH, Chen KM, Yu YC. Two-photon excited fluorescence of nitrogen-vacancy centers in proton-irradiated type ib diamond†. *The Journal of Physical Chemistry A*, 111(38):9379–9386, 2007.

38. Chen K, Tian Z, Kong L. Advances of optical miniscopes for in vivo imaging of neural activity in freely moving animals. *Frontiers in Neuroscience*, 16:994079, 2022. ISSN 1662453X.
39. Kristoffersen AS, Erga SR, Hamre1 B, Øyvind Frette. Testing fluorescence lifetime standards using two-photon excitation and time-domain instrumentation: Fluorescein, quinine sulfate and green fluorescent protein. *Journal of Fluorescence*, pages 1065–1073, 2018.
40. Nwaneshiudu1 A, Kuschal C, Sakamoto FH, Anderson RR, Schwarzenberger K, Young RC. Introduction to confocal microscopy. *Journal of Investigative Dermatology*, 132(12):1–5, 2012.
41. Pawley JB. Handbook of biological confocal microscopy, third edition. *Springer US*, 2006.
42. Pena AM, Chen X, Pence IJ, Bornschlögl T, Jeong S, Grégoire S, Luengo GS, Hallegot P, Obeidy P, Feizpour A, Chan KF, Evans CL. Imaging and quantifying drug delivery in skin – part 2: Fluorescence and vibrational spectroscopic imaging methods. *Advanced Drug Delivery Reviews*, 153:147–168, 2020. ISSN 0169-409X.
43. Lefort C. A review of biomedical multiphoton microscopy and its laser sources. *Journal of Physics D: Applied Physics*, 50(42):243001, 2017.
44. Young MD, Field JJ, Sheetz KE, Bartels RA, Squier J. A pragmatic guide to multiphoton microscope design. *Advances in Optics and Photonics*, 7(2):276–378, 2015.
45. Lee M, Kannan S, Muniraj G, Rosa V, Lu WF, Fuh JYH, Sriram G, Cao T. Two-photon fluorescence microscopy and applications in angiogenesis and related molecular events. *Tissue Eng Part B Rev*, 28(4):926–937, 2022.
46. Koenig K. Review: Clinical in vivo multiphoton flim tomography. *Methods and Applications in Fluorescence*, 8(3):034002–034002, 2020. ISSN 2050-6120.
47. Datta R, Heaster TM, Sharick JT, Gillette AA, Skala MC. Fluorescence lifetime imaging microscopy: fundamentals and advances in instrumentation, analysis, and applications. *Journal of Biomedical Optics*, 25(7):1–43, 2020.
48. Bitton A, Sambrano J, Valentino S, Houston JP. A review of new high-throughput methods designed for fluorescence lifetime sensing from cells and tissues. *Frontiers in Physics*, 9:648553, 2021.
49. Moon S, Won Y, Kim DY. Analog mean-delay method for high-speed fluorescence lifetime measurement. *Opt Express*, 17(4):2834–2849, 2009. PMID/PMCID:19219188.
50. Won YJ, Han WT, Kim DY. Precision and accuracy of the analog mean-delay method for high-speed fluorescence lifetime measurement. *J Opt Soc Am A*, 28(10):2026–2032, 2011. PMID/PMCID:21979507.
51. Bell SE, Xu Y. Infrared spectroscopy | industrial applications. *Encyclopedia of Analytical Science*, pages 124–133, 2019.
52. Griffiths PR, Haseth JAD, Winefordner JD. Fourier transform infrared spectrometry (2nd edn.). pages 131–138, 2007.
53. Carlton RA. Pharmaceutical microscopy. pages 131–138, 2011.
54. Subramanian A, Rodriguez-Saona L. Infrared spectroscopy for food quality analysis and control: Chapter 7 - Fourier transform infrared (FTIR) spectroscopy. *Academic Press*, pages 145–178, 2009.
55. Subramanian A, Rodriguez-Saona L. Fourier transform infrared (FTIR) spectroscopy. *Infrared Spectroscopy for Food Quality Analysis and Control*, pages 145–178, 2009.
56. Varughese KOG, Banerjee KK, Hradaynath R. Design of fast aplanatic IR objectives using high refractive-index materials. *Applied Optics*, 22(2):347–353, 1983.
57. Gruescu CM, Bodea RD. New advances in mechanisms, mechanical transmissions and robotics. pages 275–283, 2021.
58. Mostaço-Guidolin, Murakami LB, Nomizo LS, Auro ; Bachmann L. Fourier transform infrared spectroscopy of skin cancer cells and tissues. *Applied Spectroscopy Reviews*, 44(5):438–455, 2009.
59. Kaur H, Rana B, Tomar D, Kaur S, Jena KC. Fundamentals of ATR-FTIR spectroscopy and its role for probing in-situ molecular-level interactions. *Progress in Optical Science and Photonics*, 13:3–37, 2021. ISSN 2363510X.

60. Dazzi A, Prater CB. AFM-IR: Technology and applications in nanoscale infrared spectroscopy and chemical imaging. *Chemical Reviews*, 117:5146–5173, 2017. ISSN 15206890.
61. Lasch P. Spectral pre-processing for biomedical vibrational spectroscopy and microspectroscopic imaging. *Chemometrics and Intelligent Laboratory Systems*, 117:100–114, 2012.
62. Gautam R, Vanga S, Ariese F, Umapathy S. Review of multidimensional data processing approaches for Raman and infrared spectroscopy. *EPJ Techniques and Instrumentation*, 2:1–38, 2015.
63. Wang L, Mizaikoff B. Application of multivariate data-analysis techniques to biomedical diagnostics based on mid-infrared spectroscopy. *Analytical and bioanalytical chemistry*, 391:1641–1654, 2008.
64. Rajalahti T, Kvalheim OM. Multivariate data analysis in pharmaceutics: a tutorial review. *International Journal of Pharmaceutics*, 417(1-2):280–290, 2011.
65. Jones RR, Hooper DC, Zhang L, Wolverson D, Valev VK. Raman techniques: fundamentals and frontiers. *Nanoscale Research Letters*, 14:1–34, 2019.
66. Petry R, Schmitt M, Popp J. Raman spectroscopy—a prospective tool in the life sciences. *Chemphyschem*, 4(1):14–30, 2003.
67. Wilson R, Cossins AR, Spiller DG. Encoded microcarriers for high-throughput multiplexed detection. *Angewandte Chemie International Edition*, 45(37):6104–6117, 2006.
68. Lane LA, Qian X, Nie S. SERS nanoparticles in medicine: from label-free detection to spectroscopic tagging. *Chemical Reviews*, 115(19):10489–10529, 2015.
69. Rousseau DL. *Optical techniques in biological research*. Academic Press New York, 1984.
70. Pelletier MJ. Quantitative analysis using Raman spectrometry. *Applied Spectroscopy*, 57(1):20A–42A, 2003.
71. Paidi SK, Pandey R, Barman I. Emerging trends in biomedical imaging and disease diagnosis using Raman spectroscopy. *Molecular and Laser Spectroscopy*, pages 623–652, 2020.
72. Pence I, Mahadevan-Jansen A. Clinical instrumentation and applications of Raman spectroscopy. *Chemical Society Reviews*, 45(7):1958–1979, 2016.
73. Choe C, Schleusener J, Ri J, Choe S, Kim P, Lademann J, Darvin ME. Quantitative determination of concentration profiles of skin components and topically applied oils by tailored multivariate curve resolution-alternating least squares using in vivo confocal Raman micro-spectroscopy. *J Biophotonics*, 16(2):e202200219, 2022.
74. Iliopoulos F, Tang CF, Li Z, Rahma A, Lane ME. Confocal Raman spectroscopy for assessing bioequivalence of topical formulations. *Pharmaceutics*, 15(4):1075, 2023.
75. Kocsis D, Kichou H, Döme K, Varga-Medveczky Z, Révész Z, Antal I, Erdo F. Structural and functional analysis of excised skins and human reconstructed epidermis with confocal Raman spectroscopy and in microfluidic diffusion chambers. *Pharmaceutics*, 14(8):1689, 2022.
76. Garvie-Cook H, Hoppel M, Guy RH. Raman spectroscopic tools to probe the skin–(trans)dermal formulation interface. *Molecular Pharmaceutics*, 19(11):4010–4016, 2022.
77. Tippavajhala VK, de Oliveira Mendes T, Martin AA. In vivo human skin penetration study of sunscreens by confocal Raman spectroscopy. *AAPS PharmSciTech*, 19:753–760, 2018.
78. Zhang G, Moore DJ, Sloan KB, Flach CR, Mendelsohn R. Imaging the prodrug-to-drug transformation of a 5-fluorouracil derivative in skin by confocal Raman microscopy. *Journal of Investigative Dermatology*, 127(5):1205–1209, 2007.
79. Bonnist EY, Gorce JP, MacKay C, Pendlington RU, Pudney PD. Measuring the penetration of a skin sensitizer and its delivery vehicles simultaneously with confocal Raman spectroscopy. *Skin Pharmacology and Physiology*, 24(5):274–283, 2011.
80. dos Santos L, Tippavajhala VK, Mendes TO, da Silva MG, Fávero PP, Soto CAT, Martin AA. Evaluation of penetration process into young and elderly skin using confocal Raman spectroscopy. *Vibrational Spectroscopy*, 100:123–130, 2019.
81. Binder L, SheikhRezaei S, Baierl A, Gruber L, Wolzt M, Valenta C. Confocal Raman spectroscopy: In vivo measurement of physiological skin parameters – a pilot study. *Journal of Dermatological Science*, 88(3):280–288, 2017.

82. Choe CS, Pak GJ, Ascencio SM, Darvin ME. Quantification of skin penetration of caffeine and propylene glycol applied topically in a mixture by tailored multivariate curve resolution-alternating least squares of depth-resolved Raman spectra. *Journal of Biophotonics*, page e202300146, 2023.
83. Kichou H, Munnier E, Dancik Y, Kemel K, Byrne HJ, Tfayli A, Bertrand D, Soucé M, Chourpa I, Bonnier F. Estimating the analytical performance of Raman spectroscopy for quantification of active ingredients in human stratum corneum. *Molecules*, 27(9):2843, 2022.
84. Kang Y, Zhang F. Image of the distribution profile of targets in skin by Raman spectroscopy-based multivariate analysis. *Skin Res Technol*, 28(3):389–493, 2022.
85. Fleischli FD, Mathes S, Adlhart C. Label free non-invasive imaging of topically applied actives in reconstructed human epidermis by confocal Raman spectroscopy. *Vibrational Spectroscopy*, 68:29–33, 2013.
86. Liu Y, Lunter DJ. Profiling skin penetration using PEGylated emulsifiers as penetration enhancers via confocal Raman spectroscopy and fluorescence spectroscopy. *European Journal of Pharmaceutics and Biopharmaceutics*, 166:1–9, 2021.
87. Krombholz R, Liu Y, Lunter DJ. In-line and off-line monitoring of skin penetration profiles using confocal Raman spectroscopy. *Pharmaceutics*, 13(1):67, 2021.
88. Pyatski Y, Zhang Q, Mendelsohn R, Flach CR. Effects of permeation enhancers on flufenamic acid delivery in ex vivo human skin by confocal Raman microscopy. *International Journal of Pharmaceutics*, 505(1-2):319–328, 2016.
89. Qi Y, Zhang R, Rajarahm P, Zhang S, Ebrahim Attia AB, Bi R, Olivo M. Simultaneous dual-wavelength source Raman spectroscopy with a handheld confocal probe for analysis of the chemical composition of in vivo human skin. *Analytical Chemistry*, 95(12):5240–5247, 2023.
90. Rasmussen T. Overview of high-efficiency transmission gratings for molecular spectroscopy. *Spectroscopy*, 29(4):32–39, 2014.
91. Kumbhar K, Kshirasagar KP. Comparative study of CCD & CMOS sensors for image processing. *International Journal of Innovative Research in Electrical, Electronics, Instrumentation and Control Engineering*, 3:194–196, 2015.
92. Franzen L, Windbergs M. Applications of Raman spectroscopy in skin research—from skin physiology and diagnosis up to risk assessment and dermal drug delivery. *Advanced Drug Delivery Reviews*, 89:91–104, 2015.
93. Kong K, Kendall C, Stone N, Notingher I. Raman spectroscopy for medical diagnostics—from in-vitro biofluid assays to in-vivo cancer detection. *Advanced Drug Delivery Reviews*, 89:121–134, 2015.
94. Bielfeldt S, Bonnier F, Byrne H, Chourpa I, Dancik Y, Lane ME, Lunter DJ, Munnier E, Puppels G, Tfayli A, *et al.* Monitoring dermal penetration and permeation kinetics of topical products; the role of Raman microspectroscopy. *TrAC Trends in Analytical Chemistry*, 156:116709, 2022.
95. Caspers PJ, Bruining HA, Puppels GJ, Lucassen GW, Carter EA. In vivo confocal Raman microspectroscopy of the skin: noninvasive determination of molecular concentration profiles. *Journal of Investigative Dermatology*, 116(3):434–442, 2001.
96. Gotter B, Faubel W, Neubert R. Optical methods for measurements of skin penetration. *Skin Pharmacology and Physiology*, 21(3):156–165, 2008.
97. Strachan CJ, Rades T, Gordon KC, Rantanen J. Raman spectroscopy for quantitative analysis of pharmaceutical solids. *Journal of Pharmacy and Pharmacology*, 59(2):179–192, 2007.
98. Eckhardt G, Hellwarth RW, McClung FJ, Schwarz SE, Weiner D, Woodbury EJ. Stimulated Raman Scattering From Organic Liquids. *Physical Review Letters*, 9(11):455–457, 1962.
99. Zumbusch A, Holtom GR, Xie XS. Three-Dimensional Vibrational Imaging by Coherent Anti-Stokes Raman Scattering. *Physical Review Letters*, 82(20):4142–4145, 1999.
100. Evans CL, Xie XS. Coherent anti-stokes Raman scattering microscopy: chemical imaging for biology and medicine. *Annual Review of Analytical Chemistry*, 1(1):883, 2008.
101. Zhang C, Zhang D, Cheng JX. Coherent Raman Scattering Microscopy in Biology and Medicine. *Annual Review of Biomedical Engineering*, 17:415–445, 2015.

102. Lotem H, Lynch RT, Bloembergen N. Interference between Raman resonances in four-wave difference mixing. *Physical Review A*, 14(5):1748–1755, 1976.
103. Manifold B, Fu D. Quantitative Stimulated Raman Scattering Microscopy: Promises and Pitfalls. *Annual Review of Analytical Chemistry*, 15(1):269–289, 2022.
104. Sepp K, Lee M, Bluntzer MTJ, Helgason GV, Hulme AN, Brunton VG. Utilizing Stimulated Raman Scattering Microscopy To Study Intracellular Distribution of Label-Free Ponatinib in Live Cells. *Journal of Medicinal Chemistry*, 63(5):2028–2034, 2020.
105. Feizpour A, Marstrand T, Bastholm L, Eirefelt S, Evans CL. Label-free quantification of pharmacokinetics in skin with stimulated Raman scattering microscopy and deep learning. *Journal of Investigative Dermatology*, 141(2):395–403, 2021.
106. Francis AT, Nguyen TT, Lamm MS, Teller R, Forster SP, Xu W, Rhodes T, Smith RL, Kuiper J, Su Y, Fu D. In Situ Stimulated Raman Scattering (SRS) Microscopy Study of the Dissolution of Sustained-Release Implant Formulation. *Molecular Pharmaceutics*, 15(12):5793–5801, 2018.
107. Kuzma BA, Tu D, Goss A, Iliopoulos F, Slade JB, Wiatrowski A, Feizpour A, Evans CL. Instantaneous topical drug quantification using a 3D printed microfluidic device and coherent Raman imaging. *OpenNano*, 12:100151, 2023.
108. Boyd RW. *Nonlinear Optics*. Academic Press is an imprint of Elsevier, San Diego, 4 edition, 2019.
109. Prince RC, Frontiera RR, Potma EO. Stimulated Raman scattering: from bulk to nano. *Chemical reviews*, 117(7):5070–5094, 2017.
110. Shen YR, Bloembergen N. Theory of Stimulated Brillouin and Raman Scattering. *Physical Review*, 137(6A):A1787–A1805, 1965.
111. Rigneault H, Berto P. Tutorial: Coherent Raman light matter interaction processes. *APL Photonics*, 3(9):091101, 2018.
112. Genchi L, Laptenok SP, Liberale C. Background signals in stimulated Raman scattering microscopy and current solutions to avoid them. *Advances in Physics: X*, 8(1):2176258, 2023.
113. Goel A, Tsikritsis D, Belsey NA, Pendlington R, Glavin S, Chen T. Measurement of chemical penetration in skin using stimulated Raman scattering microscopy and multivariate curve resolution-alternating least squares. *Spectrochimica Acta Part A: Molecular and Biomolecular Spectroscopy*, 296:122639, 2023.
114. Ploetz E, Laimgruber S, Berner S, Zinth W, Gilch P. Femtosecond stimulated Raman microscopy. *Applied Physics B*, 87(3):389–393, 2007.
115. Freudiger CW, Min W, Saar BG, Lu S, Holtom GR, He C, Tsai JC, Kang JX, Xie XS. Label-Free Biomedical Imaging with High Sensitivity by Stimulated Raman Scattering Microscopy. *Science*, 322(5909):1857–1861, 2008.
116. Pence IJ, Kuzma BA, Brinkmann M, Hellwig T, Evans CL. Multi-window sparse spectral sampling stimulated Raman scattering microscopy. *Biomedical Optics Express*, 12(10):6095–6114, 2021.
117. Liao CS, Slipchenko MN, Wang P, Li J, Lee SY, Oglesbee RA, Cheng JX. Microsecond scale vibrational spectroscopic imaging by multiplex stimulated Raman scattering microscopy. *Light: Science & Applications*, 4(3):e265–e265, 2015.
118. Bae K, Zheng W, Lin K, Lim SW, Chong YK, Tang C, King NK, Ti Ang CB, Huang Z. Epi-Detected Hyperspectral Stimulated Raman Scattering Microscopy for Label-Free Molecular Subtyping of Glioblastomas. *Analytical Chemistry*, 90(17), 2018.
119. Fu D, Holtom G, Freudiger C, Zhang X, Xie XS. Hyperspectral Imaging with Stimulated Raman Scattering by Chirped Femtosecond Lasers. *The Journal of Physical Chemistry B*, 117(16):4634–4640, 2013.
120. Francis A, Berry K, Chen Y, Figueroa B, Fu D. Label-free pathology by spectrally sliced femtosecond stimulated Raman scattering (SRS) microscopy. *PLOS ONE*, 12(5):e0178750, 2017.
121. Ozeki Y, Umemura W, Sumimura K, Nishizawa N, Fukui K, Itoh K. Stimulated Raman hyperspectral imaging based on spectral filtering of broadband fiber laser pulses. *Optics Letters*, 37(3):431–433, 2012.

122. Karpf S, Eibl M, Wieser W, Klein T, Huber R. A Time-Encoded Technique for fibre-based hyperspectral broadband stimulated Raman microscopy. *Nature Communications*, 6(1):6784, 2015.
123. Figueroa B, Fu W, Nguyen T, Shin K, Manifold B, Wise F, Fu D. Broadband hyperspectral stimulated Raman scattering microscopy with a parabolic fiber amplifier source. *Biomedical Optics Express*, 9(12):6116–6131, 2018.
124. Wang P, Slipchenko MN, Zhou B, Pinal R, Cheng JX. Mechanisms of epi-detected stimulated Raman scattering microscopy. *IEEE Journal on Selected Topics in Quantum Electronics*, 18(1), 2012.
125. Slipchenko MN, Chen H, Ely DR, Jung Y, Carvajal MT, Cheng JX. Vibrational imaging of tablets by epi-detected stimulated Raman scattering microscopy. *Analyst*, 135(10), 2010.
126. Figueroa B, Nguyen T, Sotthivirat S, Xu W, Rhodes T, Lamm MS, Smith RL, John CT, Su Y, Fu D. Detecting and Quantifying Microscale Chemical Reactions in Pharmaceutical Tablets by Stimulated Raman Scattering Microscopy. *Analytical Chemistry*, 91(10):6894–6901, 2019.
127. Freudiger CW, Yang W, Holtom GR, Peyghambarian N, Xie XS, Kieu KQ. Stimulated Raman scattering microscopy with a robust fibre laser source. *Nature Photonics*, 8(2):153–159, 2014.
128. Andresen ER, Berto P, Rigneault H. Stimulated Raman scattering microscopy by spectral focusing and fiber-generated soliton as stokes pulse. *Optics Letters*, 36(13):2387–2389, 2011.
129. Liu B, Wang P, Kim JI, Zhang D, Xia Y, Chapple C, Cheng JX. Vibrational fingerprint mapping reveals spatial distribution of functional groups of lignin in plant cell wall. *Analytical chemistry*, 87(18):9436–9442, 2015.
130. Saltarelli F, Kumar V, Viola D, Crisafi F, Preda F, Cerullo G, Polli D. Broadband stimulated Raman scattering spectroscopy by a photonic time stretcher. *Optics express*, 24(19):21264–21275, 2016.
131. Liao CS, Wang P, Wang P, Li J, Lee HJ, Eakins G, Cheng JX. Spectrometer-free vibrational imaging by retrieving stimulated Raman signal from highly scattered photons. *Science advances*, 1(9):e1500738, 2015.
132. Sacré PY, De Bleye C, Chavez PF, Netchacovitch L, Hubert P, Ziemons E. Data processing of vibrational chemical imaging for pharmaceutical applications. *Journal of Pharmaceutical and Biomedical Analysis*, 101:123–140, 2014.
133. Otsuka Y, Makara K, Satoh S, Hashimoto H, Ozeki Y. On-line visualization of multicolor chemical images with stimulated Raman scattering spectral microscopy. *Analyst*, 140(9):2984–2987, 2015.
134. Masters B, So P, Gratton E. Multiphoton excitation fluorescence microscopy and spectroscopy of in vivo human skin. *Biophysical Journal*, 72(6):2405–2412, 1997. ISSN 0006-3495.
135. Masters BR, So PTC, Gratton E. Multiphoton excitation microscopy of in vivo human skin: Functional and morphological optical biopsy based on three-dimensional imaging, lifetime measurements and fluorescence spectroscopy. *Annals of the New York Academy of Sciences*, 838(1):58–67, 1998. ISSN 0077-8923.
136. Masters B, So P, Gratton E. Optical biopsy of in vivo human skin: Multi-photon excitation microscopy. *Lasers in Medical Science*, 13(3):196–203, 1998. ISSN 0268-8921.
137. Hædersdal M, Sakamoto FH, Farinelli WA, Doukas AG, Tam J, Anderson RR. Fractional co2 laser-assisted drug delivery. *Lasers in Surgery and Medicine*, 42(2):113–122, 2010. ISSN 0196-8092.
138. Hermsmeier M, Jeong S, Yamamoto A, Chen X, Nagavarapu U, Evans CL, Chan KF. Characterization of human cutaneous tissue autofluorescence: implications in topical drug delivery studies with fluorescence microscopy. *Biomedical Optics Express*, 9(11):5400–5418, 2018.
139. Evans CL, Hermsmeier M, Yamamoto A, Chan KF. Visualizing topical drug uptake with conventional fluorescence microscopy and deep learning. *Biomedical Optics Express*, 11(12):6864–6880, 2020. ISSN 2156-7085.
140. König K. *Multiphoton Microscopy and Fluorescence Lifetime Imaging*. De Gruyter, 2018. ISBN 9783110429985.

141. Chernousova S, Epple M. Silver as antibacterial agent: Ion, nanoparticle, and metal. *Angewandte Chemie (International ed)*, 52(6):1636–1653, 2013. ISSN 1433-7851.
142. Wijnhoven SW, Peijnenburg WJ, Herberts CA, Hagens WI, Oomen AG, Heugens EH, Roszek B, Bisschops J, Gosens I, Van De Meent D, Dekkers S, De Jong WH, van Zijverden M, Sips AJ, Geertsma RE. Nano-silver - a review of available data and knowledge gaps in human and environmental risk assessment. *Nanotoxicology*, 3(2):109–138, 2009. ISSN 1743-5390.
143. Koenig K, Riemann I. High-resolution multiphoton tomography of human skin with subcellular spatial resolution and picosecond time resolution. *Journal of Biomedical Optics*, 8(3):432–439, 2003. ISSN 1083-3668.
144. König K, Ehlers A, Stracke F, Riemann I. In vivo drug screening in human skin using femtosecond laser multiphoton tomography. *Skin Pharmacology and Physiology*, 19(2):78–88, 2006. ISSN 1660-5527.
145. König K, Riemann I, Ehlers A, Buckle R, Dimitrow E, Kaatz M, Fluhr J, Elsner P. In vivo multiphoton tomography of skin cancer. In *Proceedings of SPIE*, volume 6089, pages 60890R–60890R–7. SPIE, 2006. ISBN 0819461318. ISSN 1605-7422.
146. Koehler MJ, Koenig K, Elsner P, Bueckle R, Kaatz M. In vivo assessment of human skin aging by multiphoton laser scanning tomography. *Optics letters*, 31(19):2879–2881, 2006. ISSN 0146-9592.
147. Shakya BR, Shrestha P, Teppo HR, Rieppo L. The use of Fourier transform infrared (FTIR) spectroscopy in skin cancer research: a systematic review. *Applied Spectroscopy Reviews*, 56(5):347–379, 2021.
148. Mendelsohn R, Flach CR, Moore DJ. Determination of molecular conformation and permeation in skin via IR spectroscopy, microscopy, and imaging. *Biochimica et Biophysica Acta (BBA)-Biomembranes*, 1758(7):923–933, 2006.
149. Pena AM, Chen X, Pence IJ, Bornschlögl T, Jeong S, Grégoire S, Luengo GS, Hallegot P, Obeidy P, Feizpour A, *et al.* Imaging and quantifying drug delivery in skin–part 2: Fluorescence and vibrational spectroscopic imaging methods. *Advanced Drug Delivery Reviews*, 153:147–168, 2020.
150. Binder L, Kulovits EM, Petz R, Ruthofer J, Baurecht D, Klang V, Valenta C. Penetration monitoring of drugs and additives by ATR-FTIR spectroscopy/tape stripping and confocal Raman spectroscopy–a comparative study. *European Journal of Pharmaceutics and Biopharmaceutics*, 130:214–223, 2018.
151. Vater C, Hlawaty V, Werdenits P, Cichoń MA, Klang V, Elbe-Bürger A, Wirth M, Valenta C. Effects of lecithin-based nanoemulsions on skin: Short-time cytotoxicity MTT and BrdU studies, skin penetration of surfactants and additives and the delivery of curcumin. *International Journal of Pharmaceutics*, 580:119209, 2020.
152. Schwarz JC, Pagitsch E, Valenta C. Comparison of ATR–FTIR spectra of porcine vaginal and buccal mucosa with ear skin and penetration analysis of drug and vehicle components into pig ear. *European Journal of Pharmaceutical Sciences*, 50(5):595–600, 2013.
153. Nicoli S, Bunge AL, Delgado-Charro MB, Guy RH. Dermatopharmacokinetics: factors influencing drug clearance from the stratum corneum. *Pharmaceutical Research*, 26:865–871, 2009.
154. Balázs B, Sipos P, Danciu C, Avram S, Soica C, Dehelean C, Varju G, Erős G, Budai-Szűcs M, Berkó S, *et al.* ATR-FTIR and Raman spectroscopic investigation of the electroporation-mediated transdermal delivery of a nanocarrier system containing an antitumour drug. *Biomedical Optics Express*, 7(1):67–78, 2016.
155. Alberti I, Kalia YN, Naik A, Bonny JD, Guy RH. In vivo assessment of enhanced topical delivery of terbinafine to human stratum corneum. *Journal of Controlled Release*, 71(3):319–327, 2001.
156. Stani C, Vaccari L, Mitri E, Birarda G. FTIR investigation of the secondary structure of type i collagen: New insight into the amide iii band. *Spectrochimica Acta Part A: Molecular and Biomolecular Spectroscopy*, 229:118006, 2020.
157. Barba C, Alonso C, Martí M, Carrer V, Yousef I, Coderch L. Selective modification of skin barrier lipids. *Journal of pharmaceutical and biomedical analysis*, 172:94–102, 2019.

158. Dumas P, Miller L. The use of synchrotron infrared microspectroscopy in biological and biomedical investigations. *Vibrational Spectroscopy*, 32(1):3–21, 2003.
159. Nasse MJ, Walsh MJ, Mattson EC, Reininger R, Kajdacsy-Balla A, Macias V, Bhargava R, Hirschmugl CJ. High-resolution Fourier-transform infrared chemical imaging with multiple synchrotron beams. *Nature methods*, 8(5):413–416, 2011.
160. Barbosa AI, Lima SAC, Yousef I, Reis S. Evaluating the skin interactions and permeation of alginate/fucoidan hydrogels per se and associated with different essential oils. *Pharmaceutics*, 15(1):190, 2023.
161. Carrer V, Alonso C, Pont M, Zanuy M, Córdoba M, Espinosa S, Barba C, Oliver MA, Martí M, Coderch L. Effect of propylene glycol on the skin penetration of drugs. *Archives of Dermatological Research*, 312(5):337–352, 2020.
162. Marcott C, Lo M, Kjoller K, Domanov Y, Balooch G, Luengo GS. Nanoscale infrared (IR) spectroscopy and imaging of structural lipids in human stratum corneum using an atomic force microscope to directly detect absorbed light from a tunable IR laser source. *Experimental Dermatology*, 22(6):419–421, 2013.
163. Chaturvedi S, Garg A. An insight of techniques for the assessment of permeation flux across the skin for optimization of topical and transdermal drug delivery systems. *Journal of Drug Delivery Science and Technology*, 62:102355, 2021.
164. Wartewig S, Neubert RH. Pharmaceutical applications of mid-IR and Raman spectroscopy. *Advanced Drug Delivery Reviews*, 57(8):1144–1170, 2005.
165. Ruela ALM, Perissinato AG, Lino MEdS, Mudrik PS, Pereira GR. Evaluation of skin absorption of drugs from topical and transdermal formulations. *Brazilian Journal of Pharmaceutical Sciences*, 52:527–544, 2016.
166. Kopečná M, Macháček M, Nováčková A, Paraskevopoulos G, Roh J, Vávrová K. Esters of terpene alcohols as highly potent, reversible, and low toxic skin penetration enhancers. *Scientific Reports*, 9(1):14617, 2019.
167. Zhang Y, Liu C, Xu W, Quan P, Luo Z, Yang D, Fang L. An investigation on percutaneous permeation of flurbiprofen enantiomers: The role of molecular interaction between drug and skin components. *International Journal of Pharmaceutics*, 601:120503, 2021.
168. Zhang Q, Saad P, Mao G, Walters RM, Mack Correa MC, Mendelsohn R, Flach CR. Infrared spectroscopic imaging tracks lateral distribution in human stratum corneum. *Pharmaceutical Research*, 31:2762–2773, 2014.
169. Kang JH, Chon J, Kim YI, Lee HJ, Oh DW, Lee HG, Han CS, Kim DW, Park CW. Preparation and evaluation of tacrolimus-loaded thermosensitive solid lipid nanoparticles for improved dermal distribution. *International Journal of Nanomedicine*, pages 5381–5396, 2019.
170. Kemel K, Deniset-Besseau A, Baillet-Guffroy A, Faivre V, Dazzi A, Laugel C. Nanoscale investigation of human skin and study of skin penetration of Janus nanoparticles. *International Journal of Pharmaceutics*, 579:119193, 2020.
171. McAuley WJ, Chavda-Sitaram S, Mader K, Tetteh J, Lane M, Hadgraft J. The effects of esterified solvents on the diffusion of a model compound across human skin: An ATR-FTIR spectroscopic study. *International Journal of Pharmaceutics*, 447(1-2):1–6, 2013.
172. Csizmazia E, Erős G, Berkesi O, Berkó S, Szabó-Révész P, Csányi E. Ibuprofen penetration enhance by sucrose ester examined by ATR-FTIR in vivo. *Pharmaceutical Development and Technology*, 17(1):125–128, 2012.
173. Honeywell-Nguyen PL, Gooris GS, Bouwstra JA. Quantitative assessment of the transport of elastic and rigid vesicle components and a model drug from these vesicle formulations into human skin in vivo. *Journal of Investigative Dermatology*, 123(5):902–910, 2004.
174. Goh CF, Craig DQ, Hadgraft J, Lane ME. The application of ATR-FTIR spectroscopy and multivariate data analysis to study drug crystallisation in the stratum corneum. *European Journal of Pharmaceutics and Biopharmaceutics*, 111:16–25, 2017.
175. Goh CF, Moffat JG, Craig DQ, Hadgraft J, Lane ME. Monitoring drug crystallization in percutaneous penetration using localized nanothermal analysis and photothermal microspectroscopy. *Molecular Pharmaceutics*, 16(1):359–370, 2018.

176. Pirot F, Kalia YN, Stinchcomb AL, Keating G, Bunge A, Guy RH. Characterization of the permeability barrier of human skin in vivo. *Proceedings of the National Academy of Sciences*, 94(4):1562–1567, 1997.
177. Stinchcomb AL, Pirot F, Touraille GD, Bunge AL, Guy RH. Chemical uptake into human stratum corneum in vivo from volatile and non-volatile solvents. *Pharmaceutical Research*, 16:1288–1293, 1999.
178. Westfall A, Sigurdson GT, Rodriguez-Saona LE, Giusti MM. Ex vivo and in vivo assessment of the penetration of topically applied anthocyanins utilizing ATR-FTIR/PLS regression models and HPLC-PDA-MS. *Antioxidants*, 9(6):486, 2020.
179. Medendorp JP, Paudel KS, Lodder RA, Stinchcomb AL. Near infrared spectrometry for the quantification of human dermal absorption of econazole nitrate and estradiol. *Pharmaceutical Research*, 24:186–193, 2007.
180. Egawa M. In vivo simultaneous measurement of urea and water in the human stratum corneum by diffuse-reflectance near-infrared spectroscopy. *Skin Research and Technology*, 15(2):195–199, 2009.
181. Mak VH, Potts RO, Guy RH. Percutaneous penetration enhancement in vivo measured by attenuated total reflectance infrared spectroscopy. *Pharmaceutical Research*, 7:835–841, 1990.
182. Böhling A, Bielfeldt S, Himmelmann A, Keskin M, Wilhelm KP. Comparison of the stratum corneum thickness measured in vivo with confocal Raman spectroscopy and confocal reflectance microscopy. *Skin Research and Technology*, 20(1):50–57, 2014.
183. Förster M, Bolzinger MA, Ach D, Montagnac G, Briançon S. Ingredients tracking of cosmetic formulations in the skin: A confocal Raman microscopy investigation. *Pharmaceutical Research*, 28(4):858–872, 2011.
184. Krombholz R, Fressle S, Nikolić I, Pantelić I, Savić S, Sakač MC, Lunter D. ex vivo–in vivo comparison of drug penetration analysis by confocal raman microspectroscopy and tape stripping. *Experimental Dermatology*, 31(12):1908–1919, 2022.
185. Laing S, Bielfeldt S, Wilhelm KP, Obst J. Confocal raman spectroscopy as a tool to measure the prevention of skin penetration by a specifically designed topical medical device. *Skin Research and Technology*, 25(4):578–586, 2019.
186. Mateus R, Abdalghafor H, Oliveira G, Hadgraft J, Lane ME. A new paradigm in dermatopharmacokinetics – confocal Raman spectroscopy. *International Journal of Pharmaceutics*, 444(1-2):106–108, 2013.
187. Pudney PD, Mélot M, Caspers PJ, Pol AVD, Puppels GJ. An in vivo confocal Raman study of the delivery of trans-retinol to the skin. *Applied Spectroscopy*, 61(8):804–811, 2007.
188. Mélot M, Pudney PD, Williamson AM, Caspers PJ, Pol AVD, Puppels GJ. Studying the effectiveness of penetration enhancers to deliver retinol through the stratum cornum by in vivo confocal Raman spectroscopy. *Journal of Controlled Release*, 138(1):32–39, 2009.
189. Choe C, Lademann J, Darvin ME. Analysis of human and porcine skin in vivo/ex vivo for penetration of selected oils by confocal Raman microscopy. *Skin Pharmacology and Physiology*, 28(6):318–330, 2015.
190. Lohan SB, Saeidpour S, Solik A, Schanzer S, Richter H, Dong P, Darvin ME, Bodmeier R, Patzelt A, Zoubari G, Unbehauen M, Haag R, Lademann J, Teutloff C, Bittl R, Meinke MC. Investigation of the cutaneous penetration behavior of dexamethasone loaded to nano-sized lipid particles by EPR spectroscopy, and confocal Raman and laser scanning microscopy. *European Journal of Pharmaceutics and Biopharmaceutics*, 116:102–110, 2017.
191. Liu Y, Krombholz R, Lunter DJ. Critical parameters for accurate monitoring of caffeine penetration in porcine skin using confocal Raman spectroscopy. *International Journal of Pharmaceutics*, 607:121055, 2021.
192. Tosato MG, Orallo DE, Ali SM, Churio MS, Martin AA, Dicelio L. Confocal Raman spectroscopy: In vivo biochemical changes in the human skin by topical formulations under UV radiation. *Journal of Photochemistry and Photobiology B: Biology*, 153:51–58, 2015.

193. Ascencio SM, Choe C, Meinke MC, Müller RH, Maksimov GV, Wigger-Alberti W, Lademann J, Darvin ME. Confocal raman microscopy and multivariate statistical analysis for determination of different penetration abilities of caffeine and propylene glycol applied simultaneously in a mixture on porcine skin ex vivo. *European Journal of Pharmaceutics and Biopharmaceutics*, 104:51–58, 2016.
194. Yin Y, Gong Y, Zhang X, Li J, Wang K, He Q, Bai J, Qin J, Wang S. Raman spectral imaging based histopathology of human skin tissue. *Journal of Applied Spectroscopy*, 85(3):573–579, 2018.
195. Jung N, Namjoshi S, Mohammed Y, Grice JE, Benson HAE, Raney SG, Roberts MS, Windbergs M. Application of confocal Raman microscopy for the characterization of topical semisolid formulations and their penetration into human skin ex vivo. *Pharmaceutical Research*, 39:935–948, 2022.
196. Işık D, Joshi AA, Guo X, Rancan F, Klossek A, Vogt A, Rühl E, Hedtrich S, Klinger D. Sulfoxide-functionalized nanogels inspired by the skin penetration properties of DMSO. *Biomaterials Science*, 9(3):712–725, 2021.
197. Drutis DM, Hancewicz TM, Pashkovski E, Feng L, Mihalov D, Holtom G, Ananthapadmanabhan KP, Xie XS, Misra M. Three-dimensional chemical imaging of skin using stimulated Raman scattering microscopy. *Journal of Biomedical Optics*, 19(11):111604–111604, 2014.
198. Wanjiku B, Yamamoto K, Klossek A, Schumacher F, Pischon H, Mundhenk L, Rancan F, Judd MM, Ahmed M, Zoschke C, Kleuser B, Rühl E, Schäfer-Korting M. Qualifying x-ray and stimulated Raman spectromicroscopy for mapping cutaneous drug penetration. *Analytical Chemistry*, 91(11):7208–7214, 2019.
199. Belsey NA, Garrett NL, Contreras-Rojas LR, Pickup-Gerlaugh AJ, Price GJ, Moger J, Guy RH. Evaluation of drug delivery to intact and porated skin by coherent Raman scattering and fluorescence microscopies. *Journal of Controlled Release*, 174:37–42, 2014.
200. Garrett NL, Singh B, Jones A, Moger J. Imaging microscopic distribution of antifungal agents in dandruff treatments with stimulated Raman scattering microscopy. *Journal of Biomedical Optics*, 22(6):066003–066003, 2017.
201. Ito T, Obara Y, Misawa K. Single-beam phase-modulated stimulated Raman scattering microscopy with spectrally focused detection. *Journal of the Optical Society of America B*, 34(5):1004–1015, 2017.
202. Ito T, Obara Y, Misawa K. Invited article: Spectral focusing with asymmetric pulses for high-contrast pump–probe stimulated Raman scattering microscopy. *APL Photonics*, 3(9), 2018.
203. Ito T, Iguchi R, Matsuoka F, Nishi Y, Ogihara T, Misawa K. Label-free skin penetration analysis using time-resolved, phase-modulated stimulated Raman scattering microscopy. *Biomedical Optics Express*, 12(10):6545–6557, 2021.
204. Saar BG, Contreras-Rojas LR, Xie XS, Guy RH. Imaging drug delivery to skin with stimulated Raman scattering microscopy. *Molecular Pharmaceutics*, 8(3):969–975, 2011.
205. Saar BG, Freudiger CW, Reichman J, Stanley CM, Holtom GR, Xie XS. Video-rate molecular imaging in vivo with stimulated Raman scattering. *Science*, 330(6009):1368–1370, 2010.
206. Saar BG, Johnston RS, Freudiger CW, Xie XS, Seibel EJ. Coherent Raman scanning fiber endoscopy. *Optics Letters*, 36(13):2396–2398, 2011.
207. Lin P, Ni H, Li H, Vickers NA, Tan Y, Gong R, Bifano T, Cheng JX. Volumetric chemical imaging in vivo by a remote-focusing stimulated Raman scattering microscope. *Optics Express*, 28(20):30210–30221, 2020.
208. Liao CS, Wang P, Huang CY, Lin P, Eakins G, Bentley RT, Liang R, Cheng JX. In vivo and in situ spectroscopic imaging by a handheld stimulated Raman scattering microscope. *ACS Photonics*, 5(3):947–954, 2017.
209. Iliopoulos F, Tu D, Pence IJ, Li X, Ghosh P, Luke MC, Raney SG, Rantou E, Evans CL. Determining topical product bioequivalence with stimulated Raman scattering microscopy. *Journal of Controlled Release*, 367:864–76, 2024.
210. Xia Q, Li D, Yu T, Zhu J, Zhu D. In vivo skin optical clearing for improving imaging and light-induced therapy: a review. *Journal of Biomedical Optics*, 28(6):060901–060901, 2023.

Chapter 5
In Vitro Release Testing (IVRT) of Topical Products for Local Action

Isadore Kanfer and Seeprarani Rath

5.1 Background

The term "TOPICAL" is defined as "Belonging to a Place or Spot". Hence, when a topical product is applied onto the skin, several implications and consequences need to be considered. For example, following application, a portion of the active pharmaceutical ingredient (API) may reach the systemic circulation, but such a consequence is not considered to have any therapeutic benefit. Furthermore, since penetration through the *stratum corneum (SC)* layer presents a formidable challenge, the concentration of API that may subsequently enter the bloodstream is likely to be extremely low and may be undetectable [1].

There are several possible target areas following topical application to the skin [Fig. 5.1], such as "Local", where the drug is not intended to enter the systemic circulation, "Regional", where the target site is in deeper muscle tissue or "Systemic", where the product is intended to penetrate through the skin layers and enter the blood circulation (e.g. from transdermal products [1–3]).

5.2 In Vitro Release Test

An in vitro release test (IVRT) is a test designed to monitor the release of an active pharmaceutical ingredient (API) from topical dosage forms, generally semisolid

---

I. Kanfer (✉)
Leslie Dan Faculty of Pharmacy, University of Toronto, Toronto, ON, Canada

Faculty of Pharmacy, Rhodes University, Grahamstown, South Africa

S. Rath
Leslie Dan Faculty of Pharmacy, University of Toronto, Toronto, ON, Canada

S. N. Murthy (ed.), *Topical Products and Dermal Drug Delivery*,
AAPS Introductions in the Pharmaceutical Sciences 2,
https://doi.org/10.1007/978-3-031-80525-7_5

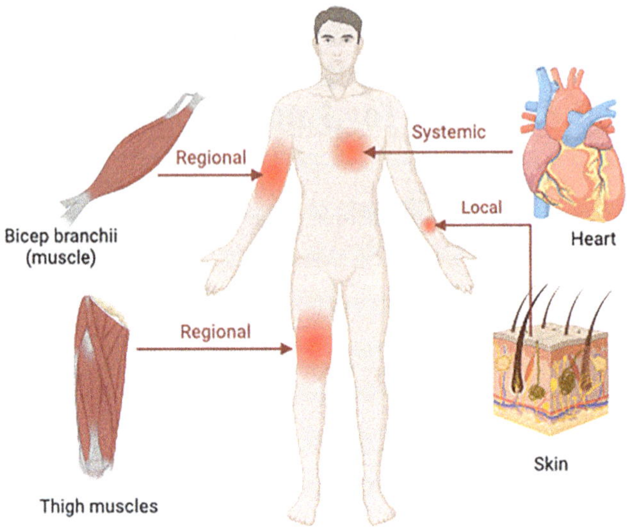

Fig. 5.1 Application sites for topical products

formulations such as creams, ointments, gels, etc., into a receptor medium across an inert membrane [4–10]. These dosage forms are generally considered to be complex products based on the definition provided by the United States Food and Drug Administration (FDA) in the Generic Drug User Fee Act (GDUFA) Reauthorization Performance Goals and Program Enhancements Fiscal Years 2018–2022 (GDUFA II Commitment Letter) [11]. Complex products are characterised according to two specific features, which involve a complex dosage form or formulation and complexity or uncertainty concerning the approval pathway or a possible alternative approach that would benefit from early scientific engagement [12].

The original primary use of IVRT was for quality control purposes, where it is well-suited to detect differences between batches. Subsequently, it has become a well-established quality control tool, widely accepted globally by regulatory authorities, including the European Medicines Agency (EMA) [13], US FDA [14] and the World Health Organization (WHO) [15]. Furthermore, IVRT can reflect numerous and combined effects of several physical and chemical parameters, including solubility, particle size of the active pharmaceutical ingredient (API) and also some rheological properties of the dosage form. Several additional applications of IVRT have been as follows:

- To quantify the amount of active pharmaceutical ingredient (API) released from semisolid dosage forms and to determine its release rate [16–18].
- To assess pharmaceutical equivalence/inequivalence involving the determination of "sameness" between pre- and post-approval product changes [14].
- For performance testing of semisolid dosage forms formulation development [19, 20].

- To detect qualitative (Q1) and quantitative (Q2) differences and also information on the microstructure and arrangement of matter between formulations (Q3) [20].
- Biowaiver to assess bioequivalence (BE), e.g. acyclovir topical ointment [21].

5.3 History

Over the years, several in vitro methods initially involved the use of excised skin for the evaluation of drug permeation across the *stratum corneum (SC)*. A VDC system is by far the most commonly used in vitro model for the study of percutaneous absorption, which depends on the interaction between the skin, the drug, and the components in the formulation vehicle. The popularity of this method is due to its simplicity, cost-effectiveness, and easily controllable experimental conditions according to the purpose of investigation [22].

Historically, in vitro diffusion cells have been employed to screen formulations, select promising candidates, elucidate the effect and mechanism of action of permeation enhancers, and establish the pharmaceutical equivalence of a formulation following post-approval changes to an approved product [23]. One of the initial studies by Treherne et al. [24] describing skin permeability involved a simple skin diffusion apparatus using rabbit skin to investigate the penetration of some non-electrolytes. Subsequently, T. Higuchi described the relationship between the amount of drug released and the square root of time [25–29] and developed a series of equations describing the principles and mechanisms involved with drug release from an ointment base (vide infra page 9). Poulsen et al. [30] studied the effect of topical vehicle compositions on the in vitro release of fluocinolone acetonide and its acetate ester. Ayres and Laskar [31] compared the release of benzocaine from some commercially available products with experimental preparations, and Bottari et al. [32–34] used a specifically designed cell to evaluate the permeation of drugs using synthetic membranes. Chowhan and Pritchard [35] also studied drug release from ointments using an apparatus consisting of a Teflon dish. Behme et al. [36] proposed a convenient and versatile technique to assess in vitro drug release from creams and ointments based on the analogy used to investigate the dissolution of solid dosage forms. Subsequently, Weng and Parrot [37] used a modification of the United States Pharmacopoeia (USP) Dissolution Apparatus 2 to study drug release from gels. Velissaratou and Papaioannou [38] studied the in vitro release of chlorpheniramine maleate from ointment bases using the Sartorius Absorption Simulator developed by Stricker [39]. A comparison of release rates of a naphthoic acid derivative from various topical formulations using an automated flow-through diffusion cell apparatus was described by Rolland et al. [40]. Sanghvi and Collins [41] used the "Enhancer Cell" developed by VanKel Industries, Inc. (Edison, NJ) and Franz cells to compare in vitro diffusion of hydrocortisone from an ointment using synthetic and biological membranes.

A simple and reproducible method using a VDC system and a synthetic membrane that could be easily adopted for the quality control of semisolid products was first developed and demonstrated by Shah and co-workers [42]. Since then, IVRT using VDCs has been regarded as one of the most widely accepted and powerful methods for evaluating the release rates of an API from semisolid dosage forms and for establishing "sameness" and the detection of differences between topical products [17]. Importantly, the rate and extent of drug release from topical products can be quantified in vitro using IVRT methods [4–6, 8–10, 17, 18, 43–47].

The release of APIs from topical dosage forms during clinical trials to establish therapeutic efficacy may be difficult due to the inherent assessment limitations of such trials. A typical example of this is the case of acyclovir ointment, where a PSG [48] using IVRT was published by the FDA in 2012. Subsequently, IVRT has become a promising tool for assessing semisolid dosage forms, including the identification of formulation factors such as Q1/Q2/Q3, which may affect the release of the API [49, 50].

5.4 Theoretical Principles of IVRT

IVRT is based on the diffusion of molecules through a matrix and involves the random movement of molecules driven by a concentration gradient, i.e. from a region of high concentration to a region of lower concentration. It involves a passive diffusion process of a permeant from a vehicle in a donor chamber through a "barrier" (semi-permeable membrane) into a receptor chamber containing a receptor fluid. The API thus partitions the membrane (permeates) and then moves (diffuses) through the membrane.

T. Higuchi described the kinetics of IVRT using a model [25, 27] that illustrates the release of an API from one side of a layer where the API is suspended in an ointment base. The process involves a consideration of the Fickian Laws of diffusion [51, 52]. The diffusion of drug molecules being released from semisolid dosage forms occurs from a region of high concentration (donor chamber) to a region of low concentration (receptor chamber) where the membrane acts as the rate-limiting step to maintain a diffusional sink, i.e. as the concentration of drug in the donor compartment decreases, the concentration of drug in the receptor medium increases.

The rate of mass transfer per unit area ("Flux") is based on Fick's first law:

$$J = -D\frac{dC}{dx} \tag{5.1}$$

where

J = flux
D = diffusion coefficient
dC = change in concentration of solute
dx = change in distance

Fick's second law states that the rate of change in concentration in a volume within a diffusional field is proportional to the rate of change in spatial concentration gradient at that point in time. The proportionality constant is the diffusion coefficient, expressed as:

$$\frac{\partial C}{\partial t} = D\left[\frac{\partial^2 c}{\partial x^2} + \frac{\partial^2 c}{\partial y^2} + \frac{\partial^2 c}{\partial z^2}\right] \tag{5.2}$$

where x, y, and z are spatial coordinates.

T. Higuchi [27] showed that for cream and certain ointment formulations, drug release follows the square-root-of-time profile and that the release rate is proportional to the square root of the drug solubility, the exposed surface area, the diffusion constant and inverse time, which is based on Fick's law of diffusion.

Using the following relationships,

$$\frac{M_t}{M_0} = K\sqrt{t} \tag{5.3}$$

where

M_{t} = amount of drug released in time t
M_0 = initial amount of drug in the formulation
k = release rate constant

A plot of the amount of API released per unit area (mg/cm^2) against the square root of time results in a straight line, and its gradient is indicative of the efficiency of the active compound.

The release of the drug from the semisolid dosage form in the VDC system is kinetically explained by the diffusion of a chemical out of a semi-infinite medium into a sink. The momentary release rate tracks the depth of penetration of the gradient within the semisolid. Beginning at the moment when the receding boundary layer's diffusional resistance assumes dominance of the kinetics of release, the amount of the drug released, M, becomes proportional to $\sqrt{t}$ (where t = time) for solution, suspension or emulsion semisolid systems alike. The momentary rate of release, *dM/dt*, becomes proportional to 1/ $\sqrt{t}$, which reflects the slowing of drug release with the passage of time. The reservoir size is kept large so that the drug is released into a medium that stays highly diluted throughout the experiment relative to the concentration of drug dissolved in the semisolid. Hence, drug release is said to take place into a diffusional sink. When a drug is totally in solution within the dosage form, the amount of drug released as a function of time can be described by:

$$M = 2.Co\sqrt{\frac{D.t}{\pi}} \tag{5.4}$$

where

M = amount of drug released into the sink per cm^2
Co = drug concentration in releasing matrix
D = drug diffusion coefficient through the matrix

A plot of M vs. $\sqrt{t}$ will be linear with a slope of $2.Co\sqrt{\frac{D}{\pi.}}$

The following equation describes drug release when the drug is in the form of a suspension within the dosage form:

$$M = \sqrt{2.Dm.Cs\left(Q - \frac{Cs}{2}\right)t} \tag{5.5}$$

where

Cs = drug solubility in the releasing matrix
Dm = drug diffusion coefficient in the semisolid matrix
Q = total amount of the drug in solution and suspended in the matrix

When $Q >> Cs$, the previous equation simplifies to:

$$M = \sqrt{2.Q.Dm.Cs.t} \tag{5.6}$$

A plot of M vs. $\sqrt{t}$ will be linear with a slope of $\sqrt{2\text{Q.Dm.Cs}}$.

Coarse particles may dissolve so slowly that the moving boundary layer recedes to some extent behind the particles. That situation introduces noticeable curvature in the $\sqrt{t}$ plot, because of a particle size effect. During release rate experiments, every attempt should be made to keep the composition of the formulation intact over the releasing period [53].

The kinetics of IVRT has been described by the Higuchi model based on the assumption that the theory is only valid if:

(a) The percentage released is less than 30% of the total drug in the vehicle
(b) Only a single drug species is present in the vehicle
(c) The diffusion coefficient does not vary with respect to time or position within the vehicle layer
(d) Only the drug diffuses out of the vehicle
(e) The drug reaching the receptor side is removed rapidly [54, 55]

Siepmann and Peppas [56] described the derivation of the Higuchi equation, its applications, and use and misuse. Importantly, Higuchi's theory was based on his studies on diffusion from ointments containing dissolved drugs, where the square-root-of-time relationship is a model based on the early time approach under perfect sink conditions. When considering Higuchi's derivation of his model [25], the derivation is a semi-infinite model since it does not consider a situation when the moving diffusion fronts have reached the centre of a film or the diffusion front from the ointment surface has reached the top of an ointment layer. However, in reality, and mathematically, once the diffusion fronts meet at the centre or reach the top, the concentration there will start to decline from the initial drug loading concentration as diffusion proceeds with time. The result is that the release will slow down and not follow a square-root-of-time relationship anymore.

W. Higuchi, in a subsequent paper in 1962 [57], presented some exact solutions to the diffusion equation to describe the drug diffusion from ointments containing dissolved drugs. After showing that these equations gave a nice fit with diffusion data reported by Patel et al. [58], Higuchi presented simplified forms of relevant equations. These simplified forms represent early time approximations to the exact solutions, that is they will fit well with the exact solutions at early times when the per cent release is not large. However, Higuchi cautioned that these approximate solutions will be more practical and easier to use and have good agreement with the exact solutions when the percentage released of the total drug in the vehicle (R) is not too large, i.e. $R < 30\%$. In the book entitled "Controlled Release of Biologically Active Agents" by R.W. Baker [59], on the subject of diffusion of drugs from monolithic devices of a slab geometry (i.e. similar to a layer of cream), it is seen that Baker's Eq. 5.11 for late-time approximation is virtually identical to Higuchi's Eq. 5.7, and Baker's Eq. 5.12 is virtually identical to Higuchi's Eq. 5.9, since the thickness of the slab in Baker's equations is for diffusion from both sides of the slab, whereas in Higuchi's equations it is for the thickness of the ointment layer releasing from one surface. Table 5.1 below provides an overview of some important equations describing the release of medicaments from semisolid dosage forms.

According to Baker, Eq. 5.12 is only valid as an approximation of Eq. 5.11 if the percentage of drug release is between 0 and 60%. Higuchi's assumption of <30% release restricts the range of applicability of the approximation equation. Therefore, several researchers have suggested increasing the limit of drug release to <60%, whereby the simplified square root of the time equation would be applicable to describe most of the drug release up to 60% release. For drug release >60%, the release rate drops off more exponentially. In such instances, either the exact solution (Higuchi's Eq. 5.7 or Baker's Eq. 5.11) may be used to analyze data or the late-time approximation given in Baker's Eq. 5.13 [60, 61].

The sources of deviation of drug release from the square-root-of-time dependence [Fig. 5.2] include boundary layer-controlled kinetics, dissolution-controlled kinetics, dimensional changes due to polymer swelling, or variable surface concentration due to polymer relaxation [61, 62].

Some of the potential reasons for deviation from linearity include:

(a) A drug release of more than 60% from its dosage form, where most of the drug has already been released within the first 2 h
(b) Improper maintenance of sink conditions in the receptor medium after 1–2 h
(c) Experimental anomalies in the setup contributing to such a deviation

Baker's Eq. 5.13 helps describe the tail end of the release. The exact solution provided by Higuchi's Eq. 5.7 or Baker's Eq. 5.11 can be used to clearly describe and fit the entire release profile from a dosage form more precisely [60, 61].

Table 5.1 Important equations describing the release of medicaments from semisolid dosage forms

Higuchis' equations (25–29)	Baker's equation (59)
$Q = hC_0\left[1 - \frac{8}{\pi^2}\sum_{m=0}^{\infty}\frac{1}{(2m+1)^2}\exp\left(-\frac{D(2m+1)^2\pi^2 t}{4h^2}\right)\right]$ (5.7)	$\frac{M_t}{M_0} = 1 - \frac{8}{\pi^2}\exp\left(\frac{-\pi^2 Dt}{h^2}\right),\ for\ 0.4 \geq \frac{M_t}{M_0} < 1.0$ (5.11)
$R = 100\left[1 - \frac{8}{\pi^2}\sum_{m=0}^{\infty}\frac{1}{(2m+1)^2}\exp\left(-\frac{D(2m+1)^2\pi^2 t}{4h^2}\right)\right]$ (5.8)	$\frac{dM_t}{dt} = \frac{8DM_0}{h^2}\exp\left(\frac{-\pi^2 Dt}{h^2}\right)$ (5.12)
$Q = 2C_0\left(\frac{Dt}{\pi}\right)^{1/2}$ (5.9)	$M_t = A\ [DKtc_s(2c_0 - c_s)]^{1/2}$ (5.13)
$R = 200\left(\frac{Dt}{\pi h^2}\right)^{1/2}$ (5.10)	$M_t \cong A\ (2DKtc_s c_0)^{1/2}, for\ c_0 \gg c_s$ (5.14)

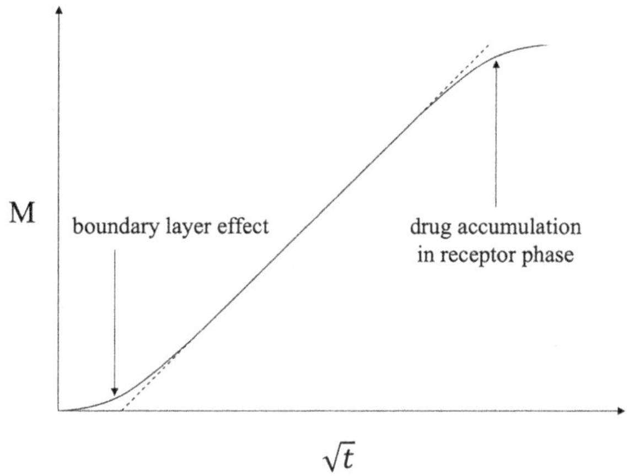

Fig. 5.2 Deviations from linearity from boundary layer effect due to the existence of a stagnant aqueous layer and/or retardation of drug release resulting from accumulation in the receptor phase [62]

While studying a delivery system, it is prudent to systematically investigate whether any specific system parameters (viz. excipient-drug interaction, dose depletion, solubility of the drug in the dosage form, influx of water due to osmosis, etc.) may cause deviation from linearity when >30% of the drug is released.

Importantly, full awareness of the underlying specific release mechanism and associated limitations is necessary to establish possible sources to explain these deviations.

5.5 Regulatory Requirements and Associated Challenges

Unlike methods to assess bioequivalence of extravascular dosage forms, where such products are formulated to ensure that the active pharmaceutical ingredient (API) efficiently enters the systemic circulation, bioequivalence assessment of topical dosage forms intended for local action, such as dermatological products, has proved to be far more challenging. Currently, apart from the use of the vasoconstrictor assay (VCA) for the bioequivalence assessment of topical corticosteroid products [63], most regulatory agencies generally require comparative clinical trials to confirm the bioequivalence of topical dosage forms intended for local action [13]. However, unlike new drug products, which require placebo-controlled clinical trials in patients to establish effectiveness on a non-inferiority basis, comparative in vivo studies to demonstrate BE for topical generic drug products are regarded as inefficient, risky, and very expensive and also considered to be the least accurate, sensitive and reproducible of the general approaches for measuring (bioavailability) BA or demonstrating BE for such products [64]. Also, clinical endpoint studies typically utilize a large sample size and are often insensitive to formulation effects. Consequently, the requirements to conduct clinical trials for such products have resulted in a dearth of generic topical products reaching the market in both developed and developing countries [65]. Hence, faster, less expensive, and more reproducible and sensitive approaches to reliably demonstrate BE for topical dermatological products are currently being investigated. Clearly, the lack of surrogate methods to conclude the BE of topical dermatological products has resulted in lower generic competition for such products, negatively impacting patients as there is limited access to affordable products.

The primary objective of the assessment of topical dermatological products is to determine the rate of release of the API from the dosage form since only once the drug is liberated from the formulation can it diffuse into the skin [66]. The application of IVRT, therefore, involves the measurement of the drug released from the dosage form into a receptor medium across an inert membrane [67]. Subsequently, the concentration of active pharmaceutical ingredient (API) released from the semisolid dosage form is measured in aliquots of receptor medium, collected over various intervals of time to determine its release rate [44].

IVRT has been established as a compendial method in the United States Pharmacopoeia (USP) General Chapter <1724> [21], where details of the apparatus, procedures for performance testing, and statistical methods to assess the "sameness" of such products are described. In this context, "sameness" refers to situations where the qualitative (Q1) and quantitative (Q2) properties, as well as the arrangement of matter (Q3) in the test and reference products, are considered to be the same. The FDA's SUPAC-SS guidance describes the application of IVRT to assess the "sameness" of semisolid dosage forms following minor changes to an approved topical dosage form, such as manufacturing site or process changes, changes in the components and composition of the drug product, or changes in batch size [14]. An in vitro release rate can reflect numerous and combined effects of several physical and chemical parameters, including solubility and particle size of the API, thereby aiding in the assessment of pharmaceutical equivalence/inequivalence between pre- and post-approval product changes [1, 2]. In 2003, the experts from the Fédération Internationale Pharmaceutique/American Association of Pharmaceutical Scientists (FIP/AAPS) pointed out that there was an absence of a standard in vitro test that could be applied to semisolid dosage forms [68]. They suggested that it is crucial to evaluate the influence of each IVRT parameter that may potentially alter the test results and to validate the operational ranges within which the IVRT remains consistent and reliable to attain the status of a generally applicable, robust, and valuable tool [69].

5.6 Application and Associated Requirements of IVRT Methods

A comprehensive validation of the IVRT method is imperative to ensure that the resulting method has the requisite attributes of sensitivity, precision, selectivity, and reproducibility necessary to detect differences relating to Q1/Q2/Q3 between products [10, 11]. Until recently, however, amongst the numerous publications involving the application of IVRT, comprehensive validation of IVRT systems and methods was conspicuously absent from the literature. Although various efforts [50, 70–72] were made by several researchers to develop a standardised method to measure in vitro drug release from a product using diffusion cells, a comprehensive validation that would be generally applicable to all topical dermatological dosage forms was first published in 2018 [16]. Appropriate validation is essential to ensure that the developed IVRT method and the associated system have the necessary discriminatory capabilities to determine "sameness" between products and also, more importantly, detect differences that may affect clinical performance.

Clinical endpoint studies have, in the past, been mandatory by most global regulatory authorities to establish the safety and efficacy of generic topical products, except in the case of topical corticosteroid products wherein a vasoconstrictor or human skin blanching assay is accepted [63], an appropriate IVRT method can provide valuable and compelling information to justify waivers of bioequivalence studies (biowaivers) for topical semisolid products intended for local action.

Interestingly, topical preparations containing nonprescription drugs, such as topical 1% clotrimazole products, do not require pharmacodynamic/clinical studies. Topical 1% clotrimazole products are marketed as over-the-counter (OTC) products based on a long history and use relating to safety and efficacy. In vivo studies are generally not required, and an FDA monograph entitled "Subpart C-Topical Antifungal Drug Products" [73] specifically mentioned that 1% topical clotrimazole products do not require clinical data for marketing authorisation since 1989. As previously mentioned, the use of clinical endpoint studies to demonstrate BE for topical dermatological drug products is associated with various limitations. Those limitations were recognised by the FDA, and the utility of IVRT was extended in 2012 when a draft guidance was published for assessing bioequivalence (BE) of acyclovir topical ointment, which was the first such publication recommending IVRT for use as a waiver of BE studies for a topical, locally acting product [74]. This particular guidance provided useful and valuable information for the future application of IVRT as a method for the sameness assessment of topical formulations intended for local action. This was the first such product-specific guidance (PSG) utilising IVRT. In view of the foregoing, since clinical endpoint studies are considered inappropriate for topical products intended for local action, several PSGs [75–79] recommending IVRT for various topical products have subsequently been published by the FDA.

IVRT has become a ubiquitous tool for assessing the release of an API from topical semisolid dosage forms, including the identification of Q1/Q2/Q3 [80]. It is interesting to note that the above-mentioned FDA guidances are product-specific and their acceptance criteria are based on those recommended in the USP [21] in accordance with the FDA's SUPAC-SS guidance [14]. In addition to PSG documents, the FDA has implemented several initiatives, such as the Drug Competition Action Plan and Generic Drug User Fee Act (GDUFA) programme to promote the entry of generic topical products on the market, thereby increasing generic competition [81]. A draft guideline on the quality and equivalence of topical products published by the European Medicines Agency (EMA) also makes provision for to use of IVRT for the approval of generic products [13]. It describes the validation conditions and requisite discriminatory power of the method to support claims of therapeutic equivalence with a reference product instead of undertaking clinical trials. Failure to comply with appropriate validation requirements can result in erroneous approvals, which may have significant implications for the clinical performance of such generic products.

In continuity with these approaches, the FDA subsequently published a draft guidance on IVRT studies for topical drug products describing the application of IVRT to support biowaivers [82]. The regulatory frameworks for the FDA and EMA differ in terms of "sameness" acceptance criteria, with the SUPAC-SS 90% confidence interval limits being 75%–133.33% [14], whereas the EMA specifies a much narrower interval of 90%–110% [13]. These guidances augur well for the speedy market approval of certain topical dosage forms intended for local action, which are likely to result in significant economic benefits and make such products more accessible to the wider population.

5.7 Current Trends and Recent Reports

The feasibility of using a non-clinical approach such as IVRT for consideration as a biowaiver to obtain market approval has recently been discussed in several publications using various model drugs. Nallagundla et al. [17] evaluated several generic creams containing 5% acyclovir in accordance with the recommendations in FDA guidances [14, 74] using different types of synthetic membranes. Four of the six synthetic membranes helped generate acceptable comparative in vitro release rate profiles, thus providing useful information for the development of regulatory guidance for a biowaiver for acyclovir creams. The release rates of acyclovir from the tested generic creams were shown to be equivalent to the innovator product.

Tiffner et al. [16] were the first to report a comprehensive approach to IVRT validation, where the research was carried out prior to the publication of the FDA draft guidance on acyclovir cream, which includes detailed information relating to IVRT validation [75]. A comprehensive characterisation of the operational parameters of an IVRT system was conducted, and an IVRT method to assess release from 5% acyclovir creams was developed and validated. In a subsequent publication, the developed IVRT method, which was subjected to a series of validation and qualification procedures, proved to be an effective tool to determine and discriminate the release rates of acyclovir among several pharmaceutically equivalent creams, as well as from a substantially different formulation, i.e. 5% acyclovir ointment, compared to the reference product [18].

Mudyahoto et al. [5] described the development and validation of an IVRT method to assess creams containing 1% hydrocortisone acetate. A positive control (Reference Listed Drug (RLD), 1% hydrocortisone acetate cream vs. itself) was included to ensure that the method had the necessary capability to confirm sameness, whereas negative controls (0.5% and 1.5% hydrocortisone acetate creams vs. RLD) were studied to ensure that the method had the requisite discriminatory power to detect significant differences. In addition, the "sameness" of two market-approved and commercially available products containing 1% hydrocortisone acetate was investigated and found to be in vitro equivalent. Furthermore, the comparative data between the different strengths of specially manufactured hydrocortisone acetate creams demonstrated that the IVRT method was able to show differences between these products that, irrefutably, are due to differences in Q2 only. The results confirmed the suitability of the IVRT method to discriminate differences in release rates of hydrocortisone acetate from topical cream formulations and provided its potential value as a useful tool in the formulation development of such products. This approach provided useful information for the future development of acceptable IVRT methods to assess topical dosage forms for local action containing different drugs.

Purazi et al. [8] utilized a comprehensive approach to address all qualification/validation considerations of an IVRT method for topical miconazole creams. In vitro evaluation of approved and marketed topical miconazole creams was carried out according to recommendations of the relevant guidances and compendial documents available at that time [14, 21]. A robust and reproducible IVRT method with adequate discriminatory power was developed and applied to confirm "sameness" and highlight differences between 2% miconazole formulations. The results indicated that the developed IVRT method could reliably detect differences relating to Q1/Q2/Q3 properties in a formulation, thereby providing compelling data for future use to obtain biowaivers.

Rath and Kanfer [4] developed and comprehensively validated an IVRT method with the necessary discriminatory power to detect "sameness" and differences between topical cream products containing 0.75% metronidazole. A novel approach involving the incorporation of positive control, whereby the reference product was compared to itself, and the inclusion of two negative controls, i.e. products containing <25% or >25% potency compared to the reference product was described. The method was applied to assess the "sameness" of a generic metronidazole cream, whereby a 0.75% metronidazole cream was compared with the reference product, Metrocreme®, 0.75% metronidazole. The positive and negative controls complied with the necessary acceptance criteria for equivalence and inequivalence, respectively, confirming the discriminatory power of the IVRT method. This is in accordance with the SUPAC-SS acceptance criteria [14] for a biowaiver where generic products containing <25% or >25% potency compared to the reference product are deemed inequivalent. The resulting data indicated that the developed IVRT method can be applied to accurately assess "sameness" and differences between metronidazole cream products and clearly indicates the utility of such an approach for use in formulation development.

In a recent publication, an IVRT method was developed by Wellington et al. [6] to measure the release rate of clotrimazole from topical products and involved an additional validation parameter, supplemental selectivity, which has generally not been included as a validation component in previously published IVRT methods. Conventional IVRT validation protocols generally only involved the assessment of selectivity to detect Q1 and Q2 differences. The inclusion of supplemental selectivity as described in a recent draft FDA guidance is intended to ensure that the IVRT method would reliably discriminate between different manufacturing methods whilst providing useful information on Q3 [82]. "Sameness" was confirmed when an extemporaneously prepared product containing identical components (Q1, Q2) was compared with the reference product. Products containing <30% or >40% clotrimazole (i.e. at a lower limit of 70% and an upper limit of 140%) served as negative controls and were included for the discriminatory power assessment, thereby ensuring appropriate discrimination to result in the requisite clinical performance. The data clearly indicate that a validated IVRT method is an extremely valuable tool and reinforces the potential for use as a biowaiver for topical products for local action and accordingly establishes IVRT as an acceptable option to assess the BE of such products.

In a subsequent publication, Wellington et al. [9] demonstrated the utility of IVRT to assess the "sameness" of topical 1% clotrimazole creams from South Africa (SA) and Canada (CA), thereby generating data to support a biowaiver based on the evaluation of qualitative (Q1) and quantitative (Q2) properties and structural arrangement of matter (Q3). None of the investigated marketed generic 1% clotrimazole creams were pharmaceutically equivalent to their respective comparator products. The IVRT method displayed the requisite ability to discriminate between the various marketed 1% clotrimazole creams, including some products with relatively insignificant formulation differences and others with substantially dissimilar attributes relating to Q1/Q2/Q3. Hence, IVRT is a valuable tool that can be used to predict possible differences in clinical performance as a result of formulation differences that result in differences in release rates. The foregoing publications again indicate the utility of IVRT to support a biowaiver for topical creams for local action, thereby obviating the need for expensive, laborious, and time-intensive clinical studies and facilitating faster entry of affordable generic products into the market.

In general, the use of IVRT methods and related publications according to the foregoing considerations clearly indicates that 'sameness' and/or differences between products can readily be confirmed.

5.8 Equipment and Methodology

IVRT has been established as a compendial method by the USP, and the compendial apparatuses and procedures for these tests are described in the USP General Chapter <1724> [21]. A schematic representation of an IVRT system for "sameness" assessment is outlined in Fig. 5.3. It depicts a 6-cell vertical diffusion cell (VDC) system used to conduct IVRT runs, where samples are collected typically over a duration of 6 h. The concentrations of drugs in the samples are analyzed using

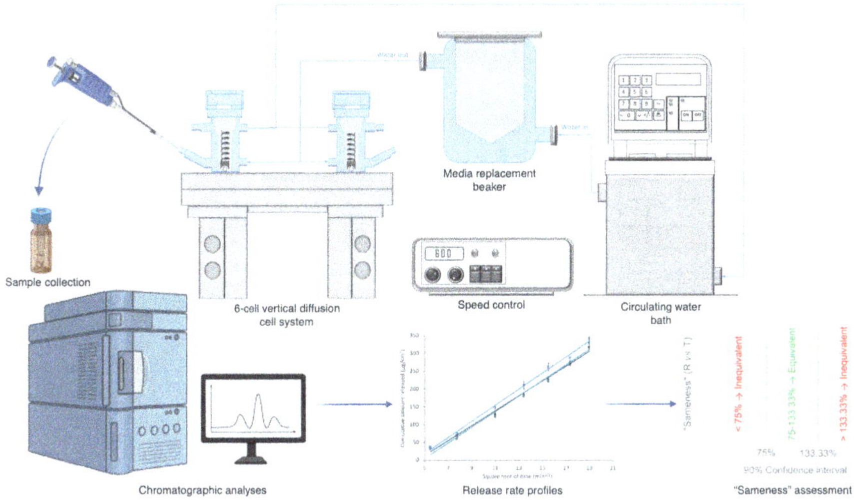

Fig. 5.3 Schematic representation of IVRT system for "sameness" assessment

chromatographic techniques. The drug amounts per unit area are calculated and plotted against the square root of time. The slopes of these plots that correspond to the release rates are estimated and subjected to "sameness" assessment using relevant statistical approaches. The "sameness" acceptance criteria laid down by the FDA [14, 82] and their interpretation are graphically represented in the bottom right corner of Fig. 5.3.

5.9 Diffusion Cells

The basic diffusion cell apparatus, known as the Franz cell, was described in 1975 [83]. The apparatus consisted of a small-volume, jacketed cell with a donor chamber for test product application, a membrane that allows drug permeation through it, and a receptor chamber containing a receptor medium into which the drug is released. Several modifications to the basic design have subsequently been made.

Diffusion cells can be classified as horizontal or vertical diffusion cells based on their physical design and may be static or flow-through. The VDC, recommended by the USP [Fig. 5.4], is one of the more widely accepted apparatuses for IVRT.

A major drawback of using a VDC is the occurrence of air bubbles at the membrane-liquid interface, which is often difficult to observe. These air bubbles can result in issues relating to the precision, accuracy, and reproducibility of the resulting data. A simple diffusion cell described by Chattaraj and Kanfer [43] involves mounting the semisolid dosage form underneath the membrane, unlike a VDC where the formulation is placed on the membrane [Fig. 5.5]. This helps

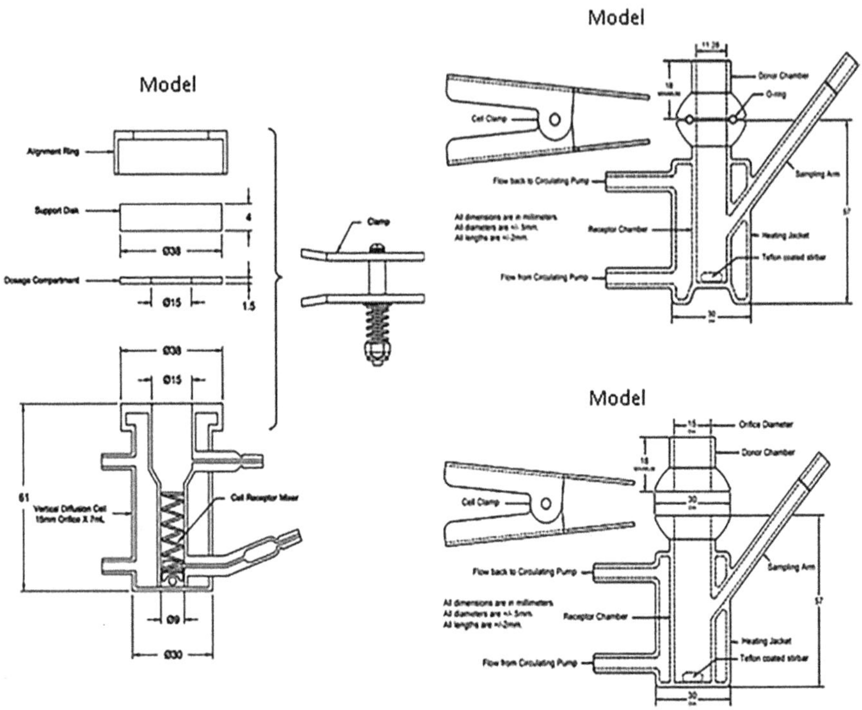

Fig. 5.4 Compendial VDCs [21]

overcome the problem of air bubble formation at the membrane/liquid interface by avoiding any discontinuity between the membrane and liquid interface.

5.10 Static Cells (Vertical or Side-by-Side)

Vertical cells, or side-by-side cells, use a fixed-volume receptor chamber, controlled temperature, a port to sample receptor fluid, and stirred receptor fluid. These side-by-side cells allow stirring of both the donor and receptor chambers [84].

5.11 Continuous Flow Cells or Flow-Through Cells

In-line cells (Type 1) have a continuous flow of the receiver media, which causes turbulence in the receptor chamber and simulates stirring. Flow-through cells (Type 2) use a fixed volume of receptor chamber, controlled temperature and adjustable flow rate [84].

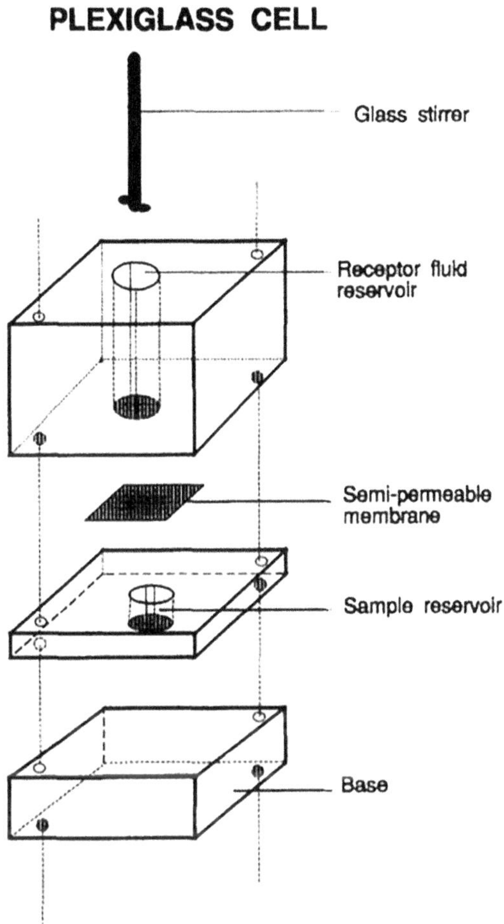

Fig. 5.5 Plexiglass cell [43]

5.12 USP Apparatus 4: Flow-Through Cell Apparatus

The use of flow-through cells to study dissolution with a continuously flowing medium was first established in 1957 for tablets [85]. The test sample is placed in a small-volume cell, and the receptor medium is pumped through the cell at a set temperature. The eluate leaving the cell is filtered and analyzed to calculate the per cent drug release [86]. Various automated and flow-through methods have been used by several researchers [40, 45, 46, 87, 88].

A method involving the use of an "insertion cell" in conjunction with a USP flow-through cell was described by Chattaraj and Kanfer [45], where the release of acyclovir from semisolid dosage forms was monitored. The drug release data were found to concur with the previously published data using different diffusion cells whilst offering several advantages such as easier usage, better adaptability, and absence of air bubble formation at the membrane/sample interface due to the continuous flow of receptor medium. Formation of air bubbles at the membrane/liquid interface often occurs in VDCs, resulting in inconsistent data.

5.13 Immersion Cells

Immersion cells are a variation of the paddle-over-disc method and can be coupled with USP Apparatus II. The four main parts of an immersion cell include the cell body containing the sample compartment, the membrane, a retaining ring used to secure the membrane to the cell body and ensure full contact with the sample, and a washer to provide a leakproof seal between the membrane, the retaining ring and the cell body. Alignment and adjustment tools are provided to allow the analyst to vary the reservoir volume within the cell. Modified flat-bottomed vessels are used to avoid the issue of void space under the cell. Typically, vessel volumes of 150–200 mL are used. The temperature is maintained at 32 ± 0.5 °C for skin preparations or at 37 ± 0.5 °C for vaginal preparations. The membranes should be thoroughly wetted with a suitable medium before use. Two models of immersion cells—models A and B—are enlisted and described in the USP Chapter <1724> [21]. Ilic et al. [80] demonstrated the utility of immersion cells in combination with USP Apparatus II to assess the effect of critical quality attributes on drug release.

5.14 VDC Systems

5.14.1 VDC Test System Model HDT 1000

The VDC test system model HDT 1000 by Copley Scientific is a compact unit designed to accommodate two rows of 5 VDCs on a heated aluminium block [Fig. 5.6]. The heated aluminium block circumvents the difficulties associated with the use of tubing present in its water-jacketed predecessors. The temperature can be adjusted and maintained from ambient +5 to 150 °C. Each test station is equipped with a powerful magnetic stirrer underneath, which can be used to vary the stirring speed from 400 to 2000 rpm. It can be manually operated or coupled with a fully automated sampler.

The cell comprises the donor chamber containing the test sample and the receptor chamber containing the receptor medium. These two chambers are separated by an inert and highly permeable synthetic membrane. The VDCs are made of borosilicate glass and possess a sidearm that acts as a sampling and media replacement port. In order to ensure homogenous distribution, the receptor medium is continuously stirred throughout the testing process by means of magnetic stirring bars [89].

Fig. 5.6 VDC test system model HDT 1000 [89]

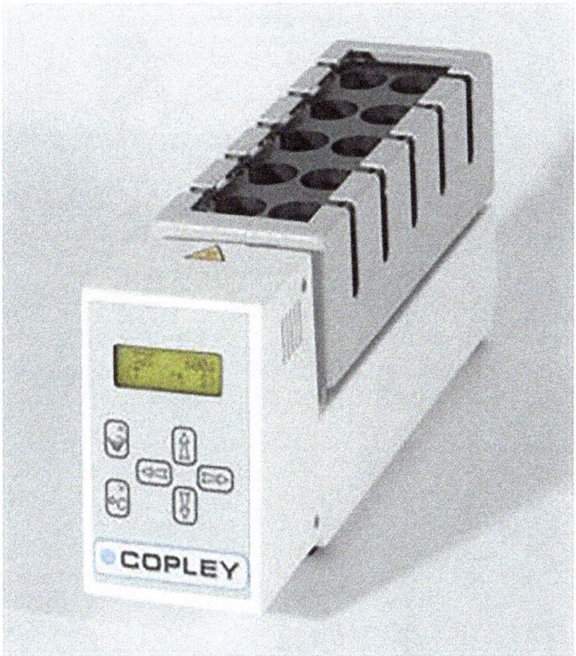

5.14.2 *PermeGear VDC System*

PermeGear vertical glass diffusion cells are classic Franz cells available in various permutations and combinations of orifice diameters, receptor cell volumes, and donor compartment types. They may be jacketed or unjacketed. Both clear glass and amber-colored glass options are available. A typical PermeGear VDC used for IVRT is shown in Fig. 5.7. These cells are placed on a 3, 6, or 9-station magnetic stirrer and connected to a circulating water bath used to mimic the temperature of the intended site of action. It is provided with a magnetic stirrer bar that facilitates homogenous mixing of the receptor chamber contents. Sampling can be achieved manually or with the help of an autosampler [90, 91].

5.14.3 *Hanson VDC System*

After over 30 years of rigorous industrial, academic, and regulatory application and collaboration, Hanson Research Corporation has developed VDCs that are designed to provide accurate results and ease of use. The cell design is similar to model A described in USP General Chapter <1724> [21]. The donor chambers are occluded with a glass cell disc to minimize any back diffusion from sampling. While VDCs of volumes 4, 7, and 12 mL are available, the 7 mL 'standard' VDC is recommended

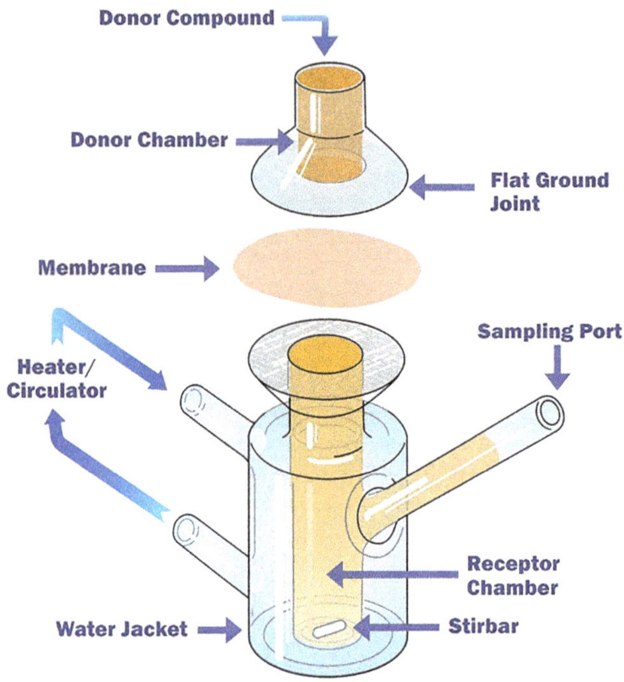

Fig. 5.7 A typical jacketed PermeGear VDC [90, 91]

by the FDA for topical products. The system is equipped with a control unit to regulate the stirring speed and a circulating water bath for temperature control. It offers the option of manual sampling or can be automated with the Hanson Vision® Microette™ diffusion test system [92–94]. A six-cell manual diffusion test system is depicted in Fig. 5.8.

5.15 Method Development

5.15.1 Temperature

The test temperature is usually maintained at 32 ± 1 °C throughout the experiment to reflect normal skin temperature [19, 69]. For vaginal preparations, a temperature of 37 ± 1 °C is used. Other temperatures may be used if justified. However, proper care should be taken while using elevated temperatures, as the formulation might undergo physical changes at higher temperatures, which may affect the release rate of the drug. A significant effect on the release of acyclovir from cream was observed when the IVRT temperature was increased from 32 °C to 37 °C, which was probably

Fig. 5.8 A six-cell manual VDC test system [94]

due to the decrease in viscosity of the formulation with an increase in temperature [43].

5.15.2 *Amount of Sample*

Typically, an "infinite" dose of 500–1000 mg of the test formulation is uniformly applied to the membrane facing the donor compartment [10, 16, 82]. This is done to ensure that the amount of drug present in the donor compartment is always greater than that in the receptor compartment.

5.15.3 *Receptor Medium*

The selection of an appropriate receptor medium is very important, and this is dependent on the solubility of the API. The most suitable receptor medium is one that mimics the physiological conditions of human skin while facilitating accurate measurement of the amount of drug released. However, when using such a medium, optimum dissolution for the product may not always be achieved. The receptor medium should provide a "diffusional sink" throughout the IVRT experiment and have a high capacity to permit accurate diffusion of the drug wherein >ten-fold maximum expected concentration of the drug is soluble [50, 95, 96]. Generally, a simple receptor medium such as saline or phosphate buffer saline solutions is preferred [16–18]. While it is desirable to have a receptor medium similar to the physiological conditions, it is also necessary to ensure that it does not suppress drug release. This can be prevented by incorporating surfactants and/or organic solvents into the receptor medium to enhance the solubility of poorly soluble drugs

[10]. When a receptor medium consisting of phosphate-buffered saline (water, 10:90 v/v, containing 7% ascorbic acid w/v) was used for terbinafine semisolid products, sink conditions were maintained and reproducible release rates with good linearity [95] were observed. In a publication by Patel, different concentrations of isopropyl alcohol (IPA), ethyl alcohol, and dimethyl sulfoxide (DMSO) with water were individually evaluated during receptor medium selection studies for dapsone gels [97]. A mixture of phosphate buffer pH 5.5 and ethanol (65:35 v/v) was used as the receiving medium for retinoic acid creams, and minor changes in alcohol concentration did not significantly affect the rate of release of retinoic acid [72].

Deaeration of the receptor medium is critical to prevent the formation of air bubbles at the interface with the membrane, as it lowers the area of contact between the receptor medium and the membrane containing the test formulation, leading to inconsistent results [50, 53].

Factors that may influence the choice of a receptor medium include the type of compound (small- or large-molecule compound), its stability when mixed in a solvent, light sensitivity, etc. [96]. The pH of the medium is an important consideration, and it should be based on the pH of the formulation, the pH-solubility profile of the API and the pH of the target membrane [44, 97, 98].

5.16 Receptor Medium Screening Studies

The solubility of an API is investigated in different receptor media by adding a relatively large amount of the API to the receptor media to produce a saturated solution. The solutions are then stirred at 600 rpm at 32 °C ± 1 °C for 6 hours, allowed to stand overnight at 32 °C without agitation, and the supernatant is withdrawn, filtered and analyzed to determine the concentration of dissolved drug.

5.16.1 Stirring Speed

The receptor medium is stirred throughout the experiment at a constant stirring speed with a magnetic bar. The tolerance is set at ±10% of the nominal stirring rate (usually 600 rpm). Various rates of stirring have been reported by researchers, viz. 400, 450, and 600–800 rpm, to ensure uniform mixing of the receptor phase [53, 71, 99]. The speed of rotation also has an effect on the barrier of unstirred layers, ultimately influencing the diffusion across the membrane [70]. Extremely high stirring speeds may lead to excessive drug release into the receptor medium and result in different drug concentrations at the membrane interface and in the receptor medium, affecting the diffusion and subsequently release rates. Very low stirring speeds, on the other hand, may cause inconsistent drug concentrations in the receptor medium. These inconsistencies ultimately affect the discriminatory ability of the IVRT method.

5.16.2 Sampling

IVRT for topical semisolid products is typically carried out for 4–6 h. Specific volumes (preferably 0.1–0.2 mL) of samples are withdrawn from the receptor chamber at predetermined time intervals and replaced with fresh receptor medium [19, 69]. In some instances, longer or shorter sampling durations may be deemed necessary. However, longer sampling durations may affect the linearity at extended times due to a drug release of more than 30% [57].

5.16.3 Membranes

The FDA recommends the use of simple, non-rate-limiting, porous synthetic membranes suitable for assessing drug release from a topical formulation, which will eliminate the high degree of variability associated with the use of skin [20]. In vitro drug release from semisolid topical dosage forms has been extensively investigated using diffusion cells with synthetic membranes such as cellulose nitrate/mixed ester, Tuffryn®, Fluoropore®, Supor® (polysulphone), Acetate Plus® (cellulose acetate), Nylon, Teflon and Polycarbonate, which are used to separate the donor compartment containing the formulation to be tested and the receptor compartment containing the appropriate receptor medium [20, 22, 69]. The type of synthetic membrane used varies with the purpose of investigation [22].

A membrane used for IVRT studies should:

- Provide an inert support for the test products with the least possible diffusional resistance without being rate-controlling.
- Allow easy spreadability of the test product.
- Be devoid of any physical or chemical interaction with the product and receptor medium. This is important because of the possibility that excipients in the product or receptor medium may affect the physical integrity of the membrane. In some instances, the active ingredient may bind to the membrane.
- Be free from any leachable material that could interfere with the analytical method.

The various considerations whilst choosing a membrane are described in the foregoing sections.

5.16.3.1 Pore Size and Viscosity of Test Formulations

Typically, membranes with a pore size of 0.45 μm are used for IVRT. Membranes of larger pore size may be used for high-viscosity products to ensure minimal resistance to diffusion [72, 96, 100].

5.16.3.2 Hydrophobic and Hydrophilic Materials

To minimize the interaction of the API with the membrane, hydrophobic membranes are generally preferred for hydrophilic drugs and vice versa. Distinct types of release profiles across a membrane may be obtained depending on the hydrophilicity/hydrophobicity of the drug and the membrane [101]. Porous and hydrophilic membranes may sometimes lead to back diffusion of receptor media into the donor compartment owing to osmotic drive [102].

5.16.3.3 Quality and Ease of Accessibility

To ensure reliability, it is essential that membranes undergo quality control testing between different lots and offer good reproducibility of results. It is advisable to use membranes that are economical and readily available [72, 96, 100].

5.16.3.4 Membrane Wetting and Equilibration

It is recommended to pre-treat the membrane by soaking it in the receptor medium and allowing it to equilibrate for 30 min prior to the start of the IVRT experiment to minimize the formation of air bubbles and uneven surface area for diffusion. The membranes soaked in isopropyl myristate (IPM) prior to the IVRT mimic a lipophilic barrier such as the SC [99]. It is important to standardize the method and duration of membrane equilibration.

In addition, other factors that influence membrane selection are membrane flux, thickness, and coating [103].

5.16.3.5 Membrane Screening Studies

Membrane inertness studies involve the assessment of different synthetic membranes where a single membrane of each type is immersed in the API solution for 6 h at 32 °C. The concentration of the drug should be equivalent to its highest achievable concentration during IVRT studies. The same solution maintained under the same conditions as the solutions containing membranes serves as a control. Aliquots of these solutions are collected and analyzed. A membrane is considered inert if the drug recovery is ≥95% [16, 82]. Following this, the inertness of the selected membranes is evaluated using lower and intermediate concentrations in addition to the highest possible concentration that could be experienced during IVRT.

The suitability of the membranes for IVRT studies is assessed by performing IVRT runs using the test products to ensure that the membrane will not interfere with the diffusivity of the drug from the donor compartment into the receptor medium.

5.17 IVRT Apparatus Qualification [4–6, 8, 16]

Inadequate apparatus qualification and/or sample handling procedures may introduce variability in the results. Hence, it is crucial to carry out a thorough apparatus qualification to ascertain that the VDC system is functioning appropriately and generating reliable data. Apparatus qualification entails the investigation and evaluation of vital parameters that may influence the experimental outcomes. Therefore, environmental factors such as suitable working area, workbench levelness, and absence of direct exposure to sunlight and/or cooling vents should be investigated and documented. Failure-resistant power supply to all the electronic components is essential. Additionally, the evaluation of the orifice diameters and capacities of each of the VDCs, the temperature of the receptor medium, stirrer speed, and the dispensed sample volume should be investigated as components of the apparatus qualification.

A suitable workbench that is evenly levelled, sturdy, and provides a high inertial mass to limit vibration should be chosen. The surface should be able to withstand the disturbances due to transient vibrations produced when objects are placed or moved on the bench without affecting its level. An open space of at least 76 cm vertically above the VDC system is necessary for proper functioning and unhindered sampling. Levelling devices may be used to compensate for any influence of benchtop surface inclination on the VDC system. The levelness of the benchtop should be measured in two orthogonal directions using a digital spirit level, wherein the inclination should not be more than $\pm 1^\circ$.

The VDC capacities should be investigated gravimetrically by individually weighing the VDCs with the magnetic stirrer bars coupled with helical mixers before and after the addition of purified water until an upper meniscus is formed on the surface of the orifice. To determine the capacity of the VDCs, the differences in weights should be converted into millilitres. The cell capacity should not exceed a tolerance limit of $\pm 5\%$. It is interesting to note that specific VDC capacities sometimes do not conform to the specifications provided by the manufacturers. However, such VDCs may still be used if the measured capacity is suitable to conduct IVRT and the VDCs exhibit a low inter-cell variability. In such instances, it is important to note that the calculation of the release rates should be based on the determined cell capacity instead of that specified in the manual.

The orifice diameters of the VDCs are critical to IVRT since they determine the dosage delivery area for the test sample. These diameters may be measured using a pair of Vernier callipers and should not exceed $\pm 5\%$ of the specified value.

Another crucial component of the IVRT is the system temperature. As previously mentioned, a change in temperature may greatly affect the release rates. Hence, it is vital to ensure uniformity of temperature between the VDCs. For this purpose, the cells should be filled with purified water and the system equilibrated at 32 °C for 30 min. The temperature of each of the receptor chambers should be individually measured using a digital thermometer. The temperature of the receptor chamber should not deviate by more than ± 1 °C from the nominal value.

Similarly, to ensure uniform mixing, the stirring speed should be measured using a tachometer by continuously stirring purified water in the VDCs with magnetic stirrer bars marked with reflective tapes. A deviation of not more than ±10% from the nominal speed is acceptable.

To verify the dispensed sampling volume, 200 μL aliquots of purified water should be withdrawn using the sampling syringe to be used during the studies and then weighed accurately to confirm the volume. A tolerance of ±10% from the nominal value is accepted.

All physical parameters should be measured in triplicate and compared to the pre-defined acceptance criteria laid down in the *stimuli* article published by Ueda et al. [53]. The means and ranges of variation of triplicate measurements should be calculated to determine the variability between individual VDCs.

5.18 Performance Verification Test (PVT)

Due to the high variability associated with IVRT studies, a performance verification test (PVT) has been proposed [53] to assess the suitability of the performance of the system used for IVRT [14]. It is conducted to ensure minimum variability and maximum reproducibility and reliability of the resultant data. A successful PVT also confirms that appropriate measures have been undertaken to control the inconsistencies arising from the IVRT system components, analyst, and/or analytical procedures. Typically, a reference product that has demonstrated consistent results across various laboratories is set as a standard for PVT of any given system. Owing to the consistent in vitro release profiles produced by 1% hydrocortisone creams using VDC systems in several studies since the late 1970s [41, 70, 88, 104–106], it is considered to be a suitable candidate for the PVT of a VDC system used for determining release rates from topical semisolid dosage forms. The acceptance criteria for a PVT have been described by Ueda et al. [53], and any deviations may be indicative of improper functioning of the VDC system, deviation from the procedure, or analyst/operator error in conducting the test [41, 70, 88, 104–106].

The PVT involves performing two IVRT runs using a 1% hydrocortisone cream as the reference standard. The IVRT runs are conducted on different days (preferably two consecutive days), in the same laboratory, using the same apparatus, and under the same experimental conditions. A pseudo infinite dose of the reference standard (~ 0.5–1 g) is uniformly applied onto a suitable membrane (e.g. Tuffryn®, polyethersulfone). The receptor medium is generally a hydroalcoholic solution (e.g. ethanol: water 30:70% v/v), which is continuously stirred at 600 ± 60 rpm and maintained at a temperature of 32 °C ± 1 °C. Aliquots of the receptor medium are sampled over a period of 6 h and analyzed to determine the amount of drug that has diffused into the receptor medium. The equivalence of two IVRT runs is evaluated by comparing the hydrocortisone release rates between the IVRT runs. Intra- and inter-run variabilities are determined by calculating the coefficient of variation (CV) of the release rates within each run and between the two runs. In order to demonstrate good reproducibility, a CV of less than 15% is desirable. The first stage of the statistical approach described in the USP General Chapter <1724> [21] is applied for sameness assessment [4–6, 8–10, 16].

5.19 Calculation of Release Rates and "Sameness" Assessment

The cumulative amount of drug released per unit area is plotted against the square root of time. The slopes of these plots correspond to their release rates, and a total of six release rates per formulation should be obtained. If two products are being tested, three VDCs are designated as references and three as tests in the first run and switched during the second run, such that at the end of the two runs there are six tests and references in each of the six VDCs. Pairwise comparison is carried out by computing the ratios of release rates for the test and reference products (T/R ratios) and establishing the 90% confidence interval (CI) using the Mann–Whitney U Test, where the 90% CI should lie within the limits of 75–133.33% to confirm "sameness" between products, in accordance with the relevant FDA guidances [14, 82] and USP General Chapter <1724> [21]. Briefly, the 90% CI is determined from the 36 T/R ratios obtained from one set of comparative IVRT runs by arranging them in an ascending order, where the eighth and twenty-ninth ratios are the lower and the upper limits, respectively.

5.20 Method Validation

5.20.1 Linearity, Precision, Reproducibility

Typically, three IVRT runs are conducted on three different days (preferably three consecutive days) using the reference product to investigate the linearity, precision, and reproducibility of the IVRT method. The 18 resultant release rates corresponding to the slopes of the release profiles are estimated.

If the drug release complies with Higuchi's Equation [25, 26], the amount released per unit area should be linear with respect to the square root of time. Linearity is confirmed if the coefficient of determination (R^2) is greater than 0.9 [107].

Intra- and inter-run variabilities between and within the three runs are estimated using a random effect model to determine the precision and reproducibility of the IVRT method. The coefficients of variance (CV) for the intra- and inter-run variabilities are calculated using the estimated mean release rates, and the variance components are estimated by a random effect model using the three runs performed as random effect variables. A CV < 15% generally confirms the precision and reproducibility of the method [16].

5.20.2 Sensitivity, Specificity, Selectivity

To investigate the sensitivity, specificity and selectivity, 3 IVRT runs are conducted using test creams containing varying strengths of the drug, i.e. half strength (50%), full strength (100%), and 1.5× strength (150%).

Sensitivity refers to the ability of an IVRT method to qualitatively detect differences in the release rates of the formulations containing different strengths of the drug [107]. It is assessed by evaluating the change in drug release rate as a function of the formulation strength.

Specificity is a quantitative metric and refers to the ability of the method to accurately measure the proportionality of the release rate with respect to the drug concentration in the test creams. A linear regression is constructed, and proportionality is confirmed if a plot of the mean release rates vs. drug concentration in the products results in an R^2 value of ≥ 0.9.

An IVRT method is selective when products containing different strengths of API can be accurately identified as inequivalent when assessed using the statistical approach for "sameness" assessment described in USP Chapter <1724> [21].

5.20.3 Recovery

Recovery relates to the percentage of drug released from a semisolid dosage form with respect to the dose applied. The average cumulative amount of drug released from the 3 IVRT runs, performed using the reference cream, is divided by the amount of drug contained in the applied dose (e.g. 500 mg of 1% drug product would contain 5 mg of the drug). A recovery of $\leq 30\%$ ensures compliance with Higuchi's assumptions [25].

5.20.3.1 Robustness

To assess the robustness of the method and rule out the effect of minor perturbations on release rates, IVRT runs with minor deviations (e.g. 10%) in temperature, stirring speed, dose, and/or receptor medium composition and/or pH from the nominal parameters are performed [82]. A deviation of <15% from the nominal mean release rate confirms the robustness of the IVRT method [16].

5.20.4 Supplemental Selectivity

The concept of IVRT supplemental selectivity was recently included in the FDA's latest draft guidance on IVRT studies for topical drug products [82]. It refers to the ability of an IVRT method to discriminate between the release rates of the reference product and an altered formulation containing the same nominal strength of the drug, such that their release rates are determined to be statistically inequivalent. This is distinct from the concept of IVRT selectivity (described earlier), which confirms the ability of the IVRT method to discriminate predictable differences in release rates of products containing different strengths of the API. On the other hand, the demonstration of IVRT supplemental selectivity validates the ability of the IVRT method to detect differences in the release rates due to formulation factors apart from the drug concentration/product strength.

The altered formulation may differ in the type and/or quantity of inactive ingredients used and/or the manufacturing process. It is important to develop or select an altered formulation with obvious differences such that they are likely to alter the drug release because not all variations may produce a significant difference in the release rate with respect to the reference product. In the event that no such differences in the release rates between the reference and altered products are observed, the supplemental selectivity results are deemed inconclusive.

Wellington et al. [6] used formulations with the same concentration of drug (1% clotrimazole) as the reference product. Whereas the test cream, which consisted of different inactive ingredients than the reference, was found to be inequivalent to the reference product, another test cream that comprised the same type and quantity of inactive ingredients as the reference passed the "sameness" acceptance criteria, indicating equivalence.

In most of the previously published papers, the validation component of supplemental selectivity has been conspicuously absent. As a result, those methods may not have been able to detect differences relating to excipient type and composition (Q1/Q2) or manufacturing processes. The inclusion of supplemental selectivity as a validation component provides useful information relating to the differences arising from Q1 and Q2 properties as well as the method of preparation.

5.20.4.1 Assessment of Discriminatory Power

Discriminatory power refers to the ability of an IVRT method to accurately identify differences between test products. This is assessed by conducting comparative IVRT runs using positive and negative controls [4–6, 8, 16].

Formulations containing more and less API than the reference product, such that they fall slightly outside the limits of acceptance (75%–133.33%), serve as negative controls. These products are compared against the reference product in three IVRT runs. In addition, the reference product is tested against itself in two IVRT runs to serve as a positive control, where three VDCs are assigned as test and 3 as a reference in each of the runs. The release rates of these runs are compared using the statistical approach described in USP Chapter <1724> [21].

This approach of using positive and negative controls is relatively new and has been applied by several researchers [4–6, 8, 16] in the recent past. Whereas Rath and Kanfer [4] used formulations containing 25% less and 26% more than the reference product as negative controls, Wellington et al. [6] used formulations containing different strengths of the API at a lower limit of 70% and an upper limit of 140% to serve as negative controls whilst investigating the discriminatory ability of IVRT methods to detect non-similarity slightly outside the limits of acceptance (75%–133.33%).

5.21 Comparative IVRT Runs

The main objective of developing and validating an IVRT method is to accurately assess the equivalence or "sameness" of test formulations. For the assessment of "sameness" between semisolid products, comparative IVRT runs are conducted with the reference and test products. The testing is carried out in accordance with the relevant guidances published by the regulatory authorities [13, 14, 82], which recommend that both products should be included in each of the runs to ensure an unbiased comparison in the event of a systematic difference between runs. The products are generally assigned to the VDCs in a systematic or random manner using an arrangement shown in Fig. 5.9. The release rate profiles, i.e. plots of cumulative amounts of drug released per unit area vs. square root of time, are constructed, and the release rates are determined. Pairwise comparison is carried out in accordance with the relevant regulatory guidance [13, 14]. The test product is considered to be equivalent if the 90% confidence interval falls within the acceptance limits of 75%–133.33% for FDA [14] and 90%–110% for EMA [13].

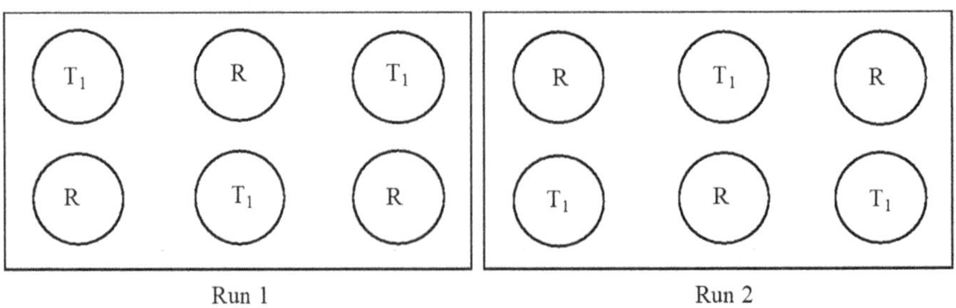

Fig. 5.9 Representative VDC assignment for IVRT runs for "sameness" assessment between the test and the reference products [4]

5.22 Concluding Remarks

IVRT has been acknowledged as a useful and valuable QC tool and has been accepted worldwide by global regulatory authorities, including the EU, US FDA and the World Health Organization (WHO), to detect similarities and differences among topical semisolid products. More recently it has been extensively proposed as an acceptable method to justify biowaivers in several FDA PSGs [82] since it can also be used to predict possible differences in product performance because differences in release rates may implicate formulation differences. The results from various studies clearly reinforce the potential for its use as a biowaiver for topical products for local action, and the published data promote the potential and possibility to establish IVRT as an acceptable option to assess BE of such products, thereby obviating the need for cost-, labour- and time-intensive clinical studies and facilitating faster entry of affordable generic products into the market. Notwithstanding, it is important to emphasise that failure to comply with appropriate validation requirements can result in erroneous approvals that may have significant implications for the clinical performance of products.

Acknowledgements The authors gratefully acknowledge and thank Dr. Ping Lee of the Leslie Dan Faculty of Pharmacy, University of Toronto, Toronto, Canada, for insightful and valuable discussions relating to the theoretical principles of drug release.

References

1. Kanfer I. Methods for the Assessment of Bioequivalence of Topical Dosage Forms: Correlations, Optimization Strategies, and Innovative Approaches. In: Shah VP, Maibach HI, Jenner J, editors. Topical Drug Bioavailability, Bioequivalence, and Penetration [Internet]. New York, NY: Springer New York; 2014. p. 113–51. https://doi.org/10.1007/978-1-4939-1289-6_9
2. Wilbur RL. The difference between topical and transdermal medications. White paper by Gensco Pharma, May. 2017
3. Flynn GL. Conceptual Differentiation of Topical and Transdermal Drug Delivery Systems. Topical Drug Bioavailability, Bioequivalence, and Penetration. 2013;369.
4. Rath S, Kanfer I. A validated IVRT method to assess topical creams containing metronidazole using a novel approach. Pharmaceutics. 2020;12(2):119.
5. Mudyahoto NA, Rath S, Ramanah A, Kanfer I. In Vitro release resting (IVRT) of topical hydrocortisone acetate creams: A novel approach using positive and negative controls. Dissolution Technol. 2020;27:6–12.
6. Wellington H, Rath S, Kanfer I. A comprehensively validated IVRT method reliably discriminates sameness and differences between several topical clotrimazole creams. European Journal of Pharmaceutical Sciences. 2024;192:106649.
7. Nallagundla S, Patnala S, Kanfer I. Application of an optimized tape stripping method for the bioequivalence assessment of topical acyclovir creams. AAPS PharmSciTech. 2018;19:1567–73.
8. Purazi P, Rath S, Ramanah A, Kanfer I. Assessment of "Sameness" and/or Differences between Marketed Creams Containing Miconazole Nitrate Using a Discriminatory in vitro Release Testing (IVRT) Method. Sci Pharm. 2020;88(1):6.

9. Wellington H, Rath S, Abboo S, Kanfer I. Utility of in vitro release testing (IVRT) to assess 'sameness' of 1% clotrimazole creams for use as a biowaiver. AAPS Open. 2023;9(1):18.
10. Kanfer I, Rath S, Purazi P, Mudyahoto NA. In vitro release testing of semi-solid dosage forms. Dissolut Technol. 2017;24(3):52–60.
11. US Food and Drug Administration. GDUFA Reauthorization Performance Goals and Program Enhancements Fiscal Years 2018–2022 [Internet]. [Cited 2024 Jan 13]. Available from: https:/ /www.fda.gov/media/101052/download
12. US Food and Drug Administration. Complex Generics News [Internet]. [Cited 2024 Jan 13]. Available from: https://www.fda.gov/drugs/generic-drugs/complex-generics-news#:~:text=Generally%2C%20complex%20generics%20are%20products,complex %20drug%2Ddevice%20combination%20products
13. Medicines Agency E. Draft guideline on quality and equivalence of topical products [Internet]. Available from: www.ema.europa.eu/contact
14. Semisolid N. Office of Training and Communications Division of Communications Management The Drug Information Branch [Internet]. Vol. 20857. Tel; 1997. Available from: http:// www.fda.gov/cder/guidance.htm
15. Barends D, Dressman J, Hubbard J, Junginger H, Patnaik R, Polli J, et al. Multisource (generic) pharmaceutical products: guidelines on registration requirements to establish interchangeability draft revision. WHO, Geneva, Switzerland. 2005;
16. Tiffner KI, Kanfer I, Augustin T, Raml R, Raney SG, Sinner F. A comprehensive approach to qualify and validate the essential parameters of an in vitro release test (IVRT) method for acyclovir cream, 5%. Int J Pharm. 2018;535(1–2):217–27.
17. Nallagundla S, Patnala S, Kanfer I. Comparison of in vitro release rates of acyclovir from cream formulations using vertical diffusion cells. AAPS PharmSciTech. 2014;15:994–9.
18. Tiffner KI, Kanfer I, Augustin T, Raml R, Raney SG, Sinner F. Comparative in vitro release testing (IVRT) of acyclovir products. Int J Pharm [Internet]. 2021;609:121186. Available from: https://www.sciencedirect.com/science/article/pii/S0378517321009923
19. Chang RK, Raw A, Lionberger R, Yu L. Generic development of topical dermatologic products: formulation development, process development, and testing of topical dermatologic products. AAPS J. 2013;15:41–52.
20. Yacobi A, Shah VP, Bashaw ED, Benfeldt E, Davit B, Ganes D, et al. Current challenges in bioequivalence, quality, and novel assessment technologies for topical products. Pharm Res. 2014;31:837–46.
21. <1724> Semisolid Drug Products—Performance Tests [Internet]. 2018. Available from: www.fda.gov/
22. Ng SF, Rouse JJ, Sanderson FD, Eccleston GM. The relevance of polymeric synthetic membranes in topical formulation assessment and drug diffusion study. Arch Pharm Res. 2012;35:579–93.
23. Narkar Y. Bioequivalence for Topical Products—An Update. Pharm Res. 2010;27:2590–601.
24. Treherne JE. The permeability of skin to some non-electrolytes. J Physiol. 1956;133(1):171.
25. Higuchi T. Rate of release of medicaments from ointment bases containing drugs in suspension. J Pharm Sci. 1961;50(10):874–5.
26. Higuchi WI. Analysis of data on the medicament release from ointments. J Pharm Sci. 1962;51(8):802–4.
27. Higuchi T. Physical chemical analysis of percutaneous absorption process from creams and ointments. J Soc Cosmet Chem. 1960;11:85–97.
28. Higuchi T. Cosm. Chem. 1960;11:85.
29. Higuchi T. Lach. JL, ibid. 43 527 (1954). 1960;49:417.
30. Poulsen BJ, Young E, Coquilla V, Katz M. Effect of topical vehicle composition on the in vitro release of fluocinolone acetonide and its acetate ester. J Pharm Sci. 1968;57(6):928–33.
31. Ayres JW, Laskar PA. Diffusion of benzocaine from ointment bases. J Pharm Sci. 1974;63(9):1402–6.
32. Bottari F, Di Colo G, Nannipieri E, Saettone MF, Serafini MF. Influence of drug concentration on in vitro release of salicylic acid from ointment bases. J Pharm Sci. 1974;63(11):1779–83.

33. Bottari F, Di Colo G, Nannipieri E, Saettone MF, Serafini MF. Evaluation of a dynamic permeation technique for studying drug–macromolecule interactions. J Pharm Sci. 1975;64(6):946–9.
34. Bottari F, Colo GDI, Nannipieri E, Saettone MF, Serafini MF. Release of drugs from ointment bases II: in vitro release of benzocaine from suspension-type aqueous gels. J Pharm Sci. 1977;66(7):926–31.
35. Chowhan ZT, Pritchard R. Release of corticoids from oleaginous ointment bases containing drug in suspension. J Pharm Sci. 1975;64(5):754–9.
36. Behme RJ, Kensler TT, Brookex D. A new technique for determining in vitro release rates of drugs from creams. J Pharm Sci. 1982;71(11):1303–5.
37. Weng HL, Parrott EL. Dissolution apparatus for gels. J Pharm Sci. 1983;72(2):186–8.
38. Velissaratou AS, Papaioannou G. In vitro release of chlorpheniramine maleate from ointment bases. Int J Pharm. 1989;52(1):83–6.
39. Stricker H. Drug absorption in the gastro-intestinal tract; in vitro investigation. Pharm Ind. 1971;33:157–60.
40. Rolland A, Demichelis G, Jamoulle JC, Shroot B. Influence of formulation, receptor fluid, and occlusion, on in vitro drug release from topical dosage forms, using an automated flow-through diffusion cell. Pharm Res. 1992;9:82–6.
41. Kumar P, Sanghvi P, Collins CC. Comparison of diffusion studies of hydrocortisone between the Franz cell and the enhancer cell. Drug Dev Ind Pharm. 1993;19(13):1573–85.
42. Shah VP, Elkins J, Lam SYY, Skelly JP. Determination of in vitro drug release from hydrocortisone creams. Int J Pharm. 1989;53(1):53–9.
43. Chattaraj SC, Swarbrick J, Kanfer I. A simple diffusion cell to monitor drug release from semi-solid dosage forms. Int J Pharm. 1995;120(1):119–24.
44. Olejnik A, Goscianska J, Nowak I. Active compounds release from semisolid dosage forms. Vol. 101, Journal of Pharmaceutical Sciences. John Wiley and Sons Inc.; 2012. p. 4032–45.
45. Chattaraj SC, Kanfer I. 'The insertion cell': A novel approach to monitor drug release from semi-solid dosage forms. Int J Pharm [Internet]. 1996;133(1):59–63. Available from: https://www.sciencedirect.com/science/article/pii/0378517395044108
46. Chattaraj SC, Kanfer I. Release of acyclovir from semi-solid dosage forms: a semi-automated procedure using a simple plexiglass flow-through cell. Int J Pharm. 1995;125(2):215–22.
47. Chattaraj SC, Das SK, Kanfer I. In vitro release of acyclovir from semisolid dosage forms: effect of cyclodextrin and polyethylene glycol. Pharm Dev Technol. 1998;3(4):565–70.
48. US Food and Drug Administration. Draft Guidance on Acyclovir. 2012 p. 1–6.
49. Klein RR, Tao JQ, Wilder S, Burchett K, Bui Q, Thakker KD. Development of an In Vitro Release Test (IVRT) for a Vaginal Microbicide Gel. Dissolut Technol. 2010;6–10.
50. Klein RR, Heckart JL, Thakker KD. In vitro release testing methodology and variability with the vertical diffusion cell (VDC). Dissolution Technol. 2018;25(3):52–61.
51. Martin A. Physical pharmacy. Physical chemical principles in the pharmaceutical sciences. 1993;126–7.
52. Troy DB. Remington: The science and practice of pharmacy. (No Title). 2005;
53. Ueda CT, Shah VP, Derdzinski K, Ewing G, Flynn G, Maibach H, et al. Topical and transdermal drug products. In: Pharmacopoeial forum. 2009. p. 750–64.
54. Chamboko BV. Evaluation of the Safety and Efficacy of. 2007; (February).
55. Rafiee-Tehrani M, Mehramizi A. In Vitro Release Studies of Piroxicam from Oil-in-Water Creams and Hydroalcoholic Gel Topical Formulations. Drug Dev Ind Pharm. 2000;26(4):409–14.
56. Siepmann J, Peppas NA. Higuchi equation: Derivation, applications, use and misuse. Vol. 418, International Journal of Pharmaceutics. 2011. p. 6–12.
57. Higuchi WI. Notes-Analysis of Data on the Medicament Release from Ointments.
58. Kanu Patel BC, Banker GS, George DeKay H. Pharmacopoeia of the United States of America. Vol. 42, J. A m. Pharm. Assoc., Pract. Phaum. Ed. McGraw-Hi11 Book Co., Inc; 1953.
59. Baker RW. Controlled release of biologically active agents. John Wiley & Sons; 1987.

60. Lee P. Discussion with Dr. Isadore Kanfer. Toronto, ON, Canada; 2018.
61. Lee P. Email to Dr. Isadore Kanfer. Toronto, ON, Canada; 2018.
62. Lee PI. Modeling of drug release from matrix systems involving moving boundaries: Approximate analytical solutions. Int J Pharm [Internet]. 2011;418(1):18–27. Available from: https://www.sciencedirect.com/science/article/pii/S0378517311000500
63. Topical Dermatologic Corticosteroids: In Vivo Bioequivalence Guidance for Industry DRAFT GUIDANCE [Internet]. Available from: https://www.fda.gov/Drugs/GuidanceComplianceRegulatoryInformation/Guidances/default.htm
64. 21 CFR Part 320.24. Vol. 5, United States of America, Federal Register—Code of Federal Regulations. 2023 [cited 2023 Sep 28]. p. 1–10 Title 21—Part 320—Bioavailability and Bioequivalence Requirements—Subpart B—Procedures for Determining the Bioavailability or Bioequivalence of Drug Products. Available from: https://www.accessdata.fda.gov/scripts/cdrh/cfdocs/cfcfr/CFRSearch.cfm?fr=320.24
65. Raney SG, Luke MC. A new paradigm for topical generic drug products: Impact on therapeutic access. J Am Acad Dermatol. 2020;82(6):1570–1.
66. Shah VP, Elkins JS, Williams RL. Evaluation of the test system used for in vitro release of drugs for topical dermatological drug products. Pharm Dev Technol. 1999;4(3):377–85.
67. Shah VP. Progress in methodologies for evaluating bioequivalence of topical formulations. Am J Clin Dermatol. 2001;2:275–80.
68. Aiache J marc, Aoyagi N, Bashaw D, Brown C, Brown W, Burgess D, et al. FIP/AAPS Guidelines to Dissolution/in Vitro Release Testing of Novel/Special Dosage Forms. AAPS PharmSciTech. 2003;4(1):1–10.
69. Brown CK, Friedel HD, Barker AR, Buhse LF, Keitel S, Cecil TL, et al. FIP/AAPS joint workshop report: dissolution/in vitro release testing of novel/special dosage forms. Springer; 2011.
70. Naik P, Shah SM, Heaney J, Hanson R, Nagarsenker MS. Influence of test parameters on release rate of hydrocortisone from cream: study using vertical diffusion cell. Dissolut Technol. 2016;23(3):14–20.
71. Shah VP, Elkins JS, Williams RL. Evaluation of the Test System Used for In Vitro Release of Drugs for Topical Dermatological Drug Products [Internet]. Vol. 4, Pharmaceutical Development and Technology. 1999. Available from: www.dekker.com
72. Thakker KD, Chern WH. Development and validation of in vitro release tests for semisolid dosage forms—case study. Vol. 10, Dissolution Technologies. Dissolution Technologies Inc; 2003. p. 10–5.
73. Electronic Code of Federal Regulations [Internet]. 2019 [cited 2019 Mar 12]. Title 21, Chapter 1 D, Part 314 A, 314.3. Available from: https://www.ecfr.gov/cgi-bin/text-idx?SID=6d3150c5705356f68d738caaa8248003&mc=true&node=se21.5.314_13&rgn=div8
74. USFDA. US Food and Drug Administration. 2012. Draft Guidance on Acyclovir (Topical Ointment). Available from: https://www.accessdata.fda.gov/drugsatfda_docs/psg/Acyclovir_topicalcream_RLD 21478_RV12-16.pdf
75. FDA Office of Generic Drugs. Draft Guidance on Acyclovir Cream. 2014.
76. US Food and Drug Administration—Office of Generic Drugs. Draft Guidance on Ivermectin. 2018;2(December 2016):1–5.
77. FDA Office of Generic Drugs. Draft Guidance on Penciclovir Cream. 2018.
78. FDA Office of Generic Drugs. Draft Guidance on Docosanol Cream. 2017.
79. FDA Office of Generic Drugs. Draft Guidance on Dapsone Gel. 2017.
80. Ilić T, Pantelić I, Lunter D, Đorđević S, Marković B, Ranković D, et al. Critical quality attributes, in vitro release and correlated in vitro skin permeation—in vivo tape stripping collective data for demonstrating therapeutic (non)equivalence of topical semisolids: A case study of "ready-to-use" vehicles. Int J Pharm. 2017;528(1–2):253–67.
81. Food and Drug Administration Safety and Innovation Act. The Senate and House of Representatives of the United States of America in Congress; 2012.

82. Fda, Cder. In Vitro Release Test Studies for Topical Drug Products Submitted in ANDAs Guidance for Industry DRAFT GUIDANCE [Internet]. Available from: https://www.fda.gov/drugs/guidance-compliance-regulatory-information/guidances-drugs
83. Franz TJ. Percutaneous absorption. On the relevance of in vitro data. Journal of Investigative Dermatology. 1975;
84. PermeGear. Diffusion Cell Basics [Internet]. [Cited 2024 Jan 21]. Available from: https://permegear.com/wp-content/uploads/2019/11/Diffusion-Cell-Basics.pdf
85. Looney TJ. USP apparatus 4 [Flow-through method] primer. Dissolut Technol. 1996;3(4):10–2.
86. Product Center. American Pharmaceutical Review Web site. [Internet]. [Cited 2024 Jan 21]. CE 7smart USP Apparatus 4—Flow Through Cell Dissolution from SOTAX. Available from: http://www.americanpharmaceuticalreview.com/25255-Dissolution-Testing-Analysis-Equipment/5821525-CE-7-i-smart-i-USP-Apparatus-4-Flow-Through-Cell-Dissolution/
87. Shah VP, Elkins J, Shaw S, Hanson R. In vitro release: Comparative evaluation of vertical diffusion cell system and automated procedure. Pharm Dev Technol. 2003;8(1):97–102.
88. Shah VP, Elkins J, Hanus J, Noorizadeh C, Skelly JP. In vitro release of hydrocortisone from topical preparations and automated procedure. Pharm Res. 1991;8:55–9.
89. Copley Scientific [Internet]. 2019 [cited 2019 Apr 14]. Vertical Diffusion Cell Test System. Available from: https://www.copleyscientific.com/home/pharmaceutical-testing/semisolids-testing/vertical-diffusion-cell-test-system
90. Diffusion testing fundamentals [Internet]. [cited 2016 Oct 3]. p. 1–8. Available from: www.permegear.com
91. PermeGear, Inc. [Internet]. [Cited 2019 Apr 13]. Franz cell—the original. Available from: https://permegear.com/franz-cells/
92. Hanson R, Heaney J. A primer on automating the vertical diffusion cell (VDC). Dissolut Technol. 2013;20(2):40–3.
93. Hanson Research Corporation [Internet]. [cited 2019 Mar 13]. MicroettePlus Multi-Group: Operation Manual. Available from: https://files.hansonresearch.com/wp-content/uploads/2015/03/Diffusion-MicroettePlus-Multi-Group-Operation-Manual.pdf
94. Teledyne Hanson Research [Internet]. 2019 [cited 2019 Apr 14]. 6-Cell manual diffusion test system. Available from: https://hansonresearch.com/diffusion-testing/manual-diffusion/
95. Upadhyay Y, Singh AK, Mishra S, Gurule SJ, Khuroo AH, Verma S, et al. Comparison of Release Rates Derived from Cutaneous Versus Film-Forming Solution Containing Terbinafine: Approach Towards Qualification and Validation of In Vitro Parameters. Dissolut Technol. 2022 Feb 1;29(1):28–35.
96. Su PH, Manager S, Wang K. IVRT insights from CMC experts of frontage labs. 2020.
97. Patel A. IVRT Method Development and Validation of Dapsone Gel.
98. Zatz JL, Segers JD. Techniques for measuring in vitro release from semisolids. Dissol Technol. 1998;5:3–13.
99. Pénzes T, Blazsó G, Aigner Z, Falkay G, Erős I. Topical absorption of piroxicam from organogels—in vitro and in vivo correlations. Int J Pharm. 2005;298(1):47–54.
100. Derel HG. Method Development and Validation of in Vitro Release Test for Fluorometholone Ophthalmic Suspension using Franz Diffusion Cell System. Global Journal of Pharmacy & Pharmaceutical Sciences [Internet]. 2021 Nov 18;9(2). Available from: https://juniperpublishers.com/gjpps/GJPPS.MS.ID.555758.php
101. Leal LB, Cordery SF, Delgado-Charro MB, Bunge AL, Guy RH. Bioequivalence methodologies for topical drug products: in vitro and ex vivo studies with a corticosteroid and an anti-fungal drug. Pharm Res. 2017;34:730–7.
102. Bhagurkar AM, Angamuthu M, Patil H, Tiwari R V, Maurya A, Hashemnejad SM, et al. Development of an ointment formulation using hot-melt extrusion technology. AAPS PharmSciTech. 2016;17:158–66.
103. Ng SF, Rouse J, Sanderson D, Eccleston G. A comparative study of transmembrane diffusion and permeation of ibuprofen across synthetic membranes using Franz diffusion cells. Pharmaceutics. 2010;2(2):209–23.

104. Shah VP, Elkins J, Lam SY, Skelly JP. Determination of in vitro drug release from hydrocortisone creams. Int J Pharm. 1989;53(1):53–9.
105. Klein RR, Bechtel JL, Burchett K, Thakker KD. Hydrocortisone as a Performance Verification Test Reference Standard for In Vitro Release Testing. Dissolution Technol. 2010;11:37–8.
106. Tuomi A, Turakka L, Raty T, Lehmussaari K. Release of hydrocortisone from topical semi solid preparations using three different in vitro methods. Acta Pharm Fenn. 1989;98:93–9.
107. Thakker K, Klein R. Drug release: topical products. Specification of Drug Substances and Products. 2020;505–34.

Chapter 6
Perspectives on Practical In Vitro Permeation Testing (IVPT) of Topical Products

Srinath Rangappa and S. Narasimha Murthy

6.1 Introduction

A typical permeation study apparatus consists of two compartments, one for placing the formulation and the other for the receiver medium. The barrier membrane is sandwiched between the two compartments. The receiver medium is sampled at different time intervals to analyze the Active Pharmaceutical Ingredient (API) and excipients. There are horizontal and vertical type static diffusion cells in use by researchers. The horizontal cells are not as appropriate for evaluating the semisolids and transdermal products. Unlike vertical diffusion cells, there is no provision to expose the formulation to the atmosphere. They are suitable for performing permeation/diffusion studies using solutions and liquid formulations. A wealth of information on diffusion cell types, uses, and dimensions is available in the literature [1, 2]. Their proven track record makes vertical static diffusion cells popular for testing semisolids and transdermal products. The vertical diffusion cells, recommended by the regulatory bodies, are a testament to their suitability for testing semisolids. The vertical diffusion cells have evolved, incorporating additional features such as dry heating (the water circulation jacket is eliminated), flow-through (the receiver medium is continuously perfused), automated sampling, automatic air bubble removal (by tilting the diffusion cell through an angle after refilling the fresh buffer), and high-throughput screening capabilities.

Vertical diffusion cells come in two types: static and flow-through. The choice between these types depends on the receiver compartment fluid dynamics. In static cells, the receiver fluid is stirred using a magnetic bead. The flow-through cells have a smaller receiver compartment through which the medium is continuously

S. Rangappa · S. N. Murthy (✉)
Topical Products Testing LLC, Batesville, MS, USA

S. N. Murthy (ed.), *Topical Products and Dermal Drug Delivery*,
AAPS Introductions in the Pharmaceutical Sciences 2,
https://doi.org/10.1007/978-3-031-80525-7_6

perfused. Despite these differences, the principle behind the experimental studies remains the same. When optimized appropriately, both cell types should yield similar conclusions.

Based on the study's objectives and the dose used for the investigation, the in vitro permeation testing protocol can be of two types: evaporation-disabled (ED) and evaporation-enabled (EN) setups. The primary difference between the two protocols is that in the evaporation-enabled study, the experimental setup lets the solvent evaporation from the formulation happen independently without interruption. The donor compartment is occluded in the evaporation-disabled setup to prevent solvent evaporation.

6.2 Evaporation-Disabled (ED) Permeation Studies

Evaporation is the conversion of a liquid to its vapor below the liquid's boiling temperature, and condensation is the transformation of the vapors into liquid molecules. The solvent's evaporation rate is the same as the rate of condensation in a closed chamber at equilibrium. The donor compartment is closed/occluded in an ED permeation study with a lid or a suitable solvent-impermeable membrane. The air above the formulation eventually gets saturated with the vapor phase of the liquid. It reaches a state where the evaporation rate equals the condensation rate, thus establishing a dynamic equilibrium.

In the clinical situation, when a therapeutically relevant dose is applied to the skin, solvent loss will bring about several changes in the formulation. The dynamic changes in quality attributes and microstructure of the formulation play a significant role in determining the performance of a topical product. Therefore, when the study's objective is to investigate the influence of differences in composition or quality attributes (globule size, particle size, viscosity, etc.) on the drug permeation, the evaporation-disabled setup is preferred to eliminate the confounding effects due to solvent evaporation.

In the ED permeation studies, one should maintain a lesser headspace in the donor compartment to allow faster equilibration of solvent vapors and minimal loss of solvent from the formulation so that there is no notable change in the properties of the formulation. In a study performed with metronidazole gels, the cumulative amount of drug permeated varied with the headspace volume. A hydrophobic plastic plug was inserted in one set of donors (1 cm^2 active diffusion area) touching the formulation surface barely (no headspace volume). In the other set, the plastic plug was used to close the donor (occluded) at the top. Despite the dose being the same (100 mg/cm^2), the cumulative amount of API permeated significantly differed between the two sets conducted with and without the headspace volumes (Fig. 6.1). Generally, it is preferable to maintain a lesser headspace in the donor compartment by performing studies at a higher dose (filling most of the donor volume) or by using an occlusion cap that inserts far enough down in the donor compartment to minimize

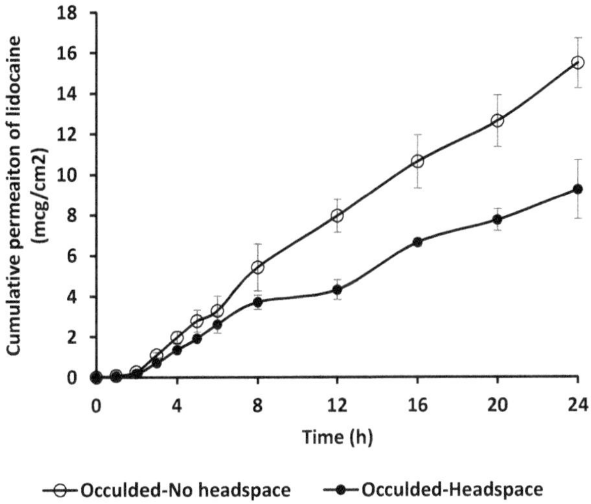

Fig. 6.1 The impact of headspace volume in the donor compartment on the permeation profile of lidocaine from topical gels

the headspace volume. Some liquid APIs, such as nitroglycerine and nicotine, could also evaporate significantly if the headspace volume is larger.

The donor compartment is often occluded using parafilm, plastic membrane/tape, or glass stoppers. One needs to evaluate the choice of membrane, as some membranes allow the evaporation of solvents. For example, parafilm is a flexible film composed of polyolefins and waxes. It is a hydrophobic membrane that completely blocks the loss of water. However, the parafilm is slightly permeable to alcohol vapors from the donor compartment. Its permeability to alcohol vapors increases if the film is stretched to be thinner. An airtight glass closure or plastic caps made of polyethylene or polypropylene would be preferable for a formulation containing alcoholic, hydroalcoholic, and highly volatile vehicles.

6.3 Evaporation-Enabled (EN) Permeation Studies

In EN studies, the donor compartment is kept open to the atmosphere to allow the evaporation of solvents from the formulation, mimicking real-world clinical conditions. When applied to the skin, the solvents from the formulation evaporate at different rates depending on their threshold kinetic energy. Liquid molecules with adequate threshold kinetic energy to escape from the skin surface at 32°C would become vapor and diffuse out of the formulation. The liquid molecules with lower kinetic energy will remain longer on the skin surface.

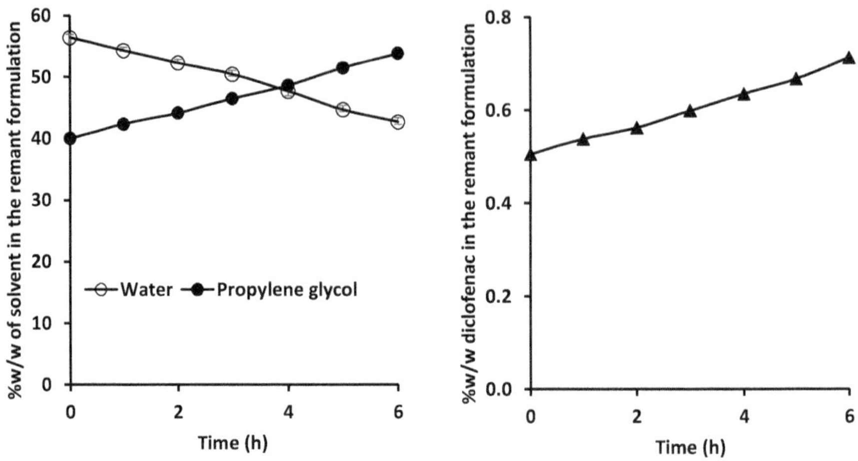

Fig. 6.2 The time course of propylene glycol, water, and diclofenac concentration due to the evaporation of solvents from the donor compartment

Figure 6.2 shows the time course of change in water and propylene glycol concentrations in a 0.5% diclofenac sodium gel containing 40% propylene glycol (bp. 188.2°C) and water (q.s.). The permeation study setup was evaporation-enabled. Please note that the dose used in this study was manyfold higher (300 mg/cm^2 of formulation to be able to sample the formulation at different time points to quantify the solvents and API) than the clinically relevant dose. The water, being relatively more volatile, evaporated rapidly from the formulation. This led to an increase in the concentration of propylene glycol in the formulation. This also resulted in an increase in the concentration of the API in the formulation due to reduced solvent volume in the remaining formulation.

6.4 Franz Diffusion Cells

Franz diffusion cells (FDC) are the vertical diffusion cells most commonly used to evaluate topical formulations [1, 3, 4–7]. The same apparatus is often used to assess formulations intended for application on mucosal, nail, and other biological barriers. The Franz diffusion cells are carefully designed, considering the hydrodynamics, sink condition, sampling, and application of the formulation. Although the dimensions of different parts of the diffusion cell remain the same across different manufacturers, some variations were also observed; nevertheless, the fundamental principles and operation are consistent [2].

6.4.1 Donor Compartment

The donor compartment is typically designed to hold the formulation in place, secure the skin to the receiver, and maintain temperature. When finite dose permeation studies are performed, donor cells maintain consistent airflow over the formulation applied. Therefore, the donor volume (in terms of the ratio of the height of the donor (h) to the skin surface area (a)) is one of the essential specifications to be paid attention to. If the ratio of the h/a of the skin is large, the solvent that evaporates from the skin surface will tend to condense on the surface of the donor during the permeation studies. The solvent condensing on the donor chamber's inner surface could impact the formulation in two ways. If the quantity of liquid is substantial, the condensate would flow down and combine with the formulation again. The second would be that the condensate will form a vapor phase and saturate the zone above the formulation, thus further affecting the solvent evaporation rate. There is nothing like a standard h/a ratio that is universally suitable for all formulation types. The extent of solvent condensation also depends on the formulation composition. Mainly, when highly volatile solvents are present in the formulation, the rapid evaporation of solvent leads to a transient decrease in the temperature, leading to condensation on the surface [8]. Therefore, the donor in a diffusion cell should be of the h/a ratio that does not hinder the natural evaporation phenomenon of the solvent. Most reputed manufacturers maintain a uniform h/a, that works for various topical formulations (Table 6.1). However, as the nature of the formulations is diverse, it would still be wise to watch for any such condensation phenomenon that could interfere with the realistic evaluation of formulation metamorphosis and performance.

The data in Fig. 6.3 shows how the drying or evaporation rate (used interchangeably) was impacted when the h/a was varied in a diffusion cell with a one sq. cm active diffusion area. A model hydrogel formulation containing 40% alcohol and 10% polyethylene glycol (PEG) 400 was made up to weight with water. The sensitivity of the gel's drying profile to h/a of the donor compartment shows how critical it is to use a donor with appropriate dimensions. Also, it underscores the importance of maintaining a uniform h/a across all the cells to minimize the variability.

Drying of Formulation Generally, the drying or weight loss profile due to the evaporation of solvents from a uniformly applied topical product occurs in two phases. In a typical graph of solvent content versus time, during the initial phase, the solvent will evaporate from the surface of the formulation. The rate of evaporation from the surface is governed by the diffusion through the stationary air film that is in contact with the surface. The duration of the initial phase is determined by the air temperature, humidity, and speed at which the solvent is transported from the bulk of the solvent to the surface. As drying proceeds, during the second phase, the solvent from the bulk reaches the surface at a decreasing rate. The diffusion coefficient in the bulk of the formulation and the progressively increasing concentration gradient will determine the loss of contents.

Table 6.1 Metrics of diffusion cell donor compartments from selected reputed manufacturers in the United States

Manufacturer	L		P					H		USP (Model B)	USP (Model C)
Active diffusion area (cm^2)	0.64	1.00	1.77	0.64	1.00	1.77	3.14	0.64	1.77	1.00	1.77
Height from the skin surface (cm)	1.7	1.9	2.5	1.7	1.9	2.5	1.8	3.6	3.6	1.8	1.8
Inner diameter (active diffusion diameter in cm)	0.9	1.13	1.5	0.9	1.13	1.5	2	0.9	1.5	1.13	1.5
h/a (height/area)	2.66	1.9	1.41	2.66	1.9	1.41	0.57	5.63	2.03	1.80	1.02

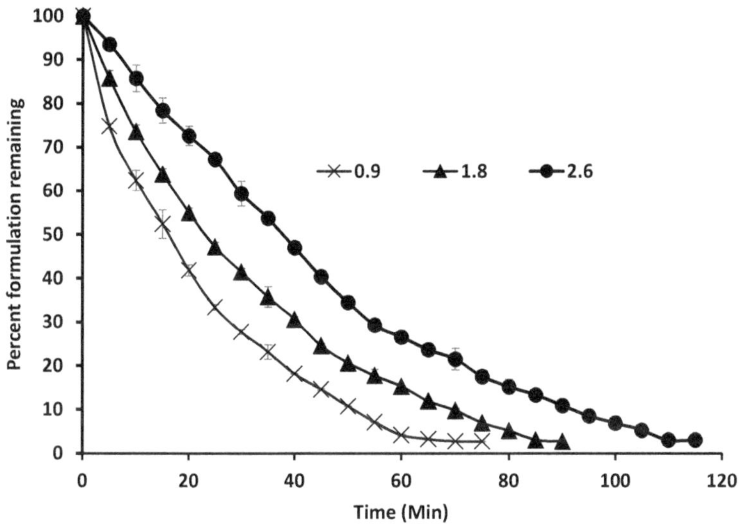

Fig. 6.3 The drying profile of a gel formulation at a 15 mg/cm^2 dose. The height of the donor was varied to get different h/a ratios (0.9, 1.8, and 2.6). Each trial ($n = 3 \pm$ SD) at 32 °C

6.4.2 Receiver Medium

The receiver medium is in contact with the dermis-epidermis layers of the tissue. Therefore, the resultant pH (pH after equilibration) of the epidermis and dermis would depend predominantly on the receiver medium's pH due to the hydrophilic nature of the tissue [9, 10]. On the other hand, the resultant pH of the stratum corneum tissue predominantly depends on the formulation pH. In most of the studies, buffers are used as receiver mediums, while some studies used saline or deionized water [11]. The choice of buffers used in any pharmaceutical evaluation generally depends on the desired pH and ionic strength. As the receiver medium should mimic the conditions of skin interstitial fluid or blood, the choice of buffers narrows down to those that can maintain a physiological pH [12–14]. The most commonly used receiver medium for performing in vitro permeation studies is phosphate-buffered saline (PBS; pH 7.4). HEPES (2-[4-(2-hydroxyethyl)piperazin-1-yl]ethanesulfonic acid) is a zwitterionic compound with a pKa value of 7.55 and has been used in several studies as well. HEPES has a reasonable buffer capacity and can maintain a stable physiological pH. Tris(Hydroxymethyl)aminomethane is also an organic compound soluble in water that can form a buffer system. It can create an efficient buffer in the pH range of 7.2–8 [15].

The formulation's ingredients, such as solvents, preservatives, antioxidants, complexing agents, penetration enhancers, and other functional excipients, can permeate across the skin and reach the receiver compartment. Therefore, the buffer capacity of the receiver medium should be high enough to resist any change in the pH due to the permeation of excipients and APIs. It is a good practice to check the potential impacts of API and excipients on the pH of the receiver medium when developing an IVPT method. It is preferable to ensure that the receiver compartment medium has an excellent tolerance to the acidic/basic characteristics of the API and the excipients present in the formulation being tested.

The solubility of the API in the receiver medium is one of the primary criteria for selecting the receiver medium. The criteria for sink condition has been arbitrarily set at 10%, meaning that the solubility in the receiver medium should be 10X of the cumulative amount of API expected to permeate. It is essential to consider the solubility at the temperature at which the experiments are performed rather than room temperature. Generally, the researchers set the circulating water temperature in the jacket to 32 °C–37°C to maintain the skin surface temperature at 32 °C [16]. The temperature of the skin surface should be measured after the tissue is completely hydrated and equilibrated. This is because the specific heat capacity of the skin tissue in the hydrated condition differs from that of partially hydrated skin. The specific heat capacity of the skin tissue also varies inversely with the thickness of the tissue [17–19].

Poor solubility of the API in the receiver medium would be flux limiting or lead to an unstirred layer phenomenon. Not many APIs are adequately soluble in the receiver medium. Some water-soluble molecules would be even less soluble in the buffers due to the salts in the buffer. The poor solubility of API in the receiver medium compels one to explore different approaches to enhance the drug solubility in the receiver medium. The choice of solubility enhancers (SEs) that can be used to improve the drug solubility in the receiver medium is limited. This is because the selection of SEs is limited by the criteria that they do not affect the skin permeability [20]. The second requirement is that, ideally, SEs should not permeate across the skin into the donor compartment and influence drug permeation. In addition, the material should not interfere with the analysis of the API and other excipients that permeate from the formulation into the receiver medium. It is also preferable not to use a SE in the receiver if it is one of the components of the formulation being evaluated. This is because the drug permeation is often influenced by the permeation of some of the excipients, too. Therefore, the presence of the same excipient in the receiver medium might impact the rate and extent of absorption of the API from the formulation.

6.4.3 Low Ionic Strength Buffer

One of the approaches to improving the drug solubility is by lowering the ionic strength. As mentioned earlier, some water-soluble drugs are less soluble in the buffers. For example, the solubility of oxybutynin hydrochloride in water is about 45 mg/ml, whereas in PBS 1X (full strength), it is about 8 mg/ml. Likewise, the solubility of diclofenac sodium is 14 mg/ml in water, whereas the solubility in PBS 1X is about 5 mg/ml. Solubility of poorly soluble drugs, like acyclovir, is generally less affected (1.57 and 1.4 mg/ml in water and PBS, respectively). On the other hand, drugs like ibuprofen would show enhanced solubility in PBS, which is favorable to generating the sink conditions [9].

The use of a lesser-ionic-strength PBS could improve the sink conditions in the case of some of the APIs. When the PBS's ionic strength was reduced by half, the solubility of oxybutynin hydrochloride increased to 20 mg/ml and diclofenac to 9 mg/ml. Minghetti and co-workers used unbuffered and degassed NaCl 0.9% w/v solution for evaluating the permeation of different diclofenac salts, as the pH of saline was close to neutral and the solubility of diclofenac salts was adequate [21]. Qvist and team used 0.05 M phosphate buffer as the medium to generate a good sink for salicylic acid, testosterone, and nicotine [22]. However, the concern associated with reducing the ionic strength of the buffer is that the buffer capacity might be significantly compromised [23, 24]. For instance, 120 μg/ml of salicylic acid and 100 μg/ml of triethanolamine altered the pH of PBS 1X by 0.1 units and 0.6 units, respectively [25]. Whereas the pH change of 0.25X PBS was 0.5 and one unit, respectively. Meanwhile, in the case of 0.5X PBS, the pH change was the same as that of PBS 1X. This implies that using a 0.5X buffer in place of 1X may be feasible if the API solubility is higher in the former without compromising the buffering ability. One should be mindful that the electrical resistivity/conductivity and pH could vary with the ionic strength of the receiver medium [26]. Besides the buffering ability, factors like sampling method, volume, and frequency of sampling of the receiver medium must also be considered while selecting a buffer system for a permeation study.

6.4.4 Surfactants

Surfactants enhance the API solubility by the micellar solubilization mechanism. Generally, nonionic surfactants have a lower critical micellar concentration value than ionic ones, and most have a negligible impact on the skin barrier at low concentrations. Poly-oxyethylene sorbitan esters (commonly known as Tweens) and Brij surfactants (poly-ethylene PEG chains with different lengths and a hydrophobic tail consisting of an alkyl chain) are the most preferred nonionic surfactants [27]. The challenge with using surfactants is that they could generate air bubbles/foam in the receiver compartment. Some surfactants at higher concentrations can increase the viscosity of the receiver medium. The surfactants might interfere with the analysis of the API, too. Therefore, factors such as surfactant solubility, molecular weight, aggregation number (the larger the better), viscosity, and hydrophilic-lipophilic balance (HLB) value are considered while choosing the surfactant to improve the solubility of the API in the receiver medium. Zambrana and co-workers used 0.1% Brij 98 in PBS to perform permeation studies of sunscreen products with 6%w/w oxybenzone [28]. A study by Kirk and coworkers explored the use of different surfactants to enhance the solubility of cannabidiol (CBD) in PBS. Tween 20, Tween 80, and sodium lauryl sulfate (SLS) enhanced the solubility up to 378.4,

388.0, and 399.8 μg/mL, respectively. Tween 20 was selected as the surfactant, likely because SLS can affect skin permeability. Tween 80 was also found to retard the degradation rate of CBD [29]. In some studies, an adequately higher surfactant concentration was used to maintain the sink condition due to practical difficulties in determining the solubility of API in the surfactant solution. For example, to evaluate the topical products containing fluoxetine, a liquid hydrophobic compound, Jung and co-workers used 6% (w/v) Brij 98 in PBS, as it was practically challenging to determine the solubility of API at different surfactant concentrations [30]. In an in-house study by the authors, the solubility of clobetasol propionate was enhanced using Brij CS 20 from 8 ug/ml in PBS to 1.8 mg/ml in a 2% surfactant solution.

6.4.5 Cosolvents

Using cosolvents is another efficient approach to enhance the solubility of APIs in aqueous solvents. However, many have been found to modulate the skin barrier significantly. For example, ethanol is one of the commonly used solvents for enhancing the solubility of APIs in formulations [20]. However, ethanol has been known to significantly enhance the skin permeability to several molecules [31, 32]. Higher concentrations (>40%) of ethanol may even disrupt the stratum corneum structure on prolonged contact [33]. Moreover, low molecular weight organic solvents pose back diffusion problems due to their volatility [34]. Likewise, polyethylene glycol 400 and propylene glycol are good cosolvents for drugs like diclofenac sodium, acyclovir, and clobetasol. However, even these glycols have been found to affect skin permeability potentially by solvation of the alpha-keratin structures [20]. Glycerol, up to 30% in PBS, was not found to impact the biophysical attributes of the skin and permeation of caffeine, a hydrophilic small molecule. Glycerol is also a cryoprotectant used to prevent ice crystals from forming when frozen. PEG 200 and hexylene glycol have been found to have a negligible impact on skin permeability up to 30% concentration in PBS (Fig. 6.4). Hexylene glycol and glycerol are also antibacterial agents that can help to preserve the skin tissue during prolonged permeation testing [35].

6.4.6 Complexing Agents

Cyclodextrins enhance the solubility of lipophilic drugs in the receiver medium, forming an inclusion complex. In most cases, as complexation with cyclodextrins is just a physical interaction between the substrate and the complexing agent, it is a generally accepted mode of enhancing the solubility of drugs in the receiver medium. The beta-cyclodextrin and hydroxypropyl beta-cyclodextrin (HP-β-CD) are the most commonly used. In one of the studies, permeation and drug release of diclofenac sodium were investigated using HP-β-CD 2.5%w/v in the receiver

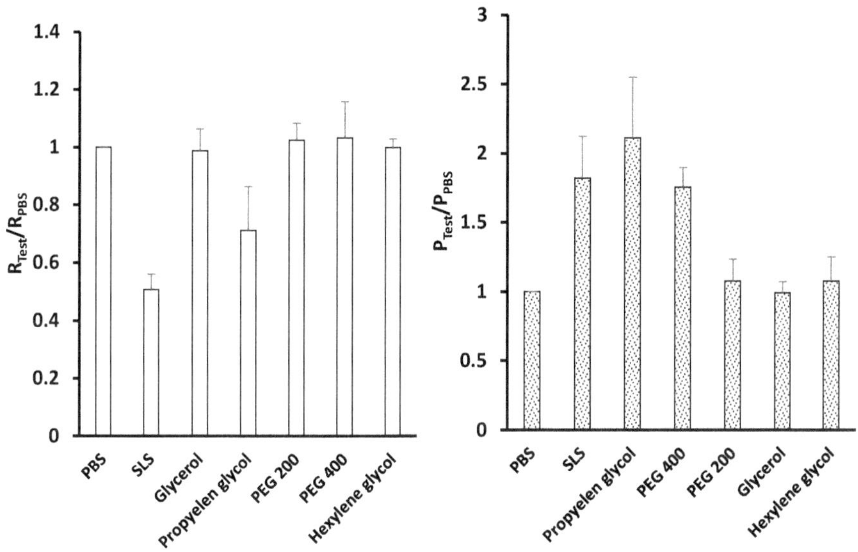

Fig. 6.4 Modulation in the skin resistivity (left) and permeability to caffeine (right) (relative to that of PBS) following 12 h of exposure to the test-receiver medium. SLS was 1%w/v in PBS, and solvents were used at 30%w/v in PBS in the receiver medium. Plain PBS was placed in the donor compartment for electrical resistivity measurements, and 2 mg/ml of caffeine was placed in the donor for permeation studies. The SLS was used as a positive experimental control. Except for propylene glycol, other glycols did not lead to a change in the electrical resistivity more significantly than the buffer did. However, the skin permeability to caffeine enhanced considerably more than the buffer in the case of PEG 400 and propylene glycol. Transepidermal electrical resistance (TEER) combined with permeation to a representative small hydrophilic molecule would be an excellent approach to screen the suitability of SEs in the receiver medium. Tools such as trans-epidermal water loss (TEWL) and permeability to marker molecules such as fluorescein, mannitol, and glucose can also be used

medium [36]. The solubility of testosterone increased in HEPES buffer from 0.02 mg/ml to 3.9 mg/ml and 3.23 mg/ml in the presence of 4%w/v HP-β-CD and Suflobutyl ether β-cyclodextrin (SE- β-CD), respectively [37].

6.4.7 *Bovine Serum Albumin (BSA)*

Bovine serum albumin renders the receiver medium biorelevant [38, 39]. Most studies reported in the literature used 4–5% BSA in the receiver medium [40]. The solubility in the receiver medium was enhanced due to the incorporation of 4%w/v BSA by 10–40 folds for lipophilic alcohols. 5%w/v BSA in PBS was used to evaluate progesterone formulations across porcine and human cadaver skin [40–42].

6.4.8 *Degradation of API and Use of Stabilizing Agents*

It's crucial to be vigilant about the stability of the API in the receiver medium. Some APIs are labile to light, while others are amenable to chemical degradation processes like oxidation or hydrolysis. The stability of the API in the receiver medium should be thoroughly investigated during the development of the IVPT protocol for the given molecule. The photolabile molecules can be protected by performing permeation studies in areas without intense lights, covering the diffusion cells with aluminum foil, or using light-resistant diffusion cells (amber-colored glass).

In the case of molecules amenable to chemical degradative reactions, one needs to investigate the rate of the degradation process primarily. If the degradation rate is significantly slower (1–2 percent in a few weeks through months), it may not affect the results of the permeation study. However, when the degradation is rapid, measures should be taken to protect the API. Some simple approaches that can be useful are sampling the entire receiver at frequent time intervals, immediate transfer of samples to the freezer until the analysis is performed, immediate extraction of API into organic solvents, or evaporation of the receiver medium solvent by freeze-drying are some approaches that might help.

Degassing the medium to remove a substantial amount of dissolved oxygen would help protect the API from undergoing oxidation to some extent. Incorporating antioxidants such as sodium sulfite, sodium bisulfite, citric acid, and chelating agents like EGTA (ethylene glycol-bis(β-aminoethyl)-tetraacetic acid) and EDTA (ethylene diamine tetraacetic acid) in the receiver medium might offer additional protection against oxidation.

In one of the in-house studies by the authors, an API repurposed for a dermal indication was found amenable to rapid hydrolytic degradation in the receiver medium. The receiver medium was incorporated with 1% Brij CS 20 to entrap the molecule in the surfactant micelle, thus minimizing its exposure to the external aqueous medium. In the presence of Brij CS 20, the API remained intact for up to 8 h, beyond which the degradation was gradual. When topical products of this API were subjected to in vitro permeation studies, the receiver medium was withdrawn at 4-hour intervals in its entirety, and the analysis was completed within the next two hours.

6.5 Degassing the Receiver Medium

It is necessary to degas the receiver medium for multiple reasons. The degassed solvent has a lesser tendency to form air bubbles in the receiver medium. The degassing also mitigates the degradation of drugs amenable to oxidation. The degassed solvent reservoir should be maintained in a container with no or minimal headspace in a water bath (to maintain 32 °C) throughout the permeation experiment. The prewarmed medium should only replace the sample drawn intermittently to minimize temperature fluctuations in the receiver compartment. Degassing can be done by heating, sonication, or negative pressure [16, 37]. The surfactants must be dissolved only after degassing, as they might generate foam in the vehicle. If cosolvents are used, it is preferable to degas the solvents in neat form first and mix them later in the required ratio.

6.6 Antimicrobial Agents

Antimicrobial agents are used in the receiver medium during long-term permeation studies. 0.01% gentamicin sulfate would mitigate most potential micro-organisms that contaminate the cadaver skin tissue [43]. Likewise, parabens could also be used. 0.01%–0.1%w/v sodium azide is one of the commonly reported antimicrobial agents in several studies reported in the literature [41, 44, 45]. Although benzalkonium chloride is an excellent antimicrobial agent, as it is a cationic surfactant, there is a potential risk of interaction with the skin. Many studies reported in the literature also used streptomycin concentrations of 100 μg/ml due to its broad-spectrum activity [46, 47]. Merwe and co-workers also evaluated the impact of streptomycin sulfate (100 μg/ml) in the receiver medium on the permeation of water. The antimicrobial agent did not affect the skin's permeability to water [48]. On the other hand, chloramphenicol sodium succinate and amphotericin B (50 μg/ml each) were found to interrupt the diffusion profile of water. The investigators recommend the use of antibiotics in the receiver medium in any study performed longer than 12 h. Indeed, they suggest the use of an antifungal drug along with an antibacterial to inhibit the microbe-mediated barrier disruption. The surfactants can affect the efficacy of antimicrobial agents. The investigator should ensure the inertness of the surfactants towards the antimicrobial agent during IVPT method development.

6.7 Sampling Procedure

The sampling tool should be made with an appropriate material that does not bind the drug. Plastic sampling tubes bind with hydrophobic APIs and APIs with a surfactant-like structure (examples: diclofenac, promethazine, propranolol). Pipette tips, long metal needles, and capillary tubes are some of the tools used to withdraw the receiver medium. The sample should be obtained each time from the middle of the receiver compartment after at least three or four push-pull mixing without causing turbulence. The sampling port should be kept closed all the time except during the sampling and replacement of the medium to avoid any solvent loss. A parafilm, cellophane membrane, or plastic cap can function efficiently to close the sampling port.

Aliquot sampling is the partial withdrawal of the receiver medium. The primary advantage of aliquot sampling is that the drug accumulates in the receiver with time, thus making the analysis less challenging. Suppose a small volume of the receiver medium is adequate for extraction and analysis; the aliquot sampling can be performed even without letting air bubbles from the skin-receiver medium interphase.

Withdrawing the entire receiver fluid is also a good approach, as one can get a large volume sample that could be subjected to freeze drying or solvent evaporation to concentrate if the amount of API permeated is below the limit of analytical quantitation. Sampling the entire receiver medium would also help to maintain a consistent sink condition, especially when the solubility of the API in the receiver medium has barely met the criteria for establishing the sink condition.

The sampling intervals are typically determined based on several factors. The expected extent of permeation is one of the very important factors to consider. In the case of partial sampling (sample volume <2 ml), it may be feasible to sample at shorter intervals as the API accumulates in the receiver medium with time. The lesser the sampling volume, the lesser the dilution when the compartment is replaced with an equal volume of prewarmed fresh buffer. When the whole receiver is withdrawn, the sampling intervals need to be long enough to accumulate an adequate amount of API in the receiver. It is preferable to frequently withdraw samples in the first few hours in finite and infinite dose permeation studies. In infinite dose studies, the initial time points' data would help calculate the lag time more precisely. In the case of finite dose studies, earlier time point data could help vividly capture the up-curve of the flux-time profile.

6.8 Skin Tissue

The skin may be harvested from patients undergoing a surgical procedure or excised from human cadavers. The source should be consistent when testing multiple products for comparison, such as bioequivalence studies. Generally, the skin is dermatomed from the posterior torso, thighs, and back regions. The dermatome skin's thickness from most skin banks is about 100–1000 micrometers. The skin isolated by plastic surgery is maintained in viable condition by keeping it in contact with a nutrition media containing antibiotics and glycerin [49, 50]. The viable skin is useful for studying the drug metabolism in the skin or for investigating the enzymatic cleavage of prodrugs in the skin during permeation. Cadaver skin is appropriate for studying the impact of different formulation variables, formulations with different quality attributes, or comparing the generic product with its respective RLD.

6.9 Skin Thickness

The thickness of the skin is one of the highly variable components of the skin obtained from skin banks. Skin thickness is variable by region and due to many other biological reasons. Although dermatome is a reliable technique to excise the tissue of the required thickness, the variability is notably high because of the viscoelastic nature of the skin and the protocol followed for excision. The dermatomed full-thickness skin (epidermis and dermis) can be used directly, or the epidermis can be separated by heat separation (split-thickness skin) [51]. After cleaning the unwanted tissue on the derm side, the skin should be cut into smaller pieces *uniformly* to fit the external diameter of the grounded surface of the donor compartment. The thickness is generally measured using a screw gauge or a vernier caliper. The thickness is also measured by microscopy [52]. One of the procedures that was found to be practical and reproducible is measuring the skin thickness using a screw gauge by sandwiching the skin between the microscopic glass slides. It is a good practice to measure the skin thickness in three or more different spots to make sure that the thickness of the active diffusion area is nearly uniform.

The impact of skin thickness variability on the permeation data depends on how lipophilic and skin-binding the drug is. For example, in an in-house study, the caffeine's ($\log P = 0.01$) permeation profile was not affected by significant differences in the thickness of the dermatomed cadaver skin. However, a two-fold difference in skin thickness significantly decreased hydrocortisone's ($\log P = 1.61$) extent of permeation by threefold. This is likely because the caffeine is hydrophilic and water-soluble, and its partitioning into the viable epidermis and dermal layers was not significantly different from its partitioning into the aqueous receiver medium, unlike the lipophilic hydrocortisone. A few studies in the literature have reported concurrent observations. The investigators in separate studies found that the impact of skin thickness on the permeation and retention of caffeine was negligible. In contrast, the permeation of testosterone ($\log P = 3.32$) and its retention in the skin were largely affected by skin thickness [53–55]. At a high level, one can generalize that when the APIs are high protein-binding types or lipophilic, the variation in the skin thickness would significantly impact the permeation profile [54, 56]. Therefore, for such APIs, it is preferable to choose skin samples as uniform as possible in thickness to minimize the inter-cell variability.

6.10 Barrier Integrity

The tritiated water permeability, the permeability of poorly permeable molecules such as fluorescein, mannitol, and glucose, has been benchmarked to check the barrier integrity and qualify it for IVPT studies [57, 58]. However, in the last two decades, trans-epidermal electrical resistance (TEER) and trans-epidermal water loss (TEWL) have become widely common for checking skin integrity. Several different instruments that measure these parameters are available in the market.

Trans-epidermal water evaporation is a physiological phenomenon of continuous and consistent outward water loss across the skin. The water activity of the skin in the inner layers is way higher than in the stratum corneum [59]. This concentration gradient in the thermodynamically active water acts as a driving force, resulting in a continuous outward water flux across the stratum corneum. The phenomenon of constant water evaporation occurs even in in vitro experimental setups. Any compromise of the stratum corneum barrier will increase the TEWL. A wealth of information is already available in books and journals regarding the types of TEWL measuring devices, principles, procedures, and the impact of barrier perturbation on the TEWL. Therefore, the scope of this chapter is limited to only practical aspects of performing in vitro permeation testing.

The threshold TEWL to qualify the skin for a permeation study across different labs typically varies from 5 to 15 g/m^2/h. The TEWL values across the cells should not vary much (preferably not more than ± 3 units) to get consistent results. The permeation data of hydrophilic molecules (examples: metronidazole, caffeine, lidocaine hydrochloride, and diclofenac) were found to be relatively more affected due to variation in the TEWL values across the cells than lipophilic molecules (clobetasol, hydrocortisone, and fentanyl). Some researchers have reported a direct correlation between the permeation of several polar drugs and TEWL values [60–62]. On the contrary, some reports in the literature also report a poor correlation.

TEER is not as harmonized as TEWL across different labs. The TEER measurement varies with the electrical settings, such as the frequency, voltage, duty cycle, and electrode dimensions. Most labs generally use an inductance (L), capacitance (C), and resistance (R) meter (LCR meter) to measure the TEER. Some labs consider 10 kOhm.cm^2 at 32°C as the threshold resistivity for qualifying the cadaver skin. The benchmark values of TEER vary (typically between 5–30 kOhm.cm^2) across labs, depending on the method, apparatus, and measurement settings [45, 63].

The skin can be stored in appropriate polyethylene bags and placed in the freezer at −70–80°C for over a year. Storage at −20°C has also been adequate and equally efficient for short-term storage. An in-house study found that the barrier integrity was intact for up to one year of storage at −70 °C. Therefore, it is suggested not to preserve the skin frozen longer. In some cases, long-term storage was also found to cause separation of the epidermis from the dermis when equilibrated at room temperature or 32°C. Another suggestion is that the skin should not be subjected to multiple freeze-thaw cycles as this could impact the barrier integrity. For example, in a study performed in-house, the permeation of metronidazole (log P -0.15) was about 2.5-fold higher in the case of the skin that was subjected to three cycles of freeze-thaw compared to that was thawed once (Fig. 6.5).

It is also suggested that the skin is not from a cadaver of a very old age person unless it closely meets all the barrier integrity testing criteria. In an in-house study, the investigators performed an "endurance challenge test" on the skin of a younger cadaver adult (43 years) (Skin A) and skin obtained from an 82-year-old cadaver

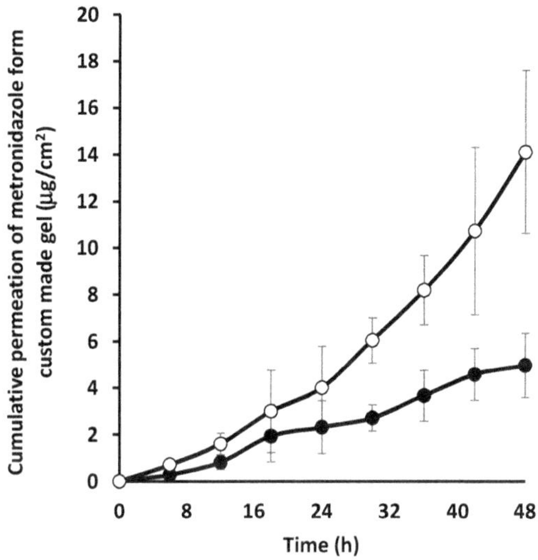

Fig. 6.5 Permeation of metronidazole across human cadaver skin subjected to one (filled) and three cycles (unfilled) of freeze-thawing ($n = 6$, SD)

(Skin B) (male Caucasian). The skin was cut into sizes suitable for the diffusion cells. The skin with comparable metrics (thickness, TEER, and TEWL) was only selected from both cadaver donors. Metronidazole (0.5%w/w) prepared in SLS solution (1%w/v) in deionized water was placed in the donor for permeation studies [60, 61]. The receiver was filled with PBS. Despite the consistency in the initial biophysical properties of the skin of the two cadaver donors, the permeability of skin B changed drastically after a few hours, while the permeation across skin A continued consistently through the 48 h of the study. This suggests that the skin of older subjects has relatively lesser endurance to harsh excipients such as SLS, Dimethyl sulfoxide (DMSO), and propylene glycol in the formulation due to aging-induced changes in the anatomical and histological attributes of the skin (Fig. 6.6).

6.11 Dosing and Application Method

Dosing is a crucial step in performing IVPT studies. Dosing the correct amount of formulation uniformly across the active diffusion area determines the extent and rate of absorption. Application of large doses is relatively less challenging than application of smaller doses. Generally, larger doses are placed directly using a positive displacement pipette or a syringe. Care should be taken not to allow air bubbles to entrap while loading the formulation, which will vary the active diffusion area. Applying a finite formulation dose uniformly on the skin requires some practice. Ideally, the dosing should be done after mounting and equilibration of the skin with the receiver medium (Mount and dose and never dose and mount). Formulation characteristics can also become the limiting factors when applying a uniform dose. Although such situations are rare, applying a highly viscous

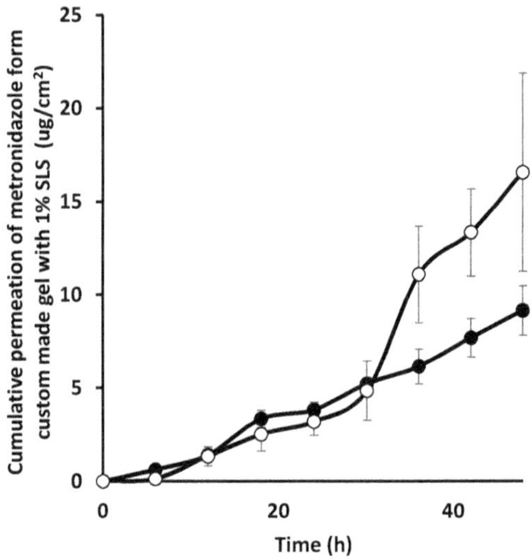

Fig. 6.6 Permeation of metronidazole across human cadaver skin from 43-year-old (filled) and 82-year-old (unfilled) males ($n = 6$, SD)

formulation with poor spreadability properties or a formulation with a high bulk density (the sample quantity at finite doses will be relatively lesser) will cause difficulty in spreading the formulation uniformly across the active diffusion area.

The investigator can develop and master a suitable application method. In an in-house study by the authors, the investigators investigated the role of the application method on the performance of a topical cream product of metronidazole. The two most common application methods were evaluated: application with a spatula and application using a glass rod. The rest of the experimental protocols were the same. The permeation flux-time profile of metronidazole from both products was found to be similar, and the dermatokinetics such as Maximum flux (Jmax), Time of maximum flux (Tmax) and Area Under the Curve 0-24h (AUC0-24) did not differ significantly. The application method did not significantly impact the API's bioavailability (Fig. 6.7; Table 6.2).

6.12 Duration of IVPT Studies

The duration of IVPT studies in infinite dose studies varies between 8 and 24 h for a moderate to highly permeable molecule. However, due to a longer lag time, the studies need to be prolonged for drugs like steroids and other highly lipophilic molecules. The feasibility of prolonged IVPT studies beyond 24 h is limited primarily by the skin's microbial stability and deteriorating barrier integrity. The antimicrobial agents are incorporated into the receiver medium to prolong the skin's life and barrier function. Using a sterile buffer (sterilized by autoclaving) incorporated with antimicrobial agents helps prolong the life of the skin [3, 64–67]. There are not many literature reports regarding permeation studies beyond 48 h.

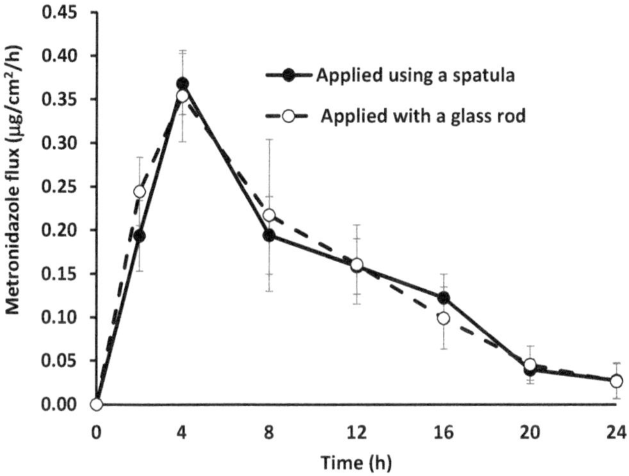

Fig. 6.7 The transdermal flux-time profile of metronidazole across human cadaver skin. The dose application method was performed with a spatula in one set and a glass rod in another ($n = 6$, SD)

Table 6.2 Dermatokinetic parameters of API—Impact of dose application method

Dermatokinetic parameters	Applied with a spatula	Applied with a glass rod
AUC ($\mu g/cm^2$)	3.60 ± 0.45	3.69 ± 0.77
Jmax ($\mu g/cm^2/h$)	0.37 ± 0.03	0.35 ± 0.05
Tmax (h)	4.00	4.00

6.13 Conclusion

The methodology for assessing skin absorption through in vitro permeation testing is well-established and relied upon. Although an extensive body of literature thoroughly covers various aspects of in vitro permeation testing of dermal products, an attempt was made to contribute an updated perspective in this chapter. The suggestions are grounded in the team's experience evaluating dermal products on industry-relevant projects over the last two decades. The aspiration is that this blend of information not only reinforces the existing knowledge but also offers valuable, real-world applications for readers. Skin washing and drug extraction protocols must be developed and validated for each formulation and API. Hence, the mass balance aspect is not touched upon in this chapter.

References

1. Raney S G, Ghosh P, Ramezanli T, Lehman P A, Franz T J. Cutaneous Pharmacokinetic Approaches to Compare Bioavailability and/or Bioequivalence for Topical Drug Products. Dermatol Clin. 2022. 40(3):319–332.

2. Ng S F, Rouse J J, Sanderson F D, Meidan V, Eccleston G M. Validation of a Static Franz Diffusion Cell System for In Vitro Permeation Studies. AAPS Pharm Sci Tech, 2010. 11(3), 1432–1441.
3. Ruela A L M, Perissinato A G, Lino M E de S, Mudrik P S, Pereira G R. Evaluation of skin absorption of drugs from topical and transdermal formulations. Brazilian J. Pharm. Sci. 2016. 52(3), 527–544.
4. Flynn G L, Shah V P, Tenjarla S N, Corbo M, DeMagistris D, Feldman T G, Franz T J, Miran D R, Pearce D M, Sequeira J A, Swarbrick J, Wang J C, Yacobi A, Zatz J L. Assessment of value and applications of in vitro testing of topical dermatological drug products. Pharm. Res. 1999. 16(9), 1325–1330.
5. Shah V P, Elkins J S, Williams R L. Evaluation of the Test System Used for In Vitro Release of Drugs for Topical Dermatological Drug Products Pharm Dev Technol. 1999. 4(3), 377–385.
6. Hauck W W, Shah V P, Shaw S W, Ueda C T. Reliability and Reproducibility of Vertical Diffusion Cells for Determining Release Rates from Semisolid Dosage Forms. Pharm. Res. 2007. 24(11), 2018–2024.
7. Franz T J. The Finite Dose Technique as a Valide in vitro Model for the Study of Percutaneous Absorption in Man. Curr. Probl., Dermatol. 1978. 7, 58–68. https://doi.org/10.1159/000401276
8. Karbanova E G. Measurement of the temperature profile during evaporation of water and ethanol. Thermophysical Basis of Energy Technologies. https://doi.org/10.1051/matecconf/20179201045
9. Chantasart D, Chootanasoontorn S, Suksiriworapong J, Li S K. Investigation of the pH influence on skin permeation behavior of weak acids using nonsteroidal anti-inflammatory drugs. J Pharm Sci. 2015. 104: 3459–3470.
10. Chantasart D, Hao J, Li S K. Evaluation of skin permeation of β-blockers for topical drug delivery. Pharm Res. 2013. 30: 866–877
11. Kim M K, Zhao H, Lee C H, Kim D D. Formulation of a reservoir-type testosterone transdermal delivery system. Int. J. Pharm. 2001. 219 (1-2), 51–59.
12. Cerdà V, Phansi P. Buffer Solutions of known Ionic Strength. Ann Adv Chem. 2023. 7. 051–056.
13. Phansi P, Mongay C, Cerda V. Buffers of formate, acetate and citrate of know ionic strength. Curr. Top. Electrochem. 2019. 21.107–117.
14. Guth K, Schäfer-Korting M, Fabian E, Landsiedel R, van Ravenzwaay B. Suitability of skin integrity tests for dermal absorption studies in vitro. Toxicology in Vitro. 2015.29(1), 113–123.
15. Franzè S, Gennari C G M, Minghetti P, Cilurzo F. Influence of chemical and structural features of low molecular weight heparins (LMWHs) on skin penetration. Int J Pharm. 2015. 481(1-2), 79–83.
16. Santos L L, Swofford N J, Santiago B G. In Vitro Permeation Test (IVPT) for Pharmacokinetic Assessment of Topical Dermatological Formulations. Curr Protoc Pharmacol. 2020.91(1).
17. Giering K, Lamprecht I, Minet O. Specific heat capacities of human and animal tissues. Laser-Tissue Interaction and Tissue Optics. SPIE. 1996. 2624:188–197
18. Webb R C, Pielak R M, Bastien P, Ayers J, Niittynen J, Kurniawan J, Rogers J A. Thermal Transport Characteristics of Human Skin Measured In Vivo Using Ultrathin Conformal Arrays of Thermal Sensors and Actuators. PLOS ONE. 2015. 10(2), e0118131.
19. Cohen M L. Measurement of the Thermal Properties of Human Skin. A Review. J. Invest. Dermatol. 1977. 69(3), 333–338.
20. Haq A, Michniak-Kohn B. Effects of solvents and penetration enhancers on transdermal delivery of thymoquinone: permeability and skin deposition study. Drug Deliv. 2018. 25(1), 1943–1949.
21. Minghetti P, Cilurzo F, Casiraghi A, Montanari L, Fini A. Ex vivo study of transdermal permeation of four diclofenac salts from different vehicles. J Pharm sci. 2007. 96(4), 814–823.
22. Qvist M H, Hoeck U, Kreilgaard B, Madsen F, Frokjaer S. Evaluation of Göttingen minipig skin for transdermal in vitro permeation studies. Eur. J. Pharm. Sci. 2000. 11(1), 59–68.

23. Aitken R L, Moody P W. The effect of valence and Ionic-strength on the measurement of pH buffer capacity. Aust. J. Soil Res. 1994. 32(5), 975–984.
24. Jordan C. Ionic strength and buffer capacity of wide-range buffers for polarography. Microchem J. 1980. 25(4), 492–499.
25. Olivella M S, Debattista N B, Pappano N B. Salicylic acid permeation: a comparative study with different vehicles and membranes. Biocell. 2006. 30(2):321–324.
26. Chaparro C V, Herrera L V, Meléndez A M, Miranda D A. Considerations on electrical impedance measurements of electrolyte solutions in a four-electrode cell. J. Phys. Conf. Ser. 2016. 687, 012101.
27. Kapoor Y, Bengani L C, Tan G, John V, Chauhan A. Aggregation and transport of Brij surfactants in hydroxyethyl methacrylate hydrogels. J. Colloid Interface Sci. 2013. 407, 390–396
28. Zambrana P N, Hammell D C, Stinchcomb A L. Advanced harmonization techniques result in accurate establishment of in vitro-in vivo correlations for oxybenzone from four complex dermal formulations with reapplication. Drug Deliv Transl Res. 2023. 13(1):275–291.
29. Kirk R D, Akanji T, Li H, Shen J, Allababidi S, Seeram N P, Bertin M J, Ma H. Evaluations of Skin Permeability of Cannabidiol and Its Topical Formulations by Skin Membrane-Based Parallel Artificial Membrane Permeability Assay and Franz Cell Diffusion Assay. Med Cannabis Cannabinoid. 2022. 10; 5(1):129–137.
30. Jung E, Kang Y P, Yoon I S, Kim J S, Kwon S W, Chung S J, Kim D D. Effect of permeation enhancers on transdermal delivery of fluoxetine: In vitro and in vivo evaluation. Int. J. Pharm. 2013. 456(2), 362–369.
31. Williams A C, Barry B W. Penetration enhancers. Adv. Drug Deliv. Rev. 2012. 64, 128–137.
32. Hewitt N J, Gregoire S, Cubberley R, Duplan H, Eilstein J, Ellison C, Lester C, Fabian E, Fernandez J, G´eni‘es C, Jacques-Jamin C, Klaric M, Rothe H, Sorrell I, Lange D, Schepky A. Measurement of the penetration of 56 cosmetic relevant chemicals into and through human skin using a standardized protocol. J. Appl. Toxicol. 2020. 40 (3), 403–415.
33. Thomas N S, Panchagnula R. Transdermal delivery of zidovudine: effect of vehicles on permeation across rat skin and their mechanism of action. Eur J Pharm Sci. 2003. 18(1), 71–79.
34. Kasting G B, Francis W R, Bowman L A, Kinnett G O. Percutaneous Absorption of Vanilloids: In Vivo and in Vitro Studies. J Pharm Sci. 1997. 86(1), 142–146.
35. Nowacka A, Douezan S, Wadso L, Topgaard D, Sparr E. Small polar molecules like glycerol and urea can preserve the fluidity of lipid bilayers under dry conditions. Soft Matter. 2012.8, 1482–1491.
36. Snorradóttir B S, Gudnason P I, Thorsteinsson F, Másson M. Experimental design for optimizing drug release from silicone elastomer matrix and investigation of transdermal drug delivery. Eur. J. Pharm. Sci. 2011. 42, 5, 559–567.
37. Schwartz D M, Kanno Y, Villarino A, Ward M, Gadina M, O'Shea J J. JAK inhibition as a therapeutic strategy for immune and inflammatory diseases. Nat. Rev. Drug Discov. 2017. 17(1), 78–78.
38. Cross S E, Anissimov Y G, Magnusson B M, Roberts M S. Bovine-Serum-Albumin-Containing Receptor Phase Better Predicts Transdermal Absorption Parameters for Lipophilic Compounds. J. Invest. Dermatol. 2003b(b). 120(4), 589–591.
39. Delange J, Vaneck P, Bruijnzeel P L B, Elliott G R. The Rate of Percutaneous Permeation of Xylene, Measured Using the "Perfused Pig Ear" Model, Is Dependent on the Effective Protein Concentration in the Perfusing Medium. Toxicol. Appl. Pharmacol. 1994. 127(2), 298–305.
40. Baert B, Roche N, Burvenich C, De Spiegeleer B. Increase of the transdermal penetration of testosterone by miconazole nitrate. Arch. Pharmacal Res. 2012.35(12), 2163–2170.
41. Heylings J R, Davies D J, Burton R. Dermal absorption of testosterone in human and pig skin in vitro. Toxicol in Vitro. 2018. 48, 71–77.

42. Van de Sandt J J M, van Burgsteden J A, Cage S, Carmichael P L, Dick I, Kenyon S, Korinth G, Larese F, Limasset J C, Maas W J M, Montomoli L, Nielsen J B, Payan J P, Robinson E, Sartorelli P, Schaller K H, Wilkinson S C, Williams F M. In vitro predictions of skin absorption of caffeine, testosterone, and benzoic acid: a multi-centre comparison study. Regul. Toxicol. Pharmacol. 2004. 39(3), 271–281.
43. Simard M, Julien P, Fradette J, Pouliot R. Modulation of the lipid profile of reconstructed skin substitutes after essential fatty acid supplementation affects testosterone permeability. Cells. 2019. 8, 1142.
44. Alinaghi A, Macedo A, Cheruvu H S, Holmes A, Roberts M S. Human epidermal *in vitro* permeation test *(IVPT)* analyses of alcohols and steroids. Int. J. Pharm. 2022. 627, 5 122114.
45. Musazzi U M, Gennari C G M, Franzè S, Minghetti P, Cilurzo F. Printing of cutaneous patches loaded with propranolol for the treatment of infantile haemangiomas. J Drug Deliv Sci Tech. 2021. 66, 102767
46. Schiemann Y, Wegmann M, Lersch P, Heisler E, Farwick M. Polar emollients in cosmetic formulations enhance the penetration and biological effects of Phytosphingosine on skin. Colloids and Surfaces A: Physicochem. Eng. Aspects. 2008.331, 103–107.
47. Cilurzo F, Minghetti P, Sinico C. Newborn pig skin as model membrane in in vitro drug permeation studies: A technical note. AAPS Pharm Sci Tech. 2007. 8(4), 97–100.
48. Merwe E V, Ackermann C, van Wyk C J. Factors affecting the permeability of water through nude mouse skin in vitro. Int J Pharm. 1988. 46(1-2), 169–171.
49. Ross A, Kearney J N. The measurement of water activity in allogeneic skin grafts preserved using high concentration glycerol or propylene glycol. Cell and Tissue Banking. 2004, 5(1), 37–44.
50. Huang Q, Pegg D E, Kearney J.N. Banking of nonviable skin allografts using high concentrations of glycerol or propylene glycol. Cell and Tissue Banking. 2004.5(1), 3–21.
51. Schreiber S, Mahmoud A, Vuia A, Rübbelke M K, Schmidt E, Schaller M, Schäfer-Korting M. Reconstructed epidermis versus human and animal skin in skin absorption studies. In Vitro Toxicol. 2005. 19(6), 813–822.
52. Therkildsen P, Haedersdal M, Lock-Andersen J, Fine Olivarius F, Poulsen T, Wulf H C. Epidermal thickness measured by light microscopy: a methodological study. Skin Res Technol. 1998. 4(4), 174–179.
53. Magnusson B M, Cross S E, Winckle G, Roberts M S. Percutaneous Absorption of Steroids: Determination of in vitro Permeability and Tissue Reservoir Characteristics in Human Skin Layers. Skin Pharmacology and Physiology. 2006. 19(6), 336–342.
54. Wilkinson S C, Maas W J M, Nielsen J B, Greaves L C, van de Sandt J J M, Williams F M. Interactions of skin thickness and physicochemical properties of test compounds in percutaneous penetration studies. Int Arch Occup Environ Health. 2006. 79(5), 405–413.
55. Cross S E, Magnusson B M, Winckle G, Anissimov Y, Roberts M S. Determination of the Effect of Lipophilicity on the in vitro Permeability and Tissue Reservoir Characteristics of Topically Applied Solutes in Human Skin Layers. J. Invest. Dermatol. 2003a(a). 120(5), 759–764.
56. Schlupp P, Weber M, Schmidts T, Geiger K, Runkel F. Development and validation of an alternative disturbed skin model by mechanical abrasion to study drug penetration. Results Pharma Sci. 2014. 4, 26–33.
57. Akomeah F, Nazir T, Martin G P, Brown M B. Effect of heat on the percutaneous absorption and skin retention of three model penetrants. Eur J Pharm Sci. 2004. 21(2-3), 337–345.
58. Akomeah F K, Martin G P, Brown M B. Variability in Human Skin Permeability In Vitro: Comparing Penetrants with Different Physicochemical Properties. J Pharm Sci. 2007. 96(4), 824–834.
59. Roskos K V, Guy R H. Assessment of skin barrier function using transepidermal water loss—effect of age. Pharm. Res. 1989. 6, 949–953.
60. Tsai J C, Hung P L, Sheu H M. Molecular weight dependence of polyethylene glycol penetration across acetone-disrupted permeability barrier. Arch. Dermatol. Res. 2001a(a). 293, 302–307.

61. Tsai J, Sheu H, Hung P, Cheng C. Effect of barrier disruption by acetone treatment on the permeability of compounds with various lipophilicities: Implications for the permeability of compromised skin. J. Pharm. Sci. 2001b(b). 90(9), 1242–1254
62. Atrux-Tallau N, Pirot F, Falson F, Roberts M S, Maibach H I. Qualitative and quantitative comparison of heat separated epidermis and dermatomed skin in percutaneous absorption studies. Arch. Dermato. Res. 2007. 299(10), 507–511.
63. Murthy S N, Sen A, Zhao Y L, Hui S W. pH influences the postpulse permeability state of skin after electroporation. J Control Release. 2003. 93(1), 49–57.
64. Janůšová B, Školová B, Tükörová K, Wojnarová L, Šimůnek T, Mladěnka P, Filipský T, Říha M, Roh J, Palát K, Hrabálek A, Vávrová K. Amino acid derivatives as transdermal permeation enhancers. J. Control. Release. 2013. 165, 2, 91–100.
65. Nishida N, Taniyama K, Sawabe T, Manome Y. Development and evaluation of a monolithic drug-inadhesive patch for valsartan. Int. J. Pharm. 2010. 402, 1-2, 103–109.
66. Novotný M, Klimentová J, Janůšová B, Palát K, Hrabálek A, Vávrová K. Ammonium carbamates as highly active transdermal permeation enhancers with a dual mechanism of action. J. Control. Release. 2011. 150, 2, 164–170.
67. Ruela A L M, Figueiredo E C, Perissinato A G, Lima A C Z, Pereira G R. In vitro evaluation of transdermal nicotine delivery systems commercially available in Brazil. Braz. J. Pharm. Sci. 2013. 49, 3, 579–588.

Chapter 7
A Statistical Approach for Establishing Bioequivalence in Topical Dermatological Drug Products Using the In Vitro Permeation Test

Elena Rantou, Nam Hee Choi, Donald Schuirmann, and Stella Grosser

7.1 Introduction

A traditional approach by which to assess the bioequivalence (BE) of topical, locally-acting dosage forms (like creams, ointments, and gels) utilizes comparative clinical endpoint BE studies. However, these studies can be relatively insensitive to distinguishing between real and placebo effects, and whilst such studies remain the only viable approach in certain cases, they are considered to be one of the least accurate, sensitive, and reproducible approaches for determining the bioavailability (BA) or BE of a drug product [1–3]. Such studies involve comparisons of how test and reference standard (RS) products heal the skin disease in patients, but the comparisons can be confounded by several issues, not the least of which is often a substantial therapeutic effect from the formulation (e.g., the cream) that is independent of the drug action [4]; the greatest challenge is not necessarily demonstrating BE between the test and RS product but also demonstrating superiority relative to the placebo vehicle. These studies can also be very costly due to study durations of several weeks or months and because they may require hundreds to thousands of patients in order to be adequately powered. For these reasons, investigators have sought to develop more efficient approaches by which to demonstrate BE for topical products [5–8].

As an alternative to comparative clinical endpoint BE studies, scientists have developed more efficient cutaneous pharmacokinetics (PK)-based BE approaches that involve measuring the rate and extent to which the drug becomes available at or near the site of action in the skin. In general, PK-based BE approaches are

E. Rantou (✉) · N. H. Choi · D. Schuirmann · S. Grosser
Office of Biostatistics, Office of Translational Sciences, Center for Drug Evaluation and Research, US Food and Drug Administration, Silver Spring, MD, USA
e-mail: Elena.Rantou@fda.hhs.gov

S. N. Murthy (ed.), *Topical Products and Dermal Drug Delivery*,
AAPS Introductions in the Pharmaceutical Sciences 2,
https://doi.org/10.1007/978-3-031-80525-7_7

considered to be one of the most accurate, sensitive, and reproducible approaches for determining the BA or BE of a drug product [3]. However, although efficient PK BE studies (e.g., plasma PK BE studies) have facilitated the demonstration of BE for most systemically-acting generic products currently marketed in the United States, these efficient approaches were challenging to adapt for locally-acting drug products (such as topical creams, ointments, and gels). The challenges were partly related to technical issues with measuring the cutaneous PK (i.e., drug concentrations) in the solid tissues of the skin and partly related to establishing suitable endpoints and efficient statistical analysis methods to adequately power PK BE studies for topical products that would be feasible to conduct [2].

Thus, although in vitro PK BE studies that utilize an in vitro permeation test (IVPT) were seriously discussed decades ago, based on evidence that they have the potential to capture clinically meaningful differences between products [1], it was not until 2022 that the US Food and Drug Administration (FDA) published general recommendations for a PK BE approach for topical drug products in a Guidance for Industry [9]. The establishment of this PK BE approach for topical products was supported by growing evidence from a substantial body of evidence from in vitro (e.g., IVPT) and in vivo (e.g., dermal open flow microperfusion) studies with multiple drugs and multiple topical dosage forms. However, the development of an appropriate statistical analysis method was central to establishing the feasibility of efficient PK BE approaches for topical products.

Traditionally, for prospective generic products, BE is assessed using the two one-sided tests (TOST) [10] applied to the difference of average log-transformed responses of the area under the curve (AUC) and maximum concentration (C_{max}). In these cases, the test and reference listed products are declared bioequivalent if the 90% confidence interval of the ratio of the population geometric means is contained within the interval [0.80, 1.25].

The recommendations in the aforementioned FDA Guidance [9] were ultimately made possible by the novel adaptation of a statistical analysis approach that provided the necessary efficiency and power to support conclusions about BE based upon data with substantial variability in drug permeation through the skin. The variability observed in cutaneous PK data arises from multiple sources, most notably inter-subject and within-subject (reference) variability. The establishment of suitable BE criteria necessitated considering the type and magnitude of variability observed, the study design, and the sample size [11]. Ultimately, a key objective of the statistical analysis was to determine whether the cutaneous PK BE study is "sufficiently discriminating and has the requisite statistical power to justify its acceptance by regulatory agencies as a surrogate for clinical testing" [12]. This work describes assessments that supported the adaptation of a reference-scaled average BE approach that involved determining BE based upon each IVPT study endpoint, as part of a set of criteria to determine BE between a test and RS topical product.

7.2 Study Design and Statistical Analysis

Although a PK BE IVPT study resembles a crossover design, there are no carryover or period effects. The data needs to be adjusted for the unique dependence of the variables, as well as due to other issues such as inter-subject variability. Under certain assumptions, the data will be treated as a special case of a paired-sample design. Earlier attempts at establishing BE for such data sets used regular average BE confidence intervals [10], which are not appropriate, given the high variability observed between different donors.

The type of IVPT data used for statistical analysis is shown as flux profiles below.

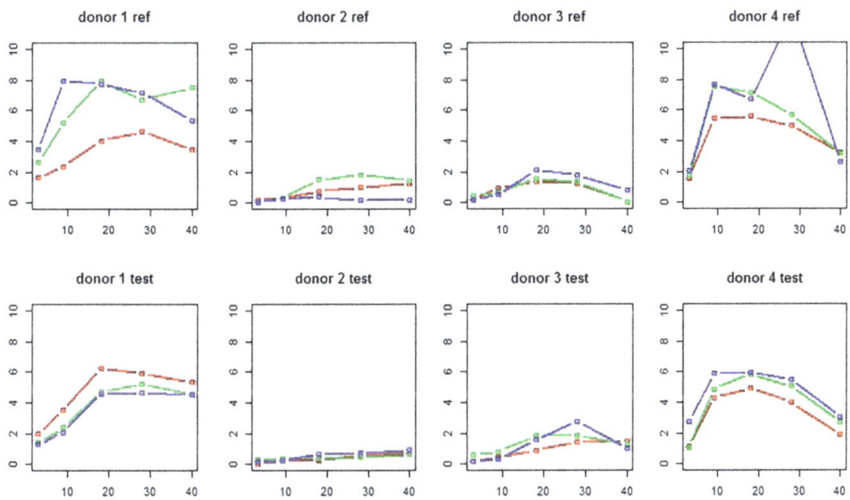

In the above figure, the first row corresponds to flux profiles from 4 donors (with 3 replicates each) dosed with the reference product, while the second row shows the same 4 donors (with 3 different replicates each), dosed with the test product. As can be observed, donors can have different behaviors, but for a good generic product, there is agreement in one donor's behavior among different products.

The response considered is the natural log-transformed total amount penetrated (AMT) or maximum flux rate ($J_{\max}$). We consider a sample of n donors, r replicates, and 2 treatments (products): prospective generic (test: T) and RS (reference: R). r replicates skin sections from each one of the n donors are collected for each formulation. Then, replicates from each donor are randomly assigned to each product. These observations can be presented in the following fashion:

Test:	Reference:
$T\ ,T\ ,\ldots,T_{n}$	$R\ ,R\ ,\ldots,R_{n}$
$T\ ,T\ ,\ldots,T_{n}$	$R\ ,R\ ,\ldots,R_{n}$
$\vdots$	$\vdots$
$T_{1},T_{2},\ldots,T$	$R_{1},R_{2},\ldots,R$

For each donor, a replicate from treatment T is randomly matched with a replicate from treatment R. Afterward, we can calculate the term $I_j = \frac{1}{r}\sum_{i=1}^{r}\left(T_{ij} - R_{ij}\right)$ and record it. This leads to the derivation of the point estimate:

$$\overline{I_.} = \frac{1}{n}\sum_{j=1}^{n} I_j \tag{7.1}$$

as well as the estimate of the inter-donor variability:

$$S_I^2 = \frac{1}{(n-1)}\sum_{j=1}^{n}\left(I_j - \overline{I_.}\right)^2$$

For different replicates R_{ij}, of the same donor, j, we define the within-reference variability as:

$$S_{WR}^2 = \frac{\sum_{j=1}^{n}\sum_{i=1}^{r}\left(R_{ij} - \overline{R_{.j}}\right)^2}{(r-1)\,n}$$

An IVPT study to evaluate BE between a test and a reference product is based on the calculation of confidence intervals (CI) for the cutaneous PK parameters corresponding to the total amount penetrated (AMT) and the maximum flux rate ($J_{\max}$) and hence necessitates the development of statistical methods to evaluate significant differences in these parameters. Under normality assumptions, the following distributional results hold:

$$\overline{I_.} \sim N\left(\mu_T - \mu_R,\ \frac{\sigma_I^2}{n}\right)$$

$$\frac{(r-1)\,n\,S_{WR}^2}{\sigma_{WR}^2} \sim \chi_{(r-1)n}^2$$

and the two quantities are statistically independent. Furthermore, we assume a balanced design and that no donor-by-formulation interaction exists. The absence of such a significant interaction effect has been systematically tested for several data sets examined [13].

It includes a mixed criterion that uses the intra (within) reference variability as a cutoff point. For smaller values than the cutoff point, the regular average BE approach is employed. For larger values, a scaled approach is used. In particular, the estimated within-reference standard deviation S_{WR} is compared to the cutoff value of 0.294, above which reference scaling is used. This cutoff value is equivalent to a coefficient of variation (CV) value of 30% as the following relationship holds [14]:

$$\sigma = \sqrt{\ln\left(1 + CV^2\right)}$$

In the case that $S_{WR} \leq 0.294$, the test and reference formulations are declared bioequivalent if the (1–2α)*100% confidence interval:

$$\overline{I}_{.} \pm t_{(n-1),1-\alpha} * \sqrt{\frac{S_I^2}{n}}$$

is contained within the limits $\left[\frac{1}{m}, m\right]$. The selection of the value of the margin m is discussed in the Results Section.

The reference-scaled criterion is accompanied by a point estimate constraint according to which the geometric mean ratio (exponential of the point estimate of the log-transformed response defined in Eq. (7.1)) has to fall within the pre-specified limits: $\left[\frac{1}{m}, m\right]$. This requirement was originally introduced to eliminate the potential that the generic product would enter the market as an apparently bioequivalent product but with a large mean difference in bioavailability [15]. A flowchart demonstrating the two possible paths to the BE decision is shown in Fig. 7.1.

The scaled BE methodology is used in the case where $S_{WR} > 0.294$ adopts the FDA's approach for the analysis of highly variable drugs [14, 16], modified for the particular design. The hypotheses to be tested are [17]:

$$H_0 : \frac{(\mu_T - \mu_R)^2}{\sigma_{WR}^2} > \theta$$

$$H_a : \frac{(\mu_T - \mu_R)^2}{\sigma_{WR}^2} \leq \theta$$

In this setting, σ_{WR}^2 is the population within-subject variance of the reference product and θ is equal to $\frac{(\ln(m))^2}{(\sigma_{W0})^2}$ [13]. The term σ_{W0} in the denominator corresponds to the regulatory constant [13]. The strategy is to construct a (1–α)*100% upper

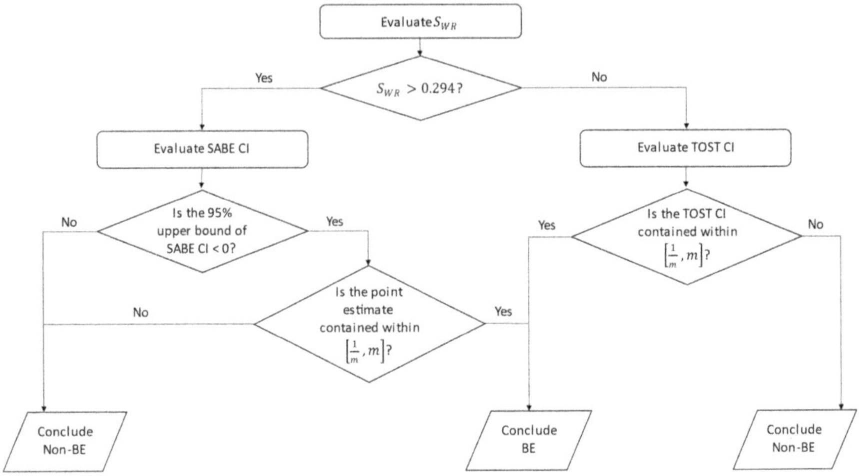

Fig. 7.1 The path to a bioequivalence decision

confidence bound for the quantity $(\mu_T - \mu_R)^2 - \theta\ \sigma_{WR}^2$[18]. If this is less than or equal to zero, the null hypothesis will be rejected. Rejection of the null hypothesis, H_0 supports BE.

Based upon Howe's approximation [19] for an upper confidence limit for $E(X + Y)$, where X and Y are two independent tabled random variables, our objective is to choose X and Y such that $E(X) = (\mu_T - \mu_R)^2$ and $E(Y) = -\theta\sigma_{WR}^2$. Additionally, we want to choose X'_β and Y'_β so that they satisfy the probability statement $P\left(E(X) \le X'_\beta\right) = P\left(E(Y) \le Y'_\beta\right) = \beta$ where $\beta = 0.95$. In this case, we choose

$$X = \overline{I}_{.}^{2} - \frac{S_I^2}{n}$$

$$Y = -\theta S_{WR}^2$$

$$X'_\beta = \left[\left|\overline{I}_{.}\right| + t_{0.95,(n-1)}\sqrt{\frac{S_I^2}{n}}\ \right]^2$$

$Y'_\beta = -\theta\frac{(r-1)n\ S_{WR}^2}{\chi_{0.95,(r-1)n}^2}$ where $t_{0.95,\,(n-1)}$ is the 95th percentile of the Student's t-distribution with $(n - 1)$ degrees of freedom and $\chi_{0.95,(r-1)n}^2$ is the 95th percentile of the Chi-Square distribution with $(r - 1) * n$ degrees of freedom.

Define the terms:

$$\tau_X = \text{sign}\left(X'_\beta - X\right)$$

$$\tau_Y = \text{sign}\left(Y'_\beta - Y\right)$$

The 95% upper confidence bound for $(\mu_T - \mu_R)^2 - \theta\ \sigma_{WR}^2$ is then constructed following Howe's approximation along with a few algebraic steps [19] as:

$$X + Y + \text{sign}\ (V) * |V|^{1/2}$$

where $= \left(X'_\beta - X\right) * \left|X'_\beta - X\right| + \left(Y'_\beta - Y\right) * |Y_\beta - Y| = \tau_X * \left(X'_\beta - X\right)^2 + \tau_Y * \left(Y'_\beta - Y\right)^2$.

The BE limits [0.80, 1.25] were utilized, and for exploratory purposes [0.75, 1.33] as well, and the statistical test performance was monitored for both cases of real data sets [11] and simulated data. The results, along with the relevant discussion, are presented in the following section.

7.3 Results

The approach described here was used for BE assessments of over 20 real data sets. Some of these cases were IVPT studies conducted comparing reference standard products with approved generic products for which BE had been demonstrated based on clinical evidence. The results obtained with these IVPT studies, assessed with the statistical test described above, were in agreement with the original conclusions of BE that led to the regulatory approval of these products. This speaks in favor of the validity of this model for providing the means of assessing BE [2]. Such cases involving approved generics as positive controls for BE can also support the selection of an appropriate margin, *m*.

The statistical test described above has also been used to assess BE with IVPT studies comparing two batches of the same reference product (as positive controls for BE) and again successfully determined the batches to be bioequivalent. The outcomes advocate the model's sensitivity to meaningful differences and its resistance to the hazard of rejecting good products.

Besides testing this statistical methodology using real data [11], large simulation studies have been used to determine the required sample size (number of donors) so that the statistical test is adequately powered. The power of the test is determined as the percentage of studies passing BE, under the assumption that the two products are indeed equivalent. For all figures below, a large number of studies has been simulated (N = 500,000). The impact of the proposed methodology on the power

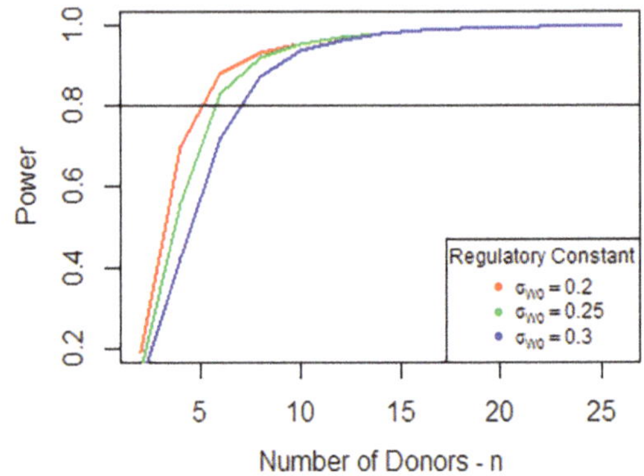

Fig. 7.2 Statistical power for different regulatory constant values and variable sample sizes

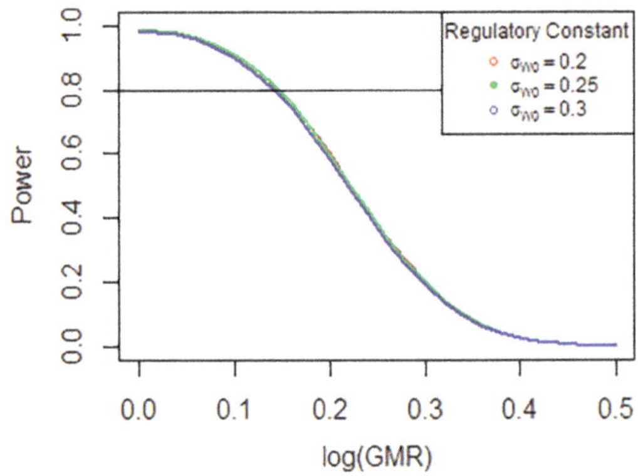

Fig. 7.3 Statistical power for different regulatory constant values and variable GMR

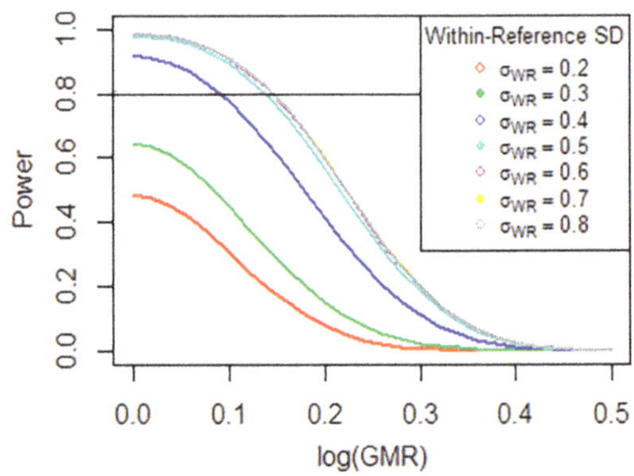

Fig. 7.4 Power for different values of σ_{WR} and variable log (GMR)

was then calculated under different conditions, such as the choice of the regulatory constant, σ_{W0} (Figs. 7.2 and 7.3), ranging the within-reference variability σ_{WR} (Fig. 7.4), ranging the values of geometric mean ratio (GMR) (Fig. 7.5), and ranging the between-donor variability $\sigma_I{}^2$ (Fig. 7.6). All simulated studies were evaluated with $r = 5$ replicate skin sections per donor.

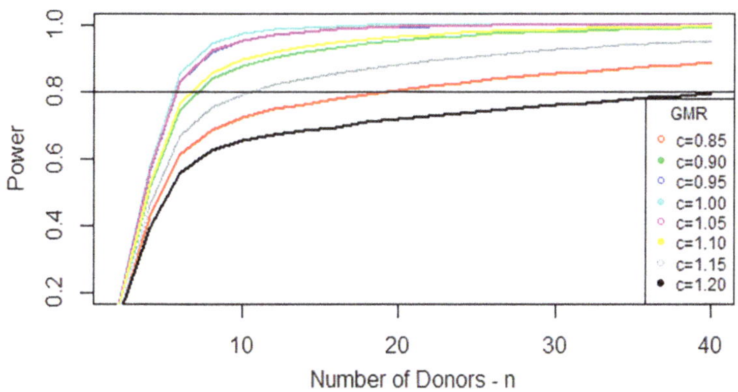

Fig. 7.5 Statistical power for different values of GMR and variable sample size

Fig. 7.6 Statistical power for different values of σ_I^2 and variable GMR

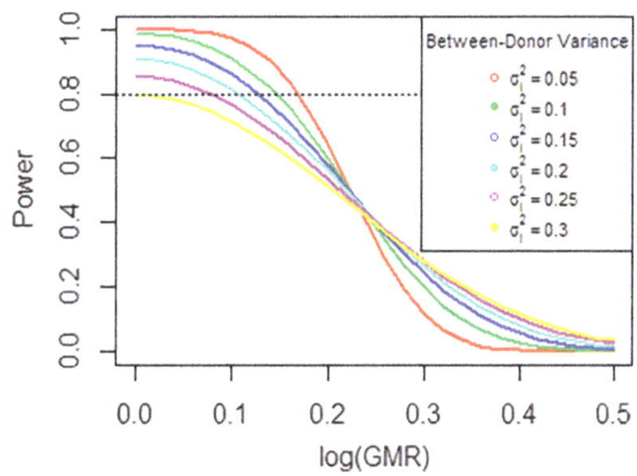

The above results refer to the determination of power through the rejection rule (i.e., the likelihood of passing BE when the null hypothesis is not assumed to be true). There is a difficulty in defining power, and consequently Type I error, due to the presence of two sets of hypotheses [18].

7.4 Extending the Methodology to the Case of Unbalanced Data

During the course of an IVPT experiment, replicate skin sections are withdrawn for a number of reasons [9, 20], such as when samples from the diffusion cell are destroyed (an anticipated experimental event) or when there is documented evidence that skin is damaged during the course of the experiment. In such cases, replicate values can be replaced so that there is no informational loss. In situations where the diffusion cell cannot be replaced, replicate values are dropped, but not uniformly

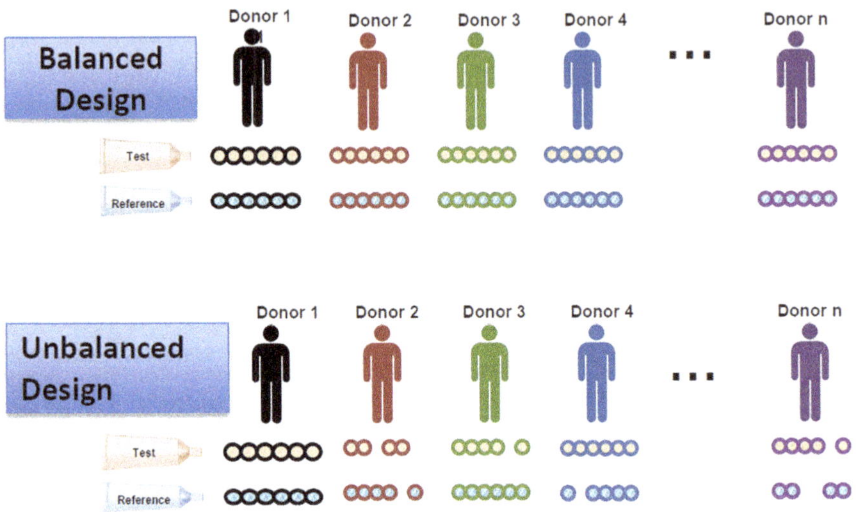

Fig. 7.7 IVPT study design

for all donors. This results in an unbalanced data set where not all donors have the same number of replicates per treatment.

The differences in the study design between balanced and unbalanced data are shown in Fig. 7.7 below.

In such a case, we consider: T_{ij} = the natural log-transformed $J_{\max}$ or AMT for the ith replicate from the jth donor $(i = 1, 2, \cdots, r_j^T,\ \ j = 1, 2, \cdots, n)$ dosed with the test product and R_{ij} = the natural log-transformed $J_{\max}$ or AMT for the ith replicate from the jth donor $(i = 1, 2, \cdots, r_j^R,\ \ j = 1, 2, \cdots, n)$ dosed with the RS product. Additionally, we define:

r_j^T = the number of skin replicates from the jth donor dosed with the test product $(j = 1, 2, \cdots, n)$,
r_j^R = the number of skin replicates from the jth donor dosed with the RS product $(j = 1, 2, \cdots, n)$,
$r^* = r_1^R + r_2^R + \cdots + r_n^R$ = the total number of skin replicates in the RS group, and
n = the number of donors.

The within-reference variance can then be estimated as $S_{WR}^2 = \frac{\sum_{j=1}^{n} \sum_{i=1}^{r_j^R} \left(R_{ij} - \overline{R}_{.j}\right)^2}{r^* - n}$ where $\overline{R}_{.j} = \frac{1}{r_j^R} \sum_{i=1}^{r_j^R} R_{ij}$ is the average of log-transformed observations across all r_j^R replicate values from the jth donor dosed with the RS product [9].

Additionally, the point estimate for the treatment mean difference $\left(\hat{I}\right)$, as well as its standard error $se\left(\hat{I}\right)$,and degrees of freedom can be approximated using a linear model with donor and treatment as factors [9].

Following the S_{WR} threshold criterion (Fig. 7.1), in the unscaled case, the scaled average bioequivalence (SABE) $(1 - 2\alpha)$*100% confidence interval for $\mu_T - \mu_R$ will be evaluated from:

$$\hat{I} \pm t_{(1-\alpha),df^*} * se\left(\hat{I}\right)$$

For the scaled case, the upper bound for the scaled CI will be constructed in the same way as in the balanced data case and will still be accompanied by the same point estimate constraint for $\hat{I}$.

7.5 Outliers in IVPT Studies

A common issue arising in IVPT study data is that of outliers. The nature of an outlying observation in this setup is a within-donor, extreme replicate value. As previously discussed, a donor's unique behavior is expected to be consistent for both the test and reference products and should be considered for the assessment of BE. Therefore, it is not meaningful to consider outlying donors.

For outlier detection, standard practices used in pharmacokinetic studies, such as those using standardized residuals, do not apply here because of the small sample size of replicate values within one donor. Small sample tests, such as the Dean-Dixon or Grubbs tests, are applicable in this case where experimental conduct anomalies are detected once the sample analysis is completed.

Figure 7.5 illustrates an example of an IVPT data set with 20 donors and its outlying replicate values per donor per treatment

To identify statistical outliers, one should be aware that varying the α value (significance level) within a specific type of outlier test seems to affect the number of outliers identified (Fig. 7.8). Additionally, we should carefully examine the impact of a combination of the choice of a specific type of outlier test and the changes in the α value.

To thoroughly investigate any potential impact of outliers on the assessment of BE, we need to consider many data sets from IVPT applications along with simulation studies under various configurations. The impact of outliers on BE assessment should then be studied through different components of the BE test. These components are:

(a) Value of S_{WR}.
(b) Value of the point estimate
(c) Value of the upper bound of the scaled confidence interval

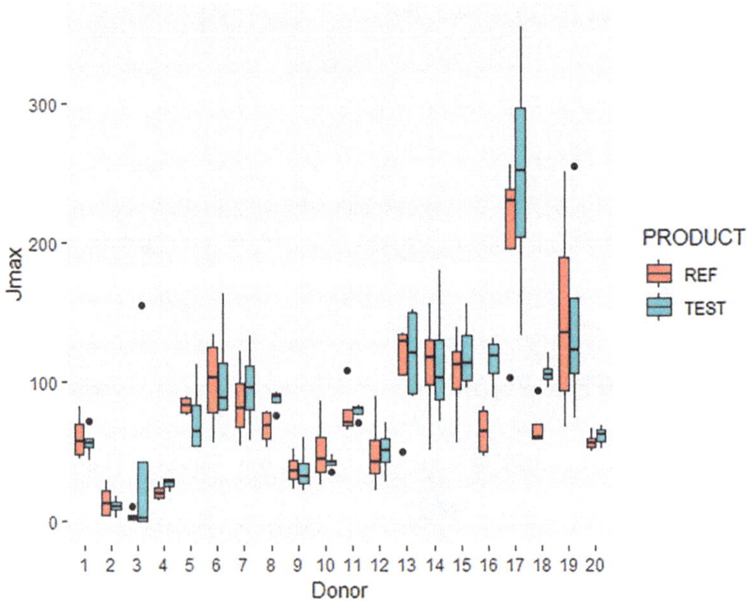

Fig. 7.8 Statistical outliers in IVPT studies

7.6 Conclusions and Discussion

The results from this study suggest that the modified, mixed-scaled approach provides a rigorous and reliable criterion for determining the BE of locally-acting topical products evaluated by cutaneous PK BE IVPT studies. This approach successfully resolved the previous absence of accepted statistical analyses with appropriate IVPT endpoints using which BE could be established more efficiently than by using comparative clinical endpoint BE studies [2].

Although IVPT studies can support a demonstration of BE as an alternative to in vivo comparative clinical endpoint BE studies, potential limitations of IVPT are that this in vitro test requires ex vivo skin from a sufficient number of human donors, and the rate and extent to which many drugs permeate into and through human skin can vary greatly (e.g., by as much as tenfold) from one donor to another. Even so, as is illustrated by the power simulation studies presented here, with the appropriate margin and regulatory constant selection, the statistical test can be adequately powered using a reasonable sample size. This is because as the simulation outcomes demonstrated, based upon the assumptions explored in this work, the statistical test can be adequately powered with approximately 10 donors, although factors impacting real-world data that may arise from anomalous biological heterogeneity in the skin may impact these assumptions and may increase the number of donors needed to power an IVPT BE study. Furthermore, the low values of the estimated level of significance advocate for the statistical test's sensitivity to meaningful differences and support the validity of IVPT BE studies utilizing a SABE statistical test to accurately, sensitively, and reproducibly assess the BE of topical drug products.

Acknowledgments The authors would like to thank Dr. Sam G. Raney from the Office of Research and Standards, Office of Generic Drugs, Center for Drug Evaluation and Research, US Food and Drug Administration for his valuable suggestions, recommendations, and discussion.

References

1. Shah VP, Flynn GL, Yacobi A, et al. Bioequivalence of topical dermatological dosage forms-methods of evaluation of bioequivalence. *Pharm Res.* 1998; 15(2): 167-171. doi: https://doi.org/10.1023/a:1011941929495
2. Grosser S, Park M, Raney SG, Rantou E. Determining Equivalence for Generic Locally Acting Drug Products. *Stat Biopharm Res.* 2015; 7(4): 337-345. doi: https://doi.org/10.1080/19466315.2015.1093541
3. United States Food and Drug Administration. Title 21 Code of Federal Regulations (CFR) Part 320, Section 24. https://www.accessdata.fda.gov/scripts/cdrh/cfdocs/cfcfr/CFRSearch.cfm?fr=320.24. Accessed 20 Mar 2023.
4. Hick J, Feldman SR. Eligibility creep: a cause for placebo group improvement in controlled trials of psoriasis treatments. *J Am Acad Dermatol.* 2007; 57(6): 972–976. doi: https://doi.org/10.1016/j.jaad.2007.08.008
5. Raney SG, Franz TJ, Lehman PA, Lionberger R, Chen M. Pharmacokinetics-Based Approaches for Bioequivalence Evaluation of Topical Dermatological Drug Products. *Clin Pharmacokinet.* 2015; 54(11): 1095-1106. https://doi.org/10.1007/s40262-015-0292-0
6. Franz, TJ, Lehman PA, Raney SG. Use of excised human skin to assess the bioequivalence of topical products. *Skin Pharmacol Physiol.* 2009; 22(5): 276-286. doi: https://doi.org/10.1159/000235828
7. Lehman PA, Franz TJ. Assessing the bioequivalence of topical retinoid products by pharmacodynamic assay. *Skin Pharmacol Physiol.* 2011; 25(5): 269-280. doi: https://doi.org/10.1159/000339899
8. Yacobi A, Vinod PS, Bashaw ED, et al. Current challenges in bioequivalence, quality, and novel assessment technologies for topical products. *Pharm Res.* 2014; 31(4): 837-846. doi: https://doi.org/10.1007/s11095-013-1259-1
9. United States Food and Drug Administration. Draft Guidance for Industry on In Vitro Permeation Test Studies for Topical Drug Products Submitted in ANDAs. https://www.fda.gov/regulatory-information/search-fda-guidance-documents/in-vitro-permeation-test-studies-topical-drug-products-submitted-andas. Published October 2022.
10. Schuirmann DJ. A comparison of the two one-sided tests procedure and the power approach for assessing the equivalence of average bioavailability. *J Pharmacokinet Biopharm.* 1987; 15(6): 657-680. doi: https://doi.org/10.1007/BF01068419
11. Shin S, Rantou E, Raney SG, Ghosh P, Hassan H, Stinchcomb A. Cutaneous Pharmacokinetics of Acyclovir Cream 5% Products: Evaluating Bioequivalence with an In Vitro Permeation Test and an Adaptation of Scaled Average Bioequivalence. *Pharm Res.* 2020; 37(210). https://doi.org/10.1007/s11095-020-02821-z
12. N'Dri-Stempfer B, Navidi WC, Guy RH, Bunge AL. Improved Bioequivalence Assessment of Topical Dermatological Drug Products Using Dermatopharmacokinetics. *Pharm Res.* 2009; 26(2): 316–328. doi: https://doi.org/10.1007/s11095-008-9742-9

13. Davit BM, Chen M, Conner DP, Haidar SH, et al. Implementation of a reference-scaled average bioequivalence approach for highly variable generic drug products by the US Food and Drug Administration. *The AAPS journal*. 2012; 14(4): 915-924. doi: https://doi.org/10.1208/s12248-012-9406-x
14. Haidar SH, Makhlouf F, Schuirmann DJ, et al. Evaluation of a scaling approach for the bioequivalence of highly variable drugs. *The AAPS journal*. 2008; 10(3): 450-454. doi: https://doi.org/10.1208/s12248-008-9053-4
15. Haidar SH, Davit B, Chen M, et al. Bioequivalence approaches for highly variable drugs and drug products. *Pharm Res*. 2008; 25(1): 237-241. doi: https://doi.org/10.1007/s11095-007-9434-x
16. United States Food and Drug Administration. Draft Guidance on Progesterone Oral Capsules. https://www.accessdata.fda.gov/drugsatfda_docs/psg/Progesterone_caps_19781_RC02-11.pdf. Recommended Apr 2010; Revised Feb 2011.
17. United States Food and Drug Administration. Draft Guidance on Acyclovir Cream. https://www.accessdata.fda.gov/drugsatfda_docs/psg/PSG_021478.pdf. Recommended Dec 2014; Revised Dec 2016; Oct 2022.
18. Lim D, Rantou E, Kim J, Choi S, Choi NH, Grosser S. Adaptive designs for IVPT data with mixed scaled average bioequivalence. Pharmaceutical Statistics. 2023 Nov;22(6):1116-34.
19. Howe WG. Approximate confidence limits on the mean of X+ Y where X and Y are two tabled independent random variables. *J Am Stat Assoc*. 1974; 69(347): 789-794. https://doi.org/10.2307/2286019
20. Rantou, E. Statistical Considerations in Assessing BE of IVPT Data. Presented at 2021 IVRT/IVPT Workshop hosted by The Center for Research on Complex Generics. https://www.complexgenerics.org/IVRTIVPT/.

Chapter 8
Cutaneous Pharmacokinetics for Topical Drug Products

Benjamin A. Kuzma and Grazia Stagni

8.1 Introduction

The application of pharmacokinetic principles and mathematical models in drug discovery and development has significantly improved therapeutic efficacy and successful use of medications. Pharmacokinetics (PK) was developed primarily to study the plasma concentrations over time as a function of an active pharmaceutical ingredient (API) dose delivered systemically. The main steps that occur after the administration of a drug product are summarized as absorption, distribution, metabolism, and excretion, also known as ADME. These processes for systemically administered drug products are generally well characterized and routinely applied in pharmaceutical research to optimize formulations and assess bioavailability. Indeed, a thorough knowledge of an API's ADME plays a vital role in the optimal development of the pharmacokinetic/pharmacodynamic (PKPD) relationship that ultimately guides the selection of the ideal dosing regimen and assures therapeutic success. Another important outcome of PK is that it provides solid principles for assessing bioequivalence (BE) at reasonable costs to the generic manufacturer and patient. This has resulted in a prosperous generic industry and made high-quality medications affordable to a broader range of patients.

However, when a drug product is applied locally, for example, to the skin, and the site of action is near the application site or is located between the application site and its uptake into the bloodstream, the traditional plasma sampling for PK

B. A. Kuzma (✉)
Arnold and Marie Schwartz College of Pharmacy, Long Island University, Brooklyn, NY, USA

Vertex Pharmaceuticals, DMPK, Boston, MA, USA

G. Stagni
Arnold and Marie Schwartz College of Pharmacy, Long Island University, Brooklyn, NY, USA

S. N. Murthy (ed.), *Topical Products and Dermal Drug Delivery*,
AAPS Introductions in the Pharmaceutical Sciences 2,
https://doi.org/10.1007/978-3-031-80525-7_8

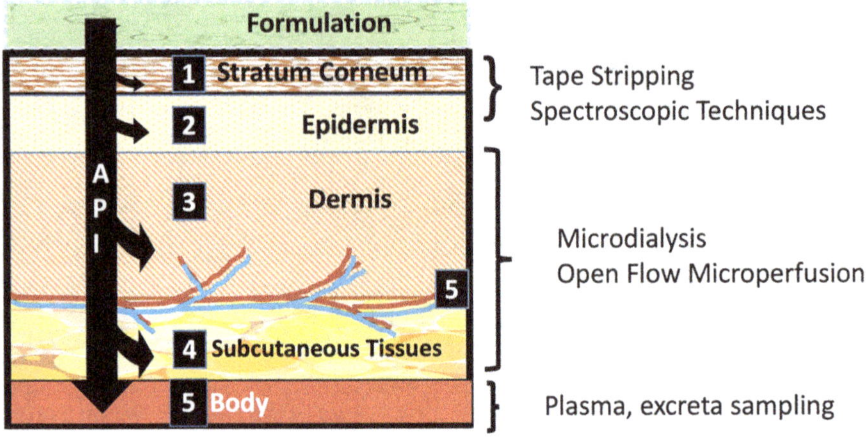

Fig. 8.1 Schematic representation of API permeation into the various skin layers. The relevance of the sampling methods depends on how close the sampling is to the site of action. Tape-stripping samples of the stratum corneum, microdialysis, and open-flow microperfusion sample dermis or subcutaneous tissue, plasma, and excreta sample APIs that are no longer in the skin and thus have reached other parts of the body. Microspectroscopic techniques can be utilized from the top down (i.e., from the formation into the skin with depths up to 200 μ) or can be used to see all skin layers from a cross-section

often fails to provide the desired beneficial results due to local sample concentration or API no longer being at the site of action. Consequently, exploring the local PK (i.e., ADME within the skin) has become paramount to optimize the therapeutic use of topical products. In this chapter, we start by reviewing the attempts to use traditional PK sampling (plasma and excreta), followed by methods that sample APIs in the skin, like dermal microdialysis, dermal open flow microperfusion, and tape stripping, in addition to various micro-spectroscopic methods. A special case is the use of a pharmacodynamic approach (blanching assay) to assess the bioequivalence of topical dermatological corticosteroids [1]. As this approach was specific to corticosteroids and is the least sensitive and robust, it will not be covered in this report. Figure 8.1 provides a schematic summary of the skin stratification and the location in which each technique can detect the API. Some of these methods are more advanced in their utility to assess cutaneous PK, bioavailability (BA), and bioequivalence (BE), while others are developing a solid foundation to be utilized in the coming years.

8.2 Plasma and Excreta Sampling

Although the application of therapeutics to the skin has been documented since the dawn of human medicine, it wasn't until the second half of the twentieth century that it became the subject of rigorous scientific investigation [2]. Initially, quantifying an API in plasma or excreta provided proof of API permeation into the body from topical or transdermal drug products to confirm that the topical/transdermal pathways were a viable way to administer medications. Given the low plasma concentration and the poor sensitivity of the analytical methods of the time, API concentrations were determined using radiolabeled compounds (carbon-14 or tritium labeled), as the concentrations were well below the detection limit of the analytical methods of the time.

Indeed, experiments focused on determining the total amount absorbed, measured from plasma or excreta samples, rather than the time course of the API in the body. For example, one of the first studies to quantify delivery from a topically applied cream used tritium-labeled diflorasone diacetate in rats, monkeys, and men [3]. Here it was found that the permeation measured using excreta (urine and feces) was highest in rats, with 88% of the dose absorbed, followed by the monkeys (49.6%), while the man had the lowest (1.1%). The suggested reasons for the stark differences between species were the increased number of hair follicles and additional anatomical differences such as skin thickness and metabolic capabilities. In another study, measurements of 3H-labeled flupamesone in blood, plasma, urine, and feces suggested lower-than-expected permeation after topical application with a total recovery of only 0.2–0.3% after 5 days [4]. Given the difficulties in measuring plasma concentration, typically, the API's extent of BA was determined from the total amount excreted into the urine [5], and the average rate of delivery was calculated as the total amount delivered divided by the application time [6–10]. However, this approach to calculating the absorption rate is suitable for devices like patches when the time of the removal is precisely known. Still, it is harder to apply to topical products like creams or gels when product metamorphosis plays a significant role in the amount absorbed.

The availability of more sensitive analytical methods such as liquid chromatography - Mass spectrometry/Mass spectrometry (LCMS/MS) has tremendously improved the study of plasma PK for transdermal delivery systems [11]. However, it is still inadequate for most topical products. For example, Arildone was topically applied to humans in either a cream or Dimethyl Sulfoxide (DMSO) formulation, with repeated applications to achieve sufficient plasma concentrations [12]. However, arildone concentrations were not detected consistently, which suggested limited percutaneous absorption or a lack of analytical sensitivity. More recently, ruxolitinib was applied to the dorsum of a mini-pig in a 1.5% cream formulation and compared to oral delivery [13]. It was found that the concentrations within the plasma after oral and topical delivery were 31-fold higher in the case of oral delivery. The plasma concentrations detected after topical application were also lower (2.7 nM) than the pharmacologically relevant concentration. Indeed, it should be noted that the concentration in the skin at the application site is much higher than that detectable in plasma, and low plasma concentrations are not uncommon after topical application. Bimiralisib was applied topically, and the plasma concentrations hovered around the lower limit of quantitation (LLOQ) of the LC-MS assay (1.0 ng/mL). However, biopsy samples indicated adequate exposure for clinical efficacy even though there was a lack of a significant clinical improvement over placebo [14].

Cai et al. evaluated the rate and extent of lidocaine released in vivo from two bioequivalent topical delivery systems (TDS) by measuring the plasma PK and residual lidocaine in the TDS following 12 h of wearing in human subjects [15]. The study included an IV bolus dose to estimate the absolute BA. They observed higher variability with the PK method of estimating the drug absorbed in vivo as compared to the residual drug method. However, the average PK profiles were similar, whereas the residual amount differed between the two bioequivalent patches. In conclusion, plasma and excreta provide proof of API permeation of a drug from the topical product as well as an assessment of potential systemic toxicity. When API concentrations are detectable, they can be utilized as PK endpoints in BE studies. They also help to understand how formulation changes impact the delivery through the skin, although this endeavor is better suited for in vitro permeation testing or in vitro release testing. However, plasma and excreta do not provide data on the ADME at the site of action. Thus, they are less likely to be helpful in the understanding of the pharmacodynamics of the API and the ensuing design of therapeutic regimens.

8.3 Dermal Microdialysis (dMD)

Microdialysis has been used for over 50 years to quantify molecules within numerous tissues, such as the brain, tumors, and skin [16]. The technique has been precious in providing target site concentrations, specifically unbound concentrations of the analyte of interest, as according to the free drug hypothesis [17], the unbound drug concentration is solely responsible for the pharmacological effect. However, one downside is the technological knowledge and expertise required to conduct these experiments and the amount of labor involved in generating high-resolution temporal data sets.

The methodology behind dermal microdialysis (dMD) will be briefly covered here, but a comprehensive review can be found in reference [18]. In dMD, a tiny hollow porous membrane (0.2 mm diameter, 1–3 cm long) has a molecular weight cut-off that prevents large analytes (such as proteins) from being collected. To obtain temporal samples, the membrane is connected to two impermeable arms, throughout which an isotonic solution (termed dialysate) is perfused at a constant, precise rate. Unbound small molecules in the dermal interstitial fluid (ISF) move into the perfusing fluid according to the concentration gradient. Because the perfusate is isotonic with the ISF and the slow perfusing rate (<3 μL/min), there is no fluid exchange between the two. Samples are collected according to the desired temporal resolution (i.e., 30 min to 1 h) and then frozen until either HPLC-UV or HPLC-MS/MS analysis. Before injection into the LC-UV or LC-MS/MS, no sample cleanup is required so that the sample can be directly injected, reducing time, labor, and resources compared to pre-processing techniques required for other methods. To accurately estimate the analyte in the ISF, a probe calibrator can be added to the perfusate or another recovery method can be applied [19, 20]. The calibrator approach has the advantage of providing a temporal readout on individual probe performance to understand if the increase in concentration is due to the local environment or if it is due to the probe itself. Typically, the probe is inserted in the lower dermis (Fig. 8.1), but it can also be placed in the subcutaneous tissues according to the study's goals.

Early cutaneous microdialysis studies laid the groundwork for gradual improvements over time. Preliminary studies by Anderson et al. evaluated the suitability of the microdialysis technique to measure ethanol [21] and histamine [22]. These were key studies that pioneered the utility of microdialysis to understand the cutaneous PK of topically applied drug products. Follow-up studies began to understand the blood flow changes after probe insertion [23] but also the changes in lymphocytes and scar tissue at the implantation site [24].

Once methodological issues had been resolved, such as probe insertion variability, experimental setup, and perfusate solution [25], researchers began to utilize microdialysis to understand the permeation pathways and processes after topical application. One issue that still needed to be resolved was the impact that probe depth had on the exposure of topically applied APIs. Müller found that for diclofenac, there was no relationship between probe depth and exposure [26], which has also been suggested for numerous other in vivo studies [27–31], yet for ex vivo studies, the probe depth is essential [32, 33]. Clearly, the dependence of recovery on probe depth depends on the characteristics of the surrounding tissue; probe recovery may differ between the dermis and subcutaneous tissue that is richer in fat. Thus, a measurement of probe depth is a vital covariate to be monitored to ensure consistency in a study.

Since then, rapid advancements have allowed dermal sampling methods to have a higher resolution temporal profile and quantify concentrations previously below the detection limit due to analytical sensitivity issues. The advancement of LC-MS/MS allows quantifying pg/mL concentrations within the skin. These levels would likely be even lower in the blood, thus rendering a plasma PK sample analysis likely not informative or requiring a sizeable topical dose per body service area to have appreciable plasma concentrations. Hence, several investigators began to apply dMD to assess cutaneous BE. Benfeldt et al. compared dMD and tape-stripping to evaluate the BE of lidocaine from an ointment and a cream product in healthy subjects [28]. They demonstrated that these methodologies deliver similar qualitative results, though dMD provided quantitative data to compare the exposure within the dermis. Applying a cream formulation and an ointment formulation containing the same amount of lidocaine (5%) resulted in higher exposure from the cream than the ointment (Fig. 8.2). However, the intra-subject accounted for 61% of the variability, allowing further improvement to reduce the number of subjects required to power a BE study.

Another example of using dMD to assess cutaneous PK and BE was utilizing ketoprofen as the API. A ketoprofen gel was applied to four sites on the same subject, and a BE assessment was conducted using a priori design to compare sites

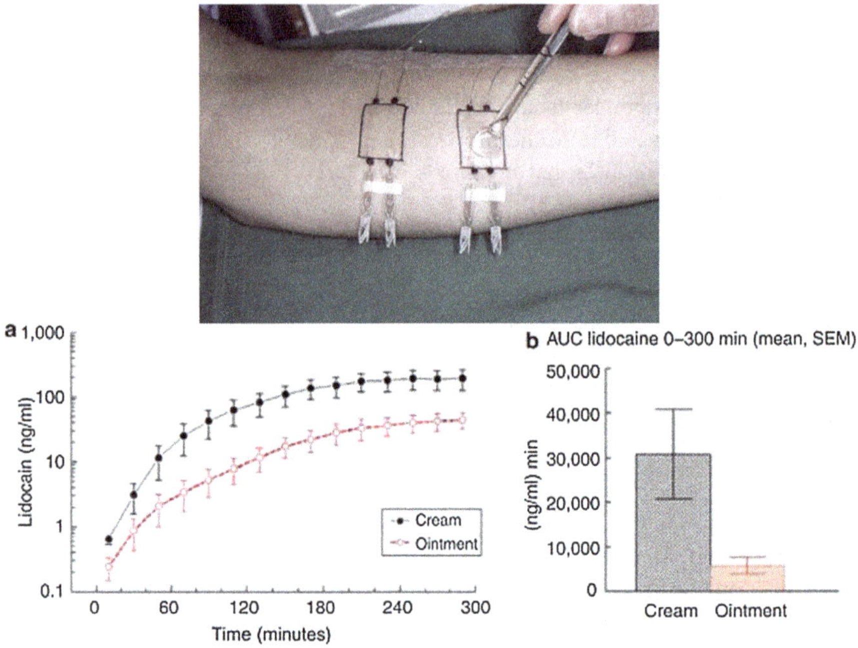

Fig. 8.2 Top: Schematic of application sites and locations of dMD probes in a human subject. Lower (**a**): concentration-time profiles of lidocaine from cream and ointment application in humans. Lower (**b**) $AUC_{0\text{-}300min}$ comparison between the cream and ointment. (Adapted from [28] with permission, copyright Journal of Investigative Dermatology (2007))

[31]. This study's intra-subject variability was approximately 10%, suggesting an improved design. However, this may not have been solely attributed to improved protocols but also because ketoprofen has a lower logP than lidocaine, and thus, it is a better substrate for dMD. A key challenge in this protocol was that the cutaneous PK profile needed to be adequately characterized. Sampling for only 5 h captures concentrations up to C_{max}, yet it is now known that topical products tend to flip-flop PK. Thus, differences in the probes/application sites may have been realized if sampling for longer.

Despite this remarkable progress, the results' high variability and poor reproducibility hampered confidence in the dMD technique. As the appeal of measuring the cutaneous PK of an API was still high in the pharmaceutical community, researchers at Long Island University, Brooklyn, in collaboration with the US FDA, developed standardized procedures to control variables that can contribute to the large variability observed. Specifically, they developed a protocol to ensure tight control of membrane length, consistent flow rate, assessment of probe stability, and sampling duration long enough to capture the entire cutaneous PK time-course adequately. In the first set of studies conducted in a Yucatan mini-pig model [34], the group demonstrated that dMD provided adequate sensitivity to quantify metronida-

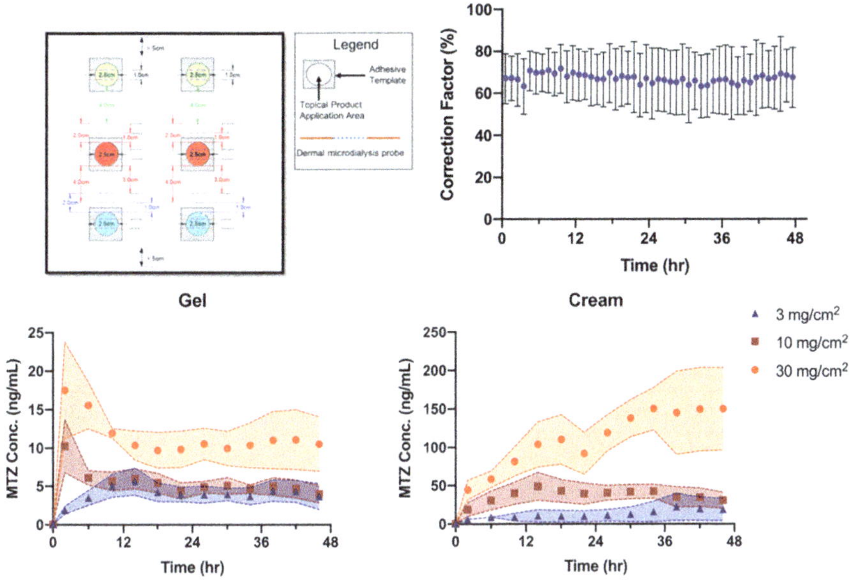

Fig. 8.3 Top left: Schematic of probe placement and application sites for the cream and gel products; Top right: Probe stability over the 48-h experimental duration. Data presented as mean +/− SD; Bottom left: Dermal concentration vs. time profiles for the gel formulation across dose levels of 3, 10, or 30 mg/cm^2; Bottom right: Dermal concentration vs. time profiles after cream application across dose levels of 3, 10, or 30 mg/cm^2 (**c**). (Figures were reprinted with permission from Elsevier. Kuzma et al. 2021 [34])

zole concentrations in escalating topically applied doses (Fig. 8.3). An additional success was utilizing a probe performance marker (deuterated metronidazole) to provide confidence that decreasing concentrations at later timepoints were due to the analyte's interstitial concentration decreasing in the periprobe and not due to technical issues such as the reduction in recovery/performance of the probe (e.g., the probe fouling [35, 36]).

Although dermal PK was monitored for 48 h, a clear elimination phase was not observed (Fig. 8.3—Bottom Panels), mainly for the cream formulation. What appeared to be a prolonged delivery into the dermis was thought to have caused increases in concentrations after 24 h. A subsequent study utilizing different dose durations (6 h, 12 h, 48 h) suggested that if a product were removed before the finalization of the product metamorphosis, 6 h in the case of the cream, the metronidazole delivery would be truncated [37]. In the same study, a few dMD probes were perfused with a known concentration of metronidazole to understand the cutaneous PK mechanistically. The half-life of the perfused metronidazole compared to the metronidazole applied topically suggested flip-flop pharmacokinetics, where the rate-limiting step is the API getting out of the formulation and permeating deeper to the skin rather than the elimination of the API. This approach can also be utilized to develop in vitro–in vivo relationships [38].

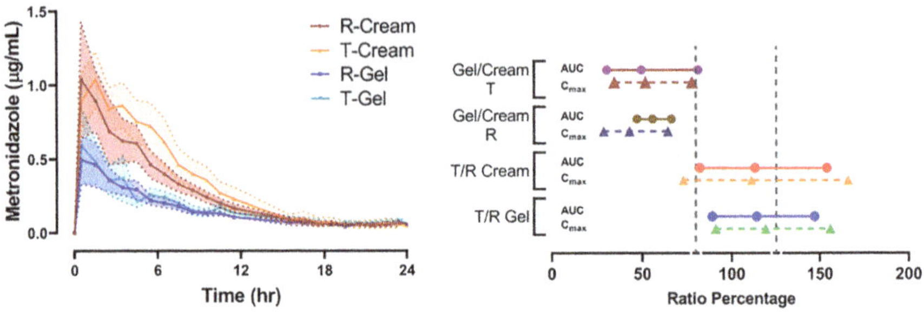

Fig. 8.4 Left panel: dermal concentration-time profile for the reference (*R*) and test (*T*) gel and cream products. Right panel: comparison of test and reference gels and creams within formulations, as well as a general comparison between the formulation times. The inset displays the application site and probe placement for the study. (Figures were reprinted with permission from Springer. Senemar et al. 2023 [39])

The same standardized dMD protocol from the previous studies was applied to conduct a pilot BE study in rabbits. They tested brand metronidazole formulations versus their corresponding FDA-approved generics. While the study was not powered to demonstrate BE, it was apparent that the increase in study duration, use of a probe performance marker, and method standardization provided consistent data to demonstrate dMD as a viable method for dermal BE assessments (Fig. 8.4) [39].

Although these studies demonstrated the suitability of dMD to assess BE of topically applied dermatological formulations when rigorous standardized procedures are applied, there are still several practical issues to be solved before translating it into a clinical setting. Mainly, the need to implant several probes in the same subject to reduce the inter-subject variability and the prolonged immobilization of the human participant, as the dMD must be perfused continuously, is a challenge. A future area for improvement is to make the microdialysis apparatus mobile so that the participant can be free to move. Alternatively, further investigation of the intermittent microdialysis approach [40] (i.e., disconnection and reconnection of probes) may be a viable alternative.

8.4 Dermal Open Flow Microperfusion

Dermal open flow microperfusion (dOFM) is like dMD in that the API can be continuously sampled in the dermis after topical drug product application for high temporal resolution concentration-time profiles. However, there are some key differences. First, dOFM, or OFM more broadly speaking, is a membrane-free probe design that allows for sampling of all molecular species within the sampling environment, such as bound or unbound molecules, lipophilic and hydrophilic APIs, and even large molecules such as monoclonal antibodies [41–46]. Another difference is that dOFM is a push-pull technique, whereas dMD is a push-only technique. Push-pull means that the pump pushes the isotonic solution at the inlet of the probe and then pulls the isotonic solution into the sampling vial at the outlet of the probe.

The lack of a membrane provides the opportunity to sample the entire dermal interstitial fluid, as previously mentioned, but this increases the complexity of sample analysis compared to dMD. The samples must be processed to remove proteins before analytical quantification, which does not dilute the API concentrations [47]. Like dMD, the pump flow rates can be modified to increase or decrease the temporal resolution required for accurate exposure estimates.

OFM has been utilized in various tissues since 1995 [48] and in the dermal tissue since 2007 [49]. One of the first studies demonstrating the utility of dOFM investigated the exposure of ketoprofen after topical application in rats [49], suggesting a similar utility to dMD, likely due to the low logP of ketoprofen, making it a good substrate for dMD recovery. In another study, dMD and dOFM were directly compared to assess the cutaneous PK of topically administered fentanyl and benzoic acid on ex vivo human skin. It was found that dOFM might be better suited for the lipophilic APIs (fentanyl) due to the lack of a membrane and non-specific binding that may occur with dMD [50].

One of the first clinical uses of dOFM was with a topical p38 inhibitor, which was applied for eight consecutive days on lesional and non-lesional skin [51]. dOFM samples were taken on Day 1 and Day 8 for 25 h to capture high-resolution concentration-time profiles, which showed an increase in concentrations from Day 1 to Day 8, which is not unexpected considering the dosing interval is more rapid than the apparent elimination of the API from the dermal interstitial fluid (Fig. 8.5).

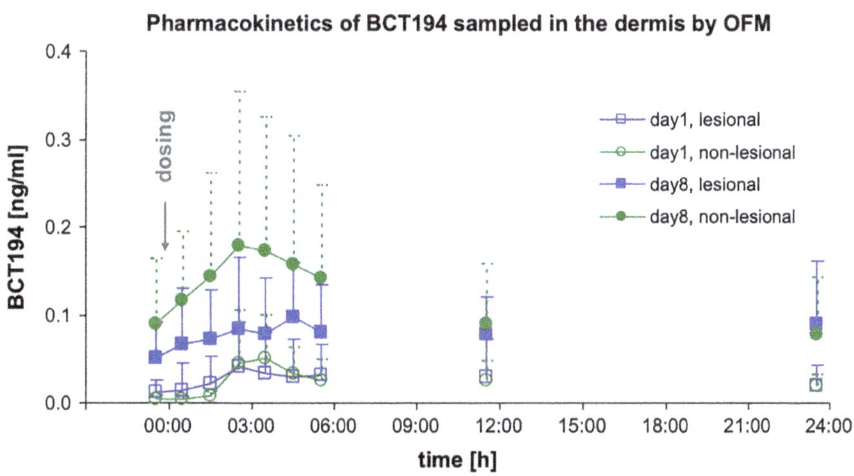

Fig. 8.5 dOFM sampling of a p-38 inhibitor in lesional and non-lesional skin after Day 1 and Day 8 of topical application. (Reprinted with Permission from Elsevier: Bodenlenz et al. 2012 [51])

More importantly, when the TNF-alpha levels were compared between Day 1 and Day 8 in lesional skin, they were significantly reduced, suggesting an effect elicited by applying the p-38 inhibitor.

This utility of dOFM in a clinical setting to evaluate PKPD relationships was only the beginning. A more recent study, in collaboration with LEO Pharma, evaluated the PKPD relationship of two topical phosphodiesterase (PDE)-4 inhibitors after topical application in fresh human skin explants [52]. There was clinical evidence that one of the PDE-4 inhibitors showed efficacy while the other did not, even though the skin total concentrations were similar after punch biopsies. The in vivo estimated TNF-alpha EC50 values were similar (13–14 nM); however, the unbound concentrations for the compound that show clinical efficacy were ~360-fold over the unbound EC50. While it is known that the unbound concentrations drive efficacy, there has been no way to measure the unbound concentrations in vivo except for dMD or dOFM, with dOFM allowing the estimation of significant biomarkers.

Understanding PKPD relationships in vivo is crucial, as is assessing the local BA, and by extension BE, of topically applied products. This is where dOFM comes into play. One potential variable in assessing topical BA/BE is the probe depth within the dermis. While this has been a minor factor in numerous dMD preclinical studies, it is essential to consistently evaluate and monitor probe depth to understand its potential impact on a BE assessment. However, with highly trained individuals, dOFM probes can be placed at a similar depth, thus eliminating this as a significant covariate. Although probe placement may cause mild discomfort, it is generally well tolerated, especially with ice, to minimize insertion trauma [41]. Another critical feature of dMD or dOFM BE assessments is the reduction in inter-subject variability by parallel application of formulations on adjacent sites on a participant, which reduces the number of subjects required to power a BE study [53].

One of the first demonstrations utilizing dOFM in the clinical setting for a BE study was utilizing acyclovir as the API [54]. Here, the group from Joanneum Research had two application sites per leg with the reference product and one with the test product, with each application site containing two dOFM probes. As a proof-of-concept study, the two reference products were compared on the same leg to confirm BE between the same products as a positive control (Fig. 8.6—Top). After 36 h of sampling, the concentrations of the reference product continued to increase while the test product concentrations began to decrease (Fig. 8.6—Bottom). This supported the utility of dOFM for BE studies, as the study design was sensitive to changes in the local BA, and the probes were stable for the entire experimental duration.

Further analysis of the acyclovir BE study investigated the sources of variability in the PK profiles. Bodenlenz et al. observed moderate variability in the overall exposure, and it was suggested that the primary source for both the test and reference products was the inter-subject variability [53]. The intra-subject variability, including probe depth, transepidermal water loss (TEWL), and site, accounted for 9–18% of the overall observed variability, demonstrating the optimization and standardization of the dOFM study design and protocol. The skin conductance and relative lactate recovery mainly explained the inter-subject variability.

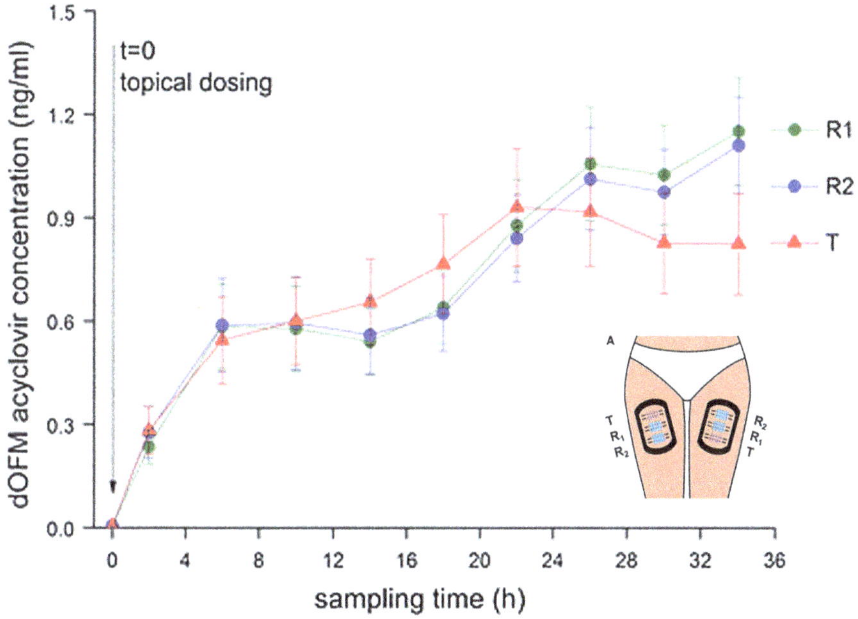

Fig. 8.6 Top Panels: (left) schematic of topical application sites and probe locations; (right) clinical setup of the schematic on the left. Bottom panel: Acyclovir concentration-time profiles from the two reference application sites per leg and the test product over time. The inset displays the application site and probe placement for the study. (Adapted with permission from Elsevier. Bodenlenz 2017 [54])

dOFM is an exciting methodology utilized in both local BA and BE assessments while also having the versatility to estimate PKPD relationships in vivo. There is a need to understand the local target site concentrations (unbound) and the relationship with biomarkers to allow for less attrition of topical innovator products. In addition, dOFM allows for the sampling of highly bound and lipophilic molecules in BA/BE assessments to allow for more market competition and reduced generic drug pricing.

8.5 Microspectroscopic Techniques

Noninvasive microspectroscopic techniques are an exciting new group of modalities to quantify chemical species within the skin. Indeed, these methodologies have recently gained traction in their utility for cutaneous PK measurements and still require investigation for optimal protocols. Like all the methodologies mentioned in this chapter, as the technologies advance, so will these methodologies as they move toward routine use in preclinical discovery or clinical utility. Several microspectroscopic techniques will be touched upon in this chapter with limited examples as they are extensively covered at another point in this book.

8.5.1 Fluorescence

Of all the techniques, fluorescence is the most challenging due to the inherent skin fluorescence and the fact that the molecule of interest must have a fluorophore. However, fluorescence tagging can alter the cPK compared to the native molecule. There are also several subdomains of fluorescence imaging, such as classical, two-photon, confocal, and fluorescence lifetime imaging microscopy (FLIM).

A specific approach to harnessing fluorescence in cutaneous PK studies involves techniques beyond conventional fluorescence. For instance, a recent study employed fluorescence imaging to investigate the permeation and biodistribution of tazarotene and minocycline in human facial skin using two-photon excitation fluorescence (TPEF) and fluorescence lifetime imaging microscopy FLIM [55]. This study successfully visualized minocycline in solution, dried minocycline, and the lipid structures of the epidermis/hair follicle. Given minocycline's inherent fluorescence, this study served as an excellent demonstration of the capacity to distinguish skin autofluorescence from that of the API.

8.5.2 Confocal Raman Spectroscopy

Spontaneous Raman spectroscopy is another non-invasive technique that can provide information about the permeation and elimination of an API after topical delivery and the relative biodistribution within the tissue. One utilized modality of spontaneous Raman spectroscopy is confocal Raman spectroscopy. The utility of this approach is obtaining concentration-time profiles as a function of depth. Nevertheless, the drawbacks are that one cannot observe the localization within the cells as in other imaging modalities and that as the entire Raman spectra are acquired, there are long acquisition times per depth.

A recent study by Caspers et al. demonstrated the ability to use confocal Raman spectroscopy as a method to noninvasively capture the concentration of an API within the skin [56]. A propylene glycol: ethanol solution with 0.3% retinol was applied to the skin, and confocal Raman was used to quantify the concentration flux over time as a function of depth. The authors demonstrated that the retinol flux was a by-product of the propylene glycol flux. It is worth noting that propylene glycol is a known permeation enhancer, which helps increase the skin's permeability to the API, thereby enhancing its absorption (Fig. 8.7).

More recently, confocal Raman spectroscopy was utilized in the BE assessment of ibuprofen. Iliopoulos et al. applied Nurofen or Ibuleve to healthy participants' skin and found no significant difference between exposure in the stratum corneum

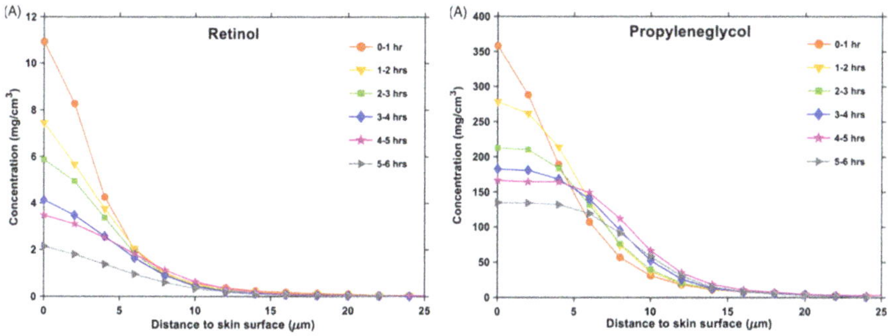

Fig. 8.7 Retinol (left) and propylene glycol (right) flux profiles as a function of depth and time. (Reused with permission from Wiley. Caspers et al., 2019 [56])

[57]. In addition, they found that the exposure did not dramatically change from 1 h post-application to 2 h post-application. This was the first demonstration of confocal Raman spectroscopy for a BE assessment, yet more probing needs to be done to compare with other in vivo cutaneous PK methods.

8.5.3 Coherent Raman

Coherent Raman scattering imaging (CRI) is a label-free, non-invasive method that is an exciting tool to quantify topically applied APIS within the skin. While there are similarities with spontaneous Raman, the critical difference is that this is a stimulated process, so the signal is amplified and provides a strong signal intensity. Within CRI, two main techniques are utilized: coherent anti-Stokes Raman scattering (CARS) and stimulated Raman scattering (SRS). CARS is typically used to inform on the skin structure, while SRS is typically utilized to quantify molecular concentrations due to its linear signal with molecular concentrations.

While there have been numerous contributions to advance the technology and the practical use of CRI [58–60], one of the first uses to quantify cPK after topical drug application was by Freudiger et al. [61]. Solution formulations of either retinoic acid or water-DMSO mixtures were applied to fresh mouse ear skin, and DMSO (670 cm^{-1}) and retinoic acid (1570 cm^{-1}) were tracked as they permeated deeper into the skin. Utilizing SRS, they were able to visualize where the compounds preferred to permeate. As these two molecules have different solubilities, they also had different permeabilities into each mouse ear skin layer—DMSO through the protein-rich region while the retinoic acid permeated through the retinoic acid region of the stratum corneum. This method was further refined by Saar et al. [62]. Here the group collected the SRS images by epi-detection for thick skin samples in vivo. As the early studies were limited to deuterated molecules or those that had vibrational modes in the high wavenumber region, they monitored trans-retinol uptake into nude

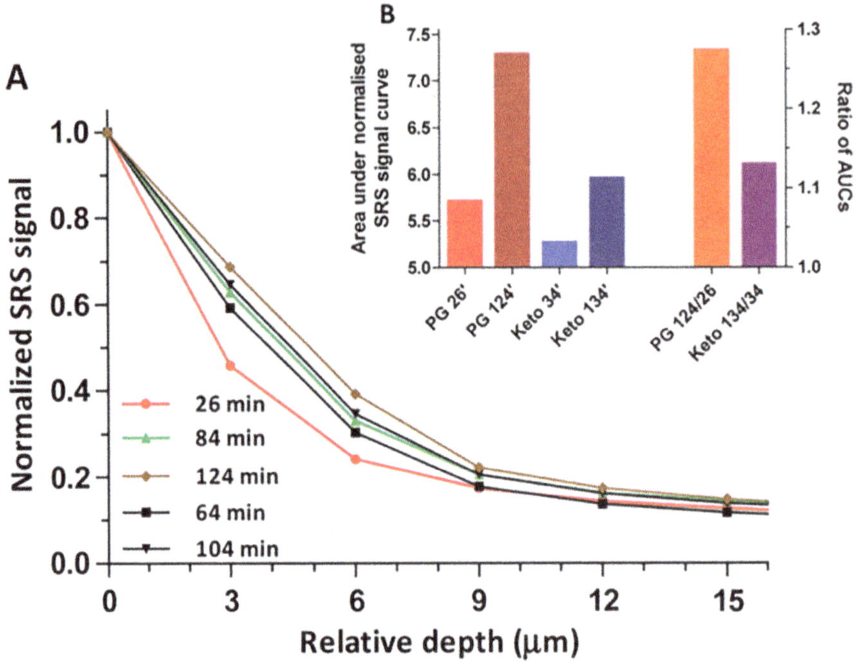

Fig. 8.8 Integrated depth-profiling analysis of percutaneous penetration. (**a**) Depth profiles of deuterated PG as a function of time. (**b**) Comparison of the areas under the normalized SRS signal versus skin depth profiles (AUCs) of PG and ketoprofen at different times (26 and 124 min, and 34 and 134 min, respectively) post-application of the formulation (left-hand y-axis). The ratios of the AUCs (right-hand y-axis) determined after the first and last measurements (at approximately 0.5 and 2 h) demonstrate the slower penetration rate of the drug compared to that of the cosolvent. (Reprinted with permission from Saar et al. Copyright (2011) American Chemical Society [63])

mouse skin and deuterated DMSO in one healthy subject. In the case of the trans-retinol, the preferred permeation route was through the hair follicles. In the case of the healthy subject, deuterated DMSO was imaged by monitoring the carbon-deuterium stretch, and it was found that the compound also localized around the hair follicles.

The proof-of-concept from Saar left more to be desired and investigated. Indeed, a follow-up study utilized two NSAIDs (ketoprofen and ibuprofen) in addition to a common solvent utilized in topical formulations—propylene glycol [63]. As propylene glycol is typically used as a permeation enhancer, it would be worthwhile to also track it utilizing deuterated propylene glycol. Propylene glycol was found to permeate more quickly than ketoprofen and ibuprofen (Fig. 8.8).

Additional advancements have been made to allow CRI, specifically SRS, to be utilized in a toolkit of approaches for cutaneous PK measurements; however, further refined protocols are required, primarily the ability to detect compounds without a deuterium band or chemically labeled and those compounds' fingerprint regions. Pence et al. recently solved this challenge through rapid wavenumber tuning and spectral unmixing to quantify various molecular species [64]. First, they could differentiate the chemical species of isotretinoin, retinoic acid, and tazarotene with rapid wavenumber switching. Utilizing this development with SRS, they could track tazarotene in a propylene glycol 200 (PEG200) solution with temporal and spatial resolution after topical application in a nude mouse ear. The PEG200 and tazarotene seemed to localize in lipid-rich regions, regardless of the skin stratification in the mouse ear.

A pressing need in microspectroscopic techniques is the ability to provide absolute molecular quantification. While relative concentration quantification may be helpful in comparing two products within the skin, innovator companies require absolute quantification. Spontaneous Raman can be used in preclinical and clinical settings, and coherent Raman is not far behind in moving to the clinic once standardized protocols/best practices are developed. Early timepoint quantification for rapidly permeating compounds is of interest, despite the long permeation of topically applied drug products. However, a significant challenge in utilizing these methodologies in the clinic is the scattering caused by the skin and many topical formulations. The turbidity in the skin and opaque formulations hinder the light from returning to the detector for accurate quantification. However, techniques such as skin clearing or adaptive optics may be implemented to increase imaging depth.

8.6 Tape Stripping

Tape stripping (TS), also known as the dermatopharmacokinetic method, is a minimally invasive technique that allows the quantification of an API at sequential layers of the stratum corneum (SC), which has been utilized throughout biology [65–67] and more relevantly in cutaneous PK [68–70]. As this has been extensively reviewed elsewhere, we will briefly overview TS [71–73]. The number of TS is correlated with a depth within the skin to obtain a concentration-depth profile. Each application site is one timepoint post-application where the application sites are evenly split into uptake and clearance phases, which provides a challenge to monitor temporal kinetics as with other techniques in this chapter. However, tape stripping is convenient when the site of action is within the stratum corneum, as methods such as dMD/dOFM cannot measure SC concentrations.

TS garnered substantial interest due to its non-invasive nature and the potential to inform about API disposition within the subsequent layers deeper in the skin. Early on, there were attempts to understand the percutaneous permeation of compounds and the amounts recovered via excretion/elimination pathways. Dupuis established a linear relationship for humans and rats with the amount in the SC and the amount penetrated over 4 days [74], suggesting a reservoir within the SC, allowing for prolonged permeation. However, the authors acknowledged species/regional differences due to the twofold increased absorption of benzoic acid in the rat. While it is to be expected that there are differences between species, the differences in anatomical sites were relatively unknown at the time. However, through research, both the site of application and the duration of application influence the concentration-depth profile obtained with tape stripping [7, 75–77].

In 1998, the US FDA released a draft guidance proposing the use of the dermatopharmacokinetic approach to evaluate the bioavailability and bioequivalence of topical dermatological drug products [78]. While this provided promise to evaluate topical drugs BA and BE as opposed to systemic measures such as blood sampling, it may have been issued before the methodology had a chance to mature. In 2002, the draft guidance was withdrawn due to inconsistencies in the method between two established laboratories [79, 80]. However, later research pointed out the deficiencies in the original draft, such as the elimination of the first two tape strips [81, 82] as well as normalizing the depth as a function of the total SC depth [83–85].

The withdrawal of this draft guidance sparked a surge of research into the methodology, aiming to understand potential improvements and limitations. A significant advancement came from a collaboration between the Colorado School of Mines and the University of Bath [86]. They introduced several critical improvements to the protocol, including using TEWL to track the SC thickness, enhanced cleaning procedures to remove residual drugs at the application sites, and a new approach to quantify all drugs present on tape strips instead of discarding the first strips. These improvements addressed the previous inconsistencies and enhanced the accuracy and reliability of bioavailability measurements. The refined protocol was subsequently utilized in various studies, exploring its potential use in diseased populations, BA/BE assessments for locally acting products, and absorption kinetics [30, 70, 87–93].

8.7 Conclusions

Cutaneous PK has been an area of research for over 50 years, but there have been more questions than answers due to analytical or practical limitations. However, the field is gaining momentum with the advances in analytical procedures and the availability of practical in vivo sampling techniques such as dMD and dOFM. Indeed, the possibility of monitoring cutaneous PK for BA and BE assessments appears within reach. This may be paired with the capability to monitor a topically applied API's pharmacodynamics, which is equally essential for successfully identifying an innovator drug product.

There is also excitement for the emerging microspectroscopic techniques that are being investigated and standardized to implement into a toolkit approach to understand cutaneous PK. However, these methodologies require more investigation before implementation into cutaneous BA and BE assessments, which are currently underway. A comparison of the techniques, the skin stratification where the APIs can be monitored, and the advantages and disadvantages are laid out in Table 8.1.

Table 8.1 Comparison of methodologies used for cutaneous PK estimates in both in vitro and in vivo settings and how they compare across key metrics

Methodology	In vitro/ in vivo	High Throughput	High operator skills required	Dynamic PK profiles	Analytical ease
Plasma sampling	In vivo	+++	+	+++	++
Microdialysis	Both	+	+++	+++	+++
Open flow microperfusion	Both	+	+++	+++	+++
Spectroscopic techniques	Both	++	++	++	++
Tape stripping	Both	+	+	+	+++

+ indicates poor suitability, while +++ indicates high suitability

References

1. F. Guidance, Topical dermatologic corticosteroids: in vivo bioequivalence, (1995).
2. M.N. Pastore, Y.N. Kalia, M. Horstmann, M.S. Roberts, Transdermal patches: history, development and pharmacology, Br. J. Pharmacol. 172 (2015) 2179–2209.
3. A.J.W. Sinha, S.R. Shaw, D.J. Weber, Percutaneous Absorption and Excretion of Tritium-Labeled Diflorasone Diacetate, a New Topical Corticosteroid in the Rat, Monkey and Man, J. Invest. Dermatol. 71 (1978) 372–377. https://doi.org/10.1111/1523-1747.ep12556772.
4. V. Rimbau, J. Forn, C. Delgado, M. Roca, Percutaneous absorption of 3H-labelled flupamesone in man, Eur. J. Drug Metab. Pharmacokinet. 1 (1976) 206–207. https://doi.org/10.1007/BF03189280.
5. M. Rowland, T.N. Tozer, Clinical Pharmacokinetics and Pharmacodynamics, (1995).
6. H. Lboutounne, J.-F. Chaulet, C. Ploton, F. Falson, F. Pirot, Sustained ex vivo skin antiseptic activity of chlorhexidine in poly (ϵ-caprolactone) nanocapsule encapsulated form and as a digluconate, J. Controlled Release. 82 (2002) 319–334.
7. A. Rougier, D. Dupuis, C. Lotte, R. Roguet, The measurement of the stratum corneum reservoir. A predictive method for in vivo percutaneous absorption studies: influence of application time, J. Invest. Dermatol. 84 (1985) 66–68.
8. R.J. Feldmann, H.I. Maibach, Percutaneous penetration of steroids in man, J. Invest. Dermatol. 52 (1969) 89–94.
9. R.J. Feldmann, H.I. Maibach, Absorption of some organic compounds through the skin in man., J. Invest. Dermatol. 54 (1970).
10. R.J. Feldmann, H.I. Maibach, Percutaneous penetration of some pesticides and herbicides in man, Toxicol. Appl. Pharmacol. 28 (1974) 126–132.
11. S.H. Shin, P. Ghosh, B. Newman, D.C. Hammell, S.G. Raney, H.E. Hassan, A.L. Stinchcomb, On the road to development of an *in vitro* permeation test (IVPT) model to compare heat effects on transdermal delivery systems: exploratory studies with nicotine and fentanyl, Pharm. Res. 34 (2017) 1817–1830.
12. J.A. Bosso, S. 1. Spruance, G. Wenerstrom, Tolerance and Percutaneous Absorption of Topically Applied Arildone, J. Clin. Pharmacol. 25 (1985) 95–99. https://doi.org/10.1002/j.1552-4604.1985.tb02808.x.
13. I. Persaud, S. Diamond, R. Pan, K. Burke, J. Harris, M. Conlin, S. Yeleswaram, Plasma pharmacokinetics and distribution of ruxolitinib into skin following oral and topical administration in minipigs, Int. J. Pharm. 590 (2020) 119889. https://doi.org/10.1016/j.ijpharm.2020.119889.

14. S.S. Wind, M.A.A. Jansen, M. Rijsbergen, M.J. van Esdonk, D. Ziagkos, W.C. Cheng, T. Niemeyer-van der Kolk, J. Korsten, A. Gruszka, D. Schmitz-Rohmer, D. Bonnel, R. Legouffe, F. Barré, M.W. Bekkenk, E.R.M. de Haas, K.D. Quint, H. Schnidar, M. Rolli, H.J. Streefkerk, J. Burggraaf, M.H. Vermeer, R. Rissmann, Topical Bimiralisib Shows Meaningful Cutaneous Drug Levels in Healthy Volunteers and Mycosis Fungoides Patients but No Clinical Activity in a First-in-Human, Randomized Controlled Trial, Cancers. 14 (2022) 1510. https://doi.org/10.3390/cancers14061510.
15. Q. Cai, A. Azarpanah, J.A. Bhatt, H. Wei, J. Carr, K.R. Morris, C. Strasinger, N.K. Brogden, Examination of the Rate and Extent of Drug Released from Commercial Topical Delivery Systems During Wear: An Example with Lidocaine Topical Systems, Pharm. Res. (2023). https://doi.org/10.1007/s11095-023-03617-7.
16. M. Hammarlund-Udenaes, Microdialysis as an Important Technique in Systems Pharmacology—a Historical and Methodological Review, AAPS J. 19 (2017) 1294–1303. https://doi.org/10.1208/s12248-017-0108-2.
17. S.G. Summerfield, J.W.T. Yates, D.A. Fairman, Free Drug Theory - No Longer Just a Hypothesis?, Pharm. Res. 39 (2022) 213–222. https://doi.org/10.1007/s11095-022-03172-7.
18. C. Shukla, E.D. Bashaw, G. Stagni, E. Benfeldt, Applications of dermal microdialysis: a review, J. Drug Deliv. Sci. Technol. 24 (2014) 259–269. https://doi.org/10.1016/S1773-2247(14)50044-5.
19. Y. Song, C.E. Lunte, Calibration methods for microdialysis sampling in vivo: muscle and adipose tissue, Anal. Chim. Acta. 400 (1999) 143–152.
20. S. Menacherry, W. Hubert, J.B. Justice, In vivo calibration of microdialysis probes for exogenous compounds, Anal. Chem. 64 (1992) 577–583.
21. C. Anderson, T. Andersson, M. Molander, Ethanol absorption across human skin measured by in vivo microdialysis technique, Acta Derm. Venereol. 71 (1991) 389–393.
22. C. Anderson, T. Andersson, R.G. Andersson, In vivo microdialysis estimation of histamine in human skin, Skin Pharmacol. Off. J. Skin Pharmacol. Soc. 5 (1992) 177–183. https://doi.org/10.1159/000211035.
23. C. Anderson, T. Andersson, K. Wårdell, Changes in skin circulation after insertion of a microdialysis probe visualized by laser Doppler perfusion imaging, J. Invest. Dermatol. 102 (1994) 807–811. https://doi.org/10.1111/1523-1747.ep12378630.
24. J.M. Ault, C.M. Riley, N.M. Meltzer, C.E. Lunte, Dermal Microdialysis Sampling in Vivo, Pharm. Res. 11 (1994) 1631–1639. https://doi.org/10.1023/A:1018922123774.
25. A.L. Krogstad, P.A. Jansson, P. Gisslén, P. Lönnroth, Microdialysis methodology for the measurement of dermal interstitial fluid in humans, Br. J. Dermatol. 134 (1996) 1005–1012.
26. M. Müller, H. Mascher, C. Kikuta, S. Schäfer, M. Brunner, G. Dorner, H.G. Eichler, Diclofenac concentrations in defined tissue layers after topical administration, Clin. Pharmacol. Ther. 62 (1997) 293–299. https://doi.org/10.1016/S0009-9236(97)90032-1.
27. G. Stagni, D. O'Donnell, Y.J. Liu, D.L. Kellogg Jr, A.M. Shepherd, Iontophoretic current and intradermal microdialysis recovery in humans, J. Pharmacol. Toxicol. Methods. 41 (1999) 49–54.
28. E. Benfeldt, S.H. Hansen, A. Vølund, T. Menné, V.P. Shah, Bioequivalence of topical formulations in humans: evaluation by dermal microdialysis sampling and the dermatopharmacokinetic method, J. Invest. Dermatol. 127 (2007) 170–178. https://doi.org/10.1038/sj.jid.5700495.
29. G. Stagni, D. O'Donnell, Y.J. Liu, D.L. Kellogg Jr, T. Morgan, A.M. Shepherd, Intradermal microdialysis:: Kinetics of iontophoretically delivered propranolol in forearm dermis, J. Controlled Release. 63 (2000) 331–339.
30. P.G. Ortiz, S.H. Hansen, V.P. Shah, T. Menné, E. Benfeldt, The effect of irritant dermatitis on cutaneous bioavailability of a metronidazole formulation, investigated by microdialysis and dermatopharmacokinetic method, Contact Dermatitis. 59 (2008) 23–30. https://doi.org/10.1111/j.1600-0536.2008.01348.x.
31. R.N.O. Tettey-Amlalo, I. Kanfer, M.F. Skinner, E. Benfeldt, R.K. Verbeeck, Application of dermal microdialysis for the evaluation of bioequivalence of a ketoprofen topical gel, Eur. J. Pharm. Sci. 36 (2009) 219–225. https://doi.org/10.1016/j.ejps.2008.09.002.

32. R. Holmgaard, E. Benfeldt, N. Bangsgaard, J.A. Sorensen, K. Brosen, F. Nielsen, J.B. Nielsen, Probe depth matters in dermal microdialysis sampling of benzoic acid after topical application: an ex vivo study in human skin, Skin Pharmacol. Physiol. 25 (2012) 9–16. https://doi.org/10.1159/000330491.
33. M. Lubda, M. Zander, A. Salazar, H. Kolmar, J. von Hagen, Comparison of Membrane Depth Determination Techniques for Active Ingredient Skin Penetration Studies Using Microdialysis, Skin Pharmacol. Physiol. 34 (2021) 203–213. https://doi.org/10.1159/000515113.
34. B.A. Kuzma, S. Senemar, T. Ramezanli, P. Ghosh, S.G. Raney, G. Stagni, Evaluation of local bioavailability of metronidazole from topical formulations using dermal microdialysis: Preliminary study in a Yucatan mini-pig model, Eur. J. Pharm. Sci. 159 (2021) 105741. https://doi.org/10.1016/j.ejps.2021.105741.
35. O. Boubriak, J. Urban, Z. Cui, Monitoring of lactate and glucose levels in engineered cartilage construct by microdialysis, J. Membr. Sci. 273 (2006) 77–83.
36. N. Torto, J. Bång, S. Richardson, G.S. Nilsson, L. Gorton, T. Laurell, G. Marko-Varga, Optimal membrane choice for microdialysis sampling of oligosaccharides, J. Chromatogr. A. 806 (1998) 265–278.
37. B.A. Kuzma, S. Senemar, T. Ramezanli, P. Ghosh, S.G. Raney, G. Stagni, The dose-duration effect on cutaneous pharmacokinetics of metronidazole from topical dermatological formulations in Yucatan mini-pigs, Eur. J. Pharm. Biopharm. Off. J. Arbeitsgemeinschaft Pharm. Verfahrenstechnik EV. 175 (2022) 43–52. https://doi.org/10.1016/j.ejpb.2022.05.001.
38. B.A. Kuzma, Investigation of Percutaneous Drug Pharmacokinetics in Yucatan Mini-Pig Utilizing Microdialysis: Development of an *In Vitro*-In Vivo Relationship with Metronidazole Topical Dermatological Drug Products, (2020).
39. S. Senemar, B.A. Kuzma, T. Ramezanli, P. Ghosh, S.G. Raney, E. Rantou, G. Stagni, Bioequivalence Evaluation of Topical Metronidazole Products Using Dermal Microdialysis in New Zealand Rabbits, AAPS PharmSciTech. 24 (2023) 204.
40. R. Juluru, C. Shukla, H. Yin, G. Stagni, Skin microdialysis-based estimation of systemic bioavailability fraction, J. Pharm. Sci. 101 (2012) 405–413.
41. M. Bodenlenz, B. Aigner, C. Dragatin, L. Liebenberger, S. Zahiragic, C. Höfferer, T. Birngruber, J. Priedl, F. Feichtner, L. Schaupp, S. Korsatko, M. Ratzer, C. Magnes, T.R. Pieber, F. Sinner, Clinical applicability of dOFM devices for dermal sampling, Skin Res. Technol. Off. J. Int. Soc. Bioeng. Skin ISBS Int. Soc. Digit. Imaging Skin ISDIS Int. Soc. Skin Imaging ISSI. 19 (2013) 474–483. https://doi.org/10.1111/srt.12071.
42. T. Birngruber, A. Ghosh, V. Perez-Yarza, T. Kroath, M. Ratzer, T.R. Pieber, F. Sinner, Cerebral open flow microperfusion: a new in vivo technique for continuous measurement of substance transport across the intact blood-brain barrier, Clin. Exp. Pharmacol. Physiol. 40 (2013) 864–871. https://doi.org/10.1111/1440-1681.12174.
43. C. Höfferer, D. Tutkur, C. Fledelius, C.L. Brand, T.J. Alsted, J. Damgaard, E. Nishimura, C.B. Jeppesen, S.I. Mautner, T.R. Pieber, F. Sinner, Open flow microperfusion: pharmacokinetics of human insulin and insulin detemir in the interstitial fluid of subcutaneous adipose tissue, Diabetes Obes. Metab. 17 (2015) 121–127. https://doi.org/10.1111/dom.12394.
44. K. Tiffner, B. Boulgaropoulos, C. Höfferer, T. Birngruber, N. Porksen, H. Linnebjerg, P. Garhyan, E.C.Q. Lam, M.P. Knadler, T.R. Pieber, F. Sinner, Quantification of Basal Insulin Peglispro and Human Insulin in Adipose Tissue Interstitial Fluid by Open-Flow Microperfusion, Diabetes Technol. Ther. 19 (2017) 305–314. https://doi.org/10.1089/dia.2016.0384.
45. C. Dragatin, F. Polus, M. Bodenlenz, C. Calonder, B. Aigner, K.I. Tiffner, J.K. Mader, M. Ratzer, R. Woessner, T.R. Pieber, Y. Cheng, C. Loesche, F. Sinner, G. Bruin, Secukinumab distributes into dermal interstitial fluid of psoriasis patients as demonstrated by open flow microperfusion, Exp. Dermatol. 25 (2016) 157–159. https://doi.org/10.1111/exd.12863.
46. F. Kolbinger, C. Loesche, M.-A. Valentin, X. Jiang, Y. Cheng, P. Jarvis, T. Peters, C. Calonder, G. Bruin, F. Polus, B. Aigner, D.M. Lee, M. Bodenlenz, F. Sinner, T.R. Pieber, D.D. Patel, β-Defensin 2 is a responsive biomarker of IL-17A-driven skin pathology in patients with psoriasis. J. Allergy Clin. Immunol. 139 (2017) 923-932.e8. https://doi.org/10.1016/j.jaci.2016.06.038.

47. D. Schimek, R. Raml, K.A. Francesconi, M. Bodenlenz, F. Sinner, Quantification of acyclovir in dermal interstitial fluid and human serum by ultra-high-performance liquid-high-resolution tandem mass spectrometry for topical bioequivalence evaluation, Biomed. Chromatogr. BMC. 32 (2018) e4194. https://doi.org/10.1002/bmc.4194.
48. F. Skrabal, Z. Trajanoski, H. Kontscheider, P. Kotanko, P. Wach, Portable system for on-line continuous ex vivo monitoring of subcutaneous tissue glucose using open tissue perfusion, Med. Biol. Eng. Comput. 33 (1995) 116–118. https://doi.org/10.1007/BF02522959.
49. K.E. Pickl, C. Magnes, M. Bodenlenz, T.R. Pieber, F.M. Sinner, Rapid online-SPE-MS/MS method for ketoprofen determination in dermal interstitial fluid samples from rats obtained by microdialysis or open-flow microperfusion, J. Chromatogr. B. 850 (2007) 432–439. https://doi.org/10.1016/j.jchromb.2006.12.026.
50. R. Holmgaard, E. Benfeldt, J.B. Nielsen, C. Gatschelhofer, J.A. Sorensen, C. Höfferer, M. Bodenlenz, T.R. Pieber, F. Sinner, Comparison of open-flow microperfusion and microdialysis methodologies when sampling topically applied fentanyl and benzoic acid in human dermis ex vivo, Pharm. Res. 29 (2012) 1808–1820. https://doi.org/10.1007/s11095-012-0705-9.
51. M. Bodenlenz, C. Höfferer, C. Magnes, R. Schaller-Ammann, L. Schaupp, F. Feichtner, M. Ratzer, K. Pickl, F. Sinner, A. Wutte, S. Korsatko, G. Köhler, F.J. Legat, E.M. Benfeldt, A.M. Wright, D. Neddermann, T. Jung, T.R. Pieber, Dermal PK/PD of a lipophilic topical drug in psoriatic patients by continuous intradermal membrane-free sampling, Eur. J. Pharm. Biopharm. Off. J. Arbeitsgemeinschaft Pharm. Verfahrenstechnik EV. 81 (2012) 635–641. https://doi.org/10.1016/j.ejpb.2012.04.009.
52. S. Eirefelt, J. Hummer, L.H. Basse, M. Bertelsen, F. Johansson, T. Birngruber, F. Sinner, J. Larsen, S.F. Nielsen, M. Lambert, Evaluating Dermal Pharmacokinetics and Pharmacodymanic Effect of Soft Topical PDE4 Inhibitors: Open Flow Microperfusion and Skin Biopsies, Pharm. Res. 37 (2020) 243. https://doi.org/10.1007/s11095-020-02962-1.
53. M. Bodenlenz, T. Augustin, T. Birngruber, K.I. Tiffner, B. Boulgaropoulos, S. Schwingenschuh, S.G. Raney, E. Rantou, F. Sinner, Variability of Skin Pharmacokinetic Data: Insights from a Topical Bioequivalence Study Using Dermal Open Flow Microperfusion, Pharm. Res. 37 (2020) 204. https://doi.org/10.1007/s11095-020-02920-x.
54. M. Bodenlenz, K.I. Tiffner, R. Raml, T. Augustin, C. Dragatin, T. Birngruber, D. Schimek, G. Schwagerle, T.R. Pieber, S.G. Raney, I. Kanfer, F. Sinner, Open Flow Microperfusion as a Dermal Pharmacokinetic Approach to Evaluate Topical Bioequivalence, Clin. Pharmacokinet. 56 (2017) 91–98. https://doi.org/10.1007/s40262-016-0442-z.
55. S. Jeong, M. Hermsmeier, S. Osseiran, A. Yamamoto, U. Nagavarapu, K.F. Chan, C.L. Evans, Visualization of drug distribution of a topical minocycline gel in human facial skin, Biomed. Opt. Express. 9 (2018) 3434–3448. https://doi.org/10.1364/BOE.9.003434.
56. P.J. Caspers, C. Nico, T.C. Bakker Schut, J. de Sterke, P.D.A. Pudney, P.R. Curto, A. Illand, G.J. Puppels, Method to quantify the in vivo skin penetration of topically applied materials based on confocal Raman spectroscopy, Transl. Biophotonics. 1 (2019) e201900004. https://doi.org/10.1002/tbio.201900004.
57. F. Iliopoulos, C.F. Tang, Z. Li, A. Rahma, M.E. Lane, Confocal Raman Spectroscopy for Assessing Bioequivalence of Topical Formulations, Pharmaceutics. 15 (2023) 1075. https://doi.org/10.3390/pharmaceutics15041075.
58. C.L. Evans, E.O. Potma, M. Puoris'haag, D. Côté, C.P. Lin, X.S. Xie, Chemical imaging of tissue in vivo with video-rate coherent anti-Stokes Raman scattering microscopy, Proc. Natl. Acad. Sci. U. S. A. 102 (2005) 16807–16812. https://doi.org/10.1073/pnas.0508282102.
59. C.L. Evans, X.S. Xie, Coherent anti-stokes Raman scattering microscopy: chemical imaging for biology and medicine, Annu. Rev. Anal. Chem. Palo Alto Calif. 1 (2008) 883–909. https://doi.org/10.1146/annurev.anchem.1.031207.112754.
60. E.O. Potma, C.L. Evans, X.S. Xie, Heterodyne coherent anti-Stokes Raman scattering (CARS) imaging, Opt. Lett. 31 (2006) 241–243. https://doi.org/10.1364/ol.31.000241.
61. C.W. Freudiger, W. Min, B.G. Saar, S. Lu, G.R. Holtom, C. He, J.C. Tsai, J.X. Kang, X.S. Xie, Label-free biomedical imaging with high sensitivity by stimulated Raman scattering microscopy, Science. 322 (2008) 1857–1861. https://doi.org/10.1126/science.1165758.

62. B.G. Saar, C.W. Freudiger, J. Reichman, C.M. Stanley, G.R. Holtom, X.S. Xie, Video-Rate Molecular Imaging in Vivo with Stimulated Raman Scattering, Science. 330 (2010) 1368–1370. https://doi.org/10.1126/science.1197236.
63. B.G. Saar, L.R. Contreras-Rojas, X.S. Xie, R.H. Guy, Imaging Drug Delivery to Skin with Stimulated Raman Scattering Microscopy, Mol. Pharm. 8 (2011) 969–975. https://doi.org/10.1021/mp200122w.
64. I.J. Pence, B.A. Kuzma, M. Brinkmann, T. Hellwig, C.L. Evans, Multi-window sparse spectral sampling stimulated Raman scattering microscopy, Biomed. Opt. Express. 12 (2021) 6095–6114. https://doi.org/10.1364/BOE.432177.
65. H. He, C.M. Olesen, A.B. Pavel, M.-L. Clausen, J. Wu, Y. Estrada, N. Zhang, T. Agner, E. Guttman-Yassky, Tape-strip proteomic profiling of atopic dermatitis on dupilumab identifies minimally invasive biomarkers, Front. Immunol. 11 (2020) 1768.
66. P.G.M. Van der Valk, H.I. Maibach, A functional study of the skin barrier to evaporative water loss by means of repeated cellophane-tape stripping, Clin. Exp. Dermatol. 15 (1990) 180–182.
67. R. Van der Molen, F. Spies, J. Van t' Noordende, E. Boelsma, A. Mommaas, H. Koerten, Tape stripping of human stratum corneum yields cell layers that originate from various depths because of furrows in the skin, Arch. Dermatol. Res. 289 (1997) 514–518.
68. C. Herkenne, A. Naik, Y.N. Kalia, J. Hadgraft, R.H. Guy, Dermatopharmacokinetic prediction of topical drug bioavailability in vivo, J. Invest. Dermatol. 127 (2007) 887–894.
69. S. Nicoli, A.L. Bunge, M.B. Delgado-Charro, R.H. Guy, Dermatopharmacokinetics: factors influencing drug clearance from the stratum corneum., Pharm. Res. 26 (2009) 865–871. https://doi.org/10.1007/s11095-008-9785-y.
70. M.A. Maciel Tabosa, M. Hoppel, A.L. Bunge, R.H. Guy, M.B. Delgado-Charro, Predicting topical drug clearance from the skin., Drug Deliv. Transl. Res. 11 (2021) 729–740. https://doi.org/10.1007/s13346-020-00864-8.
71. C. Barber, S. Boiko, Tape stripping: investigational, diagnostic, and therapeutic uses in dermatology, Clin. Dermatol. 40 (2022) 355–362. https://doi.org/10.1016/j.clindermatol.2022.02.008.
72. C. Herkenne, I. Alberti, A. Naik, Y.N. Kalia, F.-X. Mathy, V. Préat, R.H. Guy, In Vivo Methods for the Assessment of Topical Drug Bioavailability, Pharm. Res. 25 (2008) 87–103. https://doi.org/10.1007/s11095-007-9429-7.
73. J. Escobar-Chávez, L. Melgoza-Contreras, M. López-Cervantes, D. Quintanar-Guerrero, A. Ganem-Quintanar, The tape stripping method: As a valuable tool for evaluating topical applied compounds, Bentham Science Publishers: Netherlands, 2009.
74. D. Dupuis, A. Rougier, R. Roguet, C. Lotte, G. Kalopissis, In vivo relationship between horny layer reservoir effect and percutaneous absorption in human and rat, J. Invest. Dermatol. 82 (1984) 353–356.
75. A. Rougier, D. Dupuis, C. Lotte, R. Roguet, H. Schaefer, In vivo correlation between stratum corneum reservoir function and percutaneous absorption., J. Invest. Dermatol. 81 (1983) 275–278. https://doi.org/10.1111/1523-1747.ep12518298.
76. A. Rougier, D. Dupuis, C. Lotte, R. Roguet, R.C. Wester, H.I. Maibach, Regional variation in percutaneous absorption in man: measurement by the stripping method, Arch. Dermatol. Res. 278 (1986) 465–469.
77. A. Rougier, C. Lotte, H.I. Maibach, In vivo percutaneous penetration of some organic compounds related to anatomic site in humans: predictive assessment by the stripping method. J. Pharm. Sci. 76 (1987) 451–454. https://doi.org/10.1002/jps.2600760608.
78. V. Shah, Topical dermatological drug product NDAs and ANDAs—in vivo bioavailability, bioequivalence, *in vitro* release and associated studies, US Dept Health Hum. Serv. Rockv. (1998) 1–19.
79. US FDA, Draft Guidance for Industry on Topical Dermatological Drug Product NDAs and ANDAs-In Vivo Bioavailability, Bioequivalence, *In Vitro* Release and Associated Studies; Withdrawal, (n.d.). https://www.federalregister.gov/documents/2002/05/17/02-12326/draft-guidance-for-industry-on-topical-dermatological-drug-product-ndas-and-andas-in-vivo (accessed November 10, 2023).

80. L.K. Pershing, J.L. Nelsona, J.L. Corlett, S.P. Shrivastava, D.B. Hare, V.P. Shah, Assessment of dermatopharmacokinetic approach in the bioequivalence determination of topical tretinoin gel products, J. Am. Acad. Dermatol. 48 (2003) 740–751. https://doi.org/10.1067/mjd.2003.175.
81. I. Alberti, Y.N. Kalia, A. Naik, R.H. Guy, Assessment and prediction of the cutaneous bioavailability of topical terbinafine, in vivo, in man, Pharm. Res. 18 (2001) 1472.
82. I. Alberti, Y.N. Kalia, A. Naik, J.-D. Bonny, R.H. Guy, Effect of ethanol and isopropyl myristate on the availability of topical terbinafine in human stratum corneum, in vivo, Int. J. Pharm. 219 (2001) 11–19. https://doi.org/10.1016/S0378-5173(01)00616-0.
83. Y.N. Kalia, I. Alberti, N. Sekkat, C. Curdy, A. Naik, R.H. Guy, Normalization of Stratum Corneum Barrier Function and Transepidermal Water Loss In Vivo, Pharm. Res. 17 (2000) 1148–1150. https://doi.org/10.1023/A:1026474200575.
84. F. Pirot, E. Berardesca, Y.N. Kalia, M. Singh, H.I. Maibach, R.H. Guy, Stratum corneum thickness and apparent water diffusivity: facile and noninvasive quantitation in vivo, Pharm. Res. 15 (1998) 492–494. https://doi.org/10.1023/a:1011996903513.
85. J. Lademann, U. Jacobi, C. Surber, H.-J. Weigmann, J. Fluhr, The Tape Stripping procedure—Evaluation of some Critical Parameters, Eur. J. Pharm. Biopharm. Off. J. Arbeitsgemeinschaft Für Pharm. Verfahrenstechnik EV. 72 (2008) 317–23. https://doi.org/10.1016/j.ejpb.2008.08.008.
86. B. N'Dri-Stempfer, W.C. Navidi, R.H. Guy, A.L. Bunge, Improved Bioequivalence Assessment of Topical Dermatological Drug Products Using Dermatopharmacokinetics, Pharm. Res. 26 (2009) 316–328. https://doi.org/10.1007/s11095-008-9742-9.
87. P. Garcia Ortiz, S.H. Hansen, V.P. Shah, T. Menné, E. Benfeldt, Impact of adult atopic dermatitis on topical drug penetration: assessment by cutaneous microdialysis and tape stripping, Acta Derm. Venereol. 89 (2009) 33–38. https://doi.org/10.2340/00015555-0562.
88. A. Pensado, W.S. Chiu, S.F. Cordery, E. Rantou, A.L. Bunge, M.B. Delgado-Charro, R.H. Guy, Stratum Corneum Sampling to Assess Bioequivalence between Topical Acyclovir Products, Pharm. Res. 36 (2019) 180. https://doi.org/10.1007/s11095-019-2707-3.
89. L.B. Leal, S.F. Cordery, M.B. Delgado-Charro, A.L. Bunge, R.H. Guy, Bioequivalence Methodologies for Topical Drug Products: *In Vitro* and Ex Vivo Studies with a Corticosteroid and an Anti-Fungal Drug, Pharm. Res. 34 (2017) 730–737. https://doi.org/10.1007/s11095-017-2099-1.
90. T. Incecayir, I. Agabeyoglu, U. Derici, S. Sindel, Assessment of Topical Bioequivalence Using Dermal Microdialysis and Tape Stripping Methods, Pharm. Res. 28 (2011) 2165–2175. https://doi.org/10.1007/s11095-011-0444-3.
91. S.F. Cordery, A. Pensado, W.S. Chiu, M.Z. Shehab, A.L. Bunge, M.B. Delgado-Charro, R.H. Guy, Topical bioavailability of diclofenac from locally-acting, dermatological formulations, Int. J. Pharm. 529 (2017) 55–64. https://doi.org/10.1016/j.ijpharm.2017.06.063.
92. A. Pensado, A. McGrogan, K.A.J. White, A.L. Bunge, R.H. Guy, M.B. Delgado-Charro, Assessment of dermal bioavailability: predicting the input function for topical glucocorticoids using stratum corneum sampling, Drug Deliv. Transl. Res. 12 (2022) 851–861. https://doi.org/10.1007/s13346-021-01064-8.
93. M.B. Reddy, A.L. Stinchcomb, R.H. Guy, A.L. Bunge, Determining dermal absorption parameters in vivo from tape strip data, Pharm. Res. 19 (2002) 292–298. https://doi.org/10.1023/a:1014443001802.

Chapter 9
Dermatological Product Sensorial Properties: Bridging Pharmaceutical and Therapeutic Equivalence

Khanh Phan, Vânia R. Leite-Silva, Sangeeta Prakash, Sarika Namjoshi, Bhavesh Panchal, Michael Roberts, and Yousuf Mohammed

9.1 Introduction

As an important part of a practising dermatologist's treatment armamentarium, topical dermatologic products are described as drug products that are applied to the skin surface for the treatment of skin diseases [1]. A dermatological dosage form, in general, is a combination of one or more active pharmaceutical ingredients (APIs) together with different inactive ingredients to produce a final topical drug product [2]. Semisolid dosage forms, such as creams, ointments, lotions and gels, which are the most commonly used topical pharmaceutical formulations [3–5, can vary greatly in their appearance, texture and smell. They are first judged, by touch, whilst a patient applies the product on the skin. Such products may add to the complexity of assessing perception by making either positive (placebo) and/or negative (nocebo, literally "harmful") feelings independent of the presence of any active pharmaceutical ingredient (API). The application procedure of a topical product could be subdivided into four sectors, including (i) removal of the sample from its container; (ii) initial sensations on the skin; (iii) sensations during the spreading of the sample on the skin; and (iv) final impressions owing to sample residue on the skin [6]. The force needed to take a sample from its container or squeeze it out of a tube is judged as per the patient's expectation. For example, the expectation for a body lotion is to be thinner than a cream [7]. Similarly, during

K. Phan · V. R. Leite-Silva · S. Namjoshi · B. Panchal · M. Roberts · Y. Mohammed (✉)
Frazer Institute, The University of Queensland, St Lucia, QLD, Australia
e-mail: y.mohammed@uq.edu.au

S. Prakash
School of Agriculture and Food Sustainability, The University of Queensland, St Lucia, QLD, Australia

S. N. Murthy (ed.), *Topical Products and Dermal Drug Delivery*,
AAPS Introductions in the Pharmaceutical Sciences 2,
https://doi.org/10.1007/978-3-031-80525-7_9

product application, the brain subconsciously appreciates the uniformity of the sample, its absorption capacity, smoothness, and general texture [8].

The role of topical products in reinforcing human feelings of good/like and bad/dislike is well noticed by industries that use different colours in tablets to improve or reinforce a therapeutic effect, for instance, blue for sedation, red and orange for stimulants, yellows to demonstrate positivity for antidepressants, green for anxiety and white for pain, with brighter colours and embossed brand names applied to enhance effects [9]. Perception thus plays a key role, as it contributes significantly to whether a product is adored and consequently used. Any variations in the compositions of a product can potentially improve or reduce the actual response made by an API. Actually, a placebo response accounted for about half of the pain relief found for a topical pain cream [10] and one-third of the best topical product functioning for inflammatory cream for acne vulgaris [11]. Dissimilarities in culture also could impact responses, so that, for example, blue is used to link to the Italian soccer team and yellow connects to the better antimalarial drugs in Africa. Besides, verbal notes and earlier experiences relatively establish the itch reaction to an applied antihistamine cream [12]. It is not just response but compliance that can be influenced by placebo and nocebo effects. For instance, it is well known that a drug's taste and palatability are key reasons why children don't complete a course of medicines [13]. Poor sensorial perceptions of the product may adversely impact patient adherence and, sequentially, lead to product failure. Accordingly, assessing the sensorial properties of topical products utilizing the senses (including sight, odour, touch, taste and hearing) to elicit, determine, quantify, analyse and understand responses to products/materials [14] is vital in topical product evaluation.

9.2 Therapeutic Performance of Topical Products: A Sensorial Experience

A "high-quality drug product" is defined by the US Food and Drug Administration (FDA) as a contamination-free product that can provide therapeutic benefits to the user, as specified by the label claim [15]. For approved generic drug products by the FDA, they are therapeutically equivalent to a reference-listed drug (RLD) product and are pharmaceutical equivalents for which bioequivalence has been demonstrated [1]. A perception of superior quality, thus, is one key to generic success for topical semisolid products that are applied on the skin, recognizing that placebo effects such as poor texture and stability issues may suggest otherwise. Here, prior experiences, inferior texture quality, appearance, and smell, reinforced by a negative opinion held by both professionals [16] and members of the public [17], may impede their take-up by the market with adverse health costs, even though generics now are being prescribed heavily in many countries in place of the reference products [18]. Also, any notion that generics are less effective than the innovator products will likely lead to reduced acceptance in healthcare provision. This can also trigger a "nocebo" effect.

Realizing that the therapeutic performance of topical products is a sensorial experience can give proper orientation to therapeutic research and outcomes. Changes in topical product composition may impact its physical and structural attributes and, in turn, a patient's perception of the therapeutic effect. Skin bioavailability and safety of active ingredients within a given product, including identifying in vitro methods, can be predictive of in vivo performance in humans. However, for most skin disorders, it is not just the activity that matters. Generally, skin conditions show in the form of dry, wet, crusty, thickened, or erythematous skin; therefore, the presence and functionalities of the inactive ingredients in a product can assist in ameliorating these skin conditions. Some common examples include emollients that provide a soothing effect for irritated/dry skin and enhance the sensory properties of the topical product. APIs may also influence the patient's perception of the texture and feel of a product. Thus, it is the final/full product (with total active and inactive ingredients) produced in a certain way, which defines sensorial/perception equivalence and should be taken into account, along with product bioavailability. This is exceptionally related to where topical products are pricey, and the consumer counts on value for money in not only the product's aesthetic quality but in its therapeutic effect also [19].

It is incumbent on generic manufacturers to make superior quality products and for regulatory bodies to convey this expectation to generic manufacturers, recognizing that this will eventually also change the prescription habits of healthcare professionals and subsequently patient attitudes [20]. A key requirement, therefore, for a generic product is to have similar organoleptic properties (look, feel, touch, perception and odour) as the originator product. This can be achieved by using similar raw materials, both in terms of quality and physicochemical properties, to the reference. Besides, an overall approach or a procedure should be set up to successfully correlate in vitro characterization results with perception data and to understand if characterization of product quality attributes in vitro is a good measure to predict functional and sensorial properties of topical products.

In order to evaluate sensory perceptions, the first step is to identify and define sensorial parameters of topical products that will be perceptible by human subjects. Secondly, there's a need to recognize product quality tests that can be used to characterize the physical attributes in topical dosage forms with different microstructures and are indicative of sensorial aspects. For instance, infinite shear rheology can be used to correlate flow properties during spreading and tribological parameters (friction) to understand the behaviour of products during rubbing. Other critical parameters that are also valuable include the influence of drying, particle size and geometry, and texture on sensorial properties.

9.3 Sensory Testing Topical Products

Topical pharmaceutical products are assessed by touch when a patient spreads a finite amount of dose on the skin; during this application process, the brain subconsciously perceives the look and texture of the product applied, softness, grittiness, spreadability, shine and any secondary sensations that the product may have, such as cooling or warming sensations [8]. Some general perceptions that are more relevant among patients are that ointments are more stable and structured than creams, whereas creams are stronger than gels, or lotions are thinner than creams [7]. Traditionally, four main sets of forces have been used to describe the processes that follow the application of a topical product on the skin [21, 22]: drug active–product, drug active–skin, product–skin and drug active–product–skin interactions. Among them, product-skin interactions play an important role in identifying the specific quality attributes of topical products that may be perceptible by subjects and/or patients. Sensory attributes that can be perceived by patients during and after the application of topical products on the skin are listed in Table 9.1.

Characterization of sensory attributes of topical products has been reported widely [23, 24]. The standard sensory method commonly used is quantitative descriptive profiling, allowing obtaining accurate sensory images of the products [25]. The descriptive method identifies and evaluates the intensity of sensory attributes of the product by a group of assessors [26].

The type of excipients such as texturizing polymers, emulsifiers, emollients, oils and waxes used in the formulation of the different dosage forms (gels, creams, ointments, etc.) will have a significant impact on the sensorial properties of the product [27, 28]. Creams are emulsion microstructures that could either be primarily oil-in-water (O/W) or water-in-oil (W/O). The amount of water in the cream will affect the skin's feel. The water will evaporate from the formulation, resulting in a cooling sensation on the skin. A well-known example is vanishing creams. Emollients and moisturizing agents increase hydration of the skin by increasing occlusivity of the skin. The type of emollient selected in a formulation usually has an impact on the degree of oiliness and the residual film left on the skin [29]. In the case of emulsions, the sensorial emulsifiers largely influence the perception of the product. It has been reported that emulsifiers have a more significant impact on the texture of the final product than emollients [30]. For an ointment, the level of greasiness and the residual film formed at the site of action (thicker film – more occlusive effect) may impact product perception by the patient. Where the ointment base is composed of suspended actives, the degree of dispersion in the base, the particle size and the hardness of the particles can determine the resulting grittiness of the product when applied on the skin. Also, the amount of humectants such as propylene glycol and glycerine and various oils and polymers can affect the stickiness of products. While stickiness is a desirable attribute in terms of efficient contact with skin, patients tend to perceive products with high levels of stickiness

Table 9.1 Exhaustive list of attributes that can be perceived by patients during and after the application of topical products on the skin [31]

Attribute	Definition
Perceivability	The point at which the patient/subject perceives the product on the skin
Hardness	Force exerted to apply the product
Grittiness	Grainy or gritty feel of the product due to the suspended nature of the active or certain excipients
Cooling sensation	This sensation is perceived due to specific excipients incorporated in the product, such as emulsifiers, emollients, and fragrances
Absorption point	Number of rubs (rotations) needed for the product to start to be absorbed by the skin
Spreadability	Ease of spreading the product on the skin
Slipperiness	Ease of sliding the finger over the skin
Stickiness	Degree to which fingers adhere to the skin
Immediate white residue	White film formed on the skin immediately after the spreading of the product
Residual white residue	White film formed on the skin 1 minute after spreading the product
Velvety/soft film	Feeling of softness
Dry touch	Skin non-sticky, non-greasy and dull.
Immediate gloss on skin	Light intensity reflected on the skin immediately after the product spreading
Residual gloss on skin	Light intensity reflected on the skin 2 minutes after spreading the product
Immediate oiliness	Sensation of oil on the skin during and immediately after the product spreading
Residual oiliness	Sensation of oil on the skin 2 minutes after spreading the product
Immediate greasy film	Sensation of grease film, formed on the skin, immediately after the product spreading
Residual greasy film	Sensation of grease film, formed on the skin, 2 minutes after spreading the product

negatively. Similarly, in gels, the texture/hardness/spreadability is determined by the type of polymer and polymer matrix formed.

Figure 9.1 shows mock compositions of a gel and a cream formulation and a representative corresponding sensorial map that highlights the subtle differences between the two products. Responses to pharmaceutical topical products may also be due to a therapeutic or placebo effect. The placebo effect may differ because of odour, spreadability, speed of absorption, creaminess, colour, sheen, etc. There are usually two dimensions of product perception on the skin: when the person sees the product (visual, smell and hearing) and when they apply it (in use).

Standard industry questionnaires for conducting sensorial testing are applicable to sensory analysis in topical bioequivalence testing. For example, a cooling sensation is generally associated with a lotion and not an ointment. Questionnaires for sensorial testing need to be designed so as to remove any sources of ambiguity that can lead to false positives. If the product to be tested is an ointment,

Excipients	Gel	Cream
Wax	-	5
Oil (Liquid paraffin)	-	20
Emulsifier		5
Polymer thickener	3	1.0
Emollient	-	3
Co-solvent	5	5
Preservative mixture	1	1
Water	To 100	To100

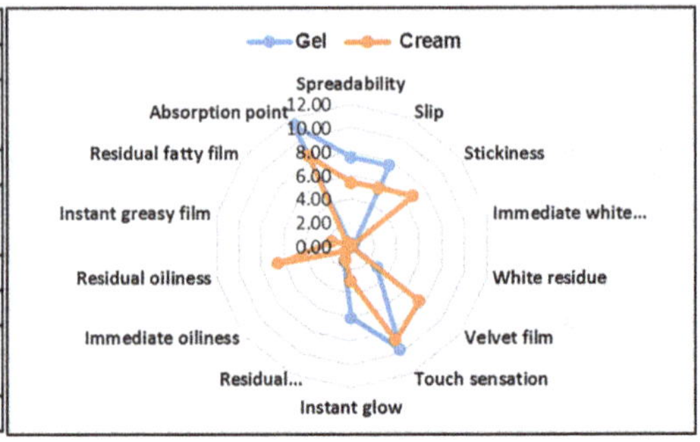

Fig. 9.1 Representative (mock) formulas for gel and cream microstructures with associated representative sensorial maps highlighting the differences in perception of the two products

Table 9.2 Outline of sensorial tests to be performed on various topical products

Sensory tests	*Cream*	*Lotion*	*Gels*	*Ointment*	*Foam*	*Suspensions*
	Emollient properties	Cooling sensation	Non-greasy	Greasy	Easily spreadable	Grittiness
	Hydration	Ease of spreading	Ease of spreading	Residual film after application	Minimal residue	Residual film
	Grittiness	Light feel	Speed of absorption	Hardness/ stiffness	Quick drying	
	Softness		Cooling sensation		Ease of application	
	Texture/ greasiness		Quick drying			

microstructure-specific questionnaires should be designed. So, an ointment may have fewer sensorial tests compared to a complex emulsion where more parameters may be required to sense sensorial differences. Another example is an ointment where spreadability is not a key parameter that needs to be assessed, as they are generally applied to a confined region of the skin to leave a greasy film of activity and to promote occlusion. Being more spreadable may be counterproductive, as this will only lead to easy wash-off and less retention. Instead of spreadability, a more relevant question would be to check for residual film formation. Table 9.2 lists some sensorial tests that will be performed on non-Q1 (qualitatively)/Q2 (quantitatively) products [1] of different microstructures. A descriptive sensory analysis can be conducted with 13–15 assessors or more (no gender bias and individuals 18 years and above), who are trained and validated to evaluate sensory attributes in topical products.

The sensory characterization of skin care products during product application also depends on the physicochemical characteristics of the film left on the skin and its interaction with the skin. These characteristics are time-dependent and vary during application owing to different simultaneous processes, such as variations in rheological properties caused by the shear exerted during the application, loss of water and volatile ingredients due to evaporation, melting of solid ingredients, fluctuations in the structure of emulsions, a mixture of the product with sebum and sweat and adsorption into the skin [32].

9.4 Critical Quality Attributes (CQAs) and Predicting Sensorial Perceptions

Characterization of physical quality related to the sensorial aspects of products can elucidate the relationships between product quality attributes and their functional properties [2, 4, 33]. The quality attribute characterization tests are critical in understanding the anticipated behaviour of products during use and application. Some of the most relevant product quality attributes, or CQAs, that are related to the perception of a product are rheology, tribology and texture. Inactive ingredients such as polymers, waxes, surfactants and rheology modifiers can alter the rheological, mechanical, textural and sensory properties of the formulation [34]. This area has received a great deal of traction in the cosmetic industry, where it is accepted that the application and acceptance (sensory) of cosmetics are related to their flow properties [35, 36].

With this knowledge, it will then be possible to potentially engineer products with desired sensory properties and to predict sensorial differences between two products, for example, a reference and a generic topical product. The sensory judgement of rheological/tribological properties is a well-studied area within food sciences [37] but not extensively studied in the area of topical product development, and there is no established set of rules that link these areas [38]. If the sensorial experience can be efficiently correlated to textural and rheological properties, these methods of analysis can be used to predict the acceptability of pharmaceutical products [39].

Investigating the applicability of characterization techniques for products that are not Q1 (qualitative) and Q2 (quantitative) similar and may have different Q3 (microstructure) attributes that may be outside the range of variation that usually exists among batches of the RLD product is also therefore essential. For example, the texture attributes that play a significant role in the perception of efficacy related to hydration and moisturizing properties of the product and the extent to which product quality attributes (rheology, texture, tribology, particle/globule size, rate of drying, etc.) may differ between non-Q1/Q2 products but still not have a significant impact on perception.

- *Rheology*

For topical pharmaceuticals, rheological characterization is most often related to the physical stability of the product and as a quality control tool for raw materials and finished products [33, 40]. Shear rates relevant to the application of topical

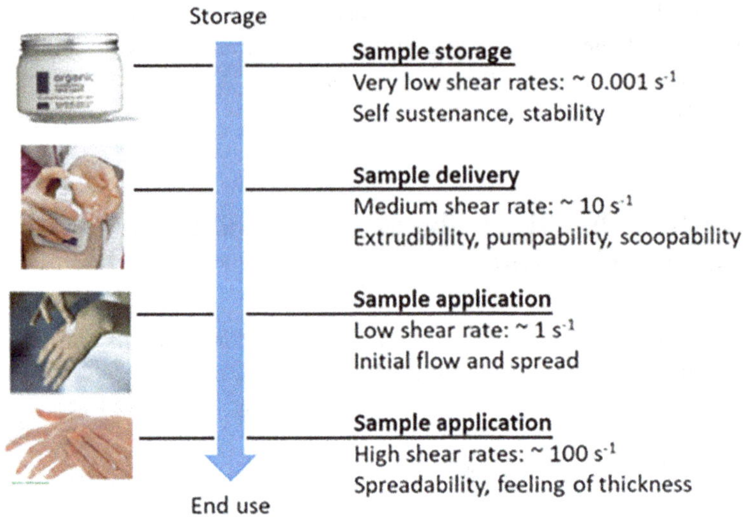

Fig. 9.2 Schematic demonstrating the shear experienced by a topical product from rest to in-use

products could be in the range of 10–100,000 s^{-1} [6, 41–43], while typical shear rates during processing may be 500–1000 s^{-1}, 50–100 s^{-1} when pouring a sample from a bottle, and 100–10,000 s^{-1} when removing a sample from a tube [6, 43]. Typical shear rates a product is subjected to from pick up to product application are illustrated in Fig. 9.2.

Rheometers incorporating both oscillatory as well as continuous shear measurements can be used to quantify viscoelastic behaviour in addition to flow. As critical parameters, rheological differences can indicate unsimilar microstructures of different formulations made by varying the type and quantity of surfactants and viscosity modifiers of products that affect product perception. The parameters of paramount importance include storage (G') and loss (G") modulus, as well as tan delta (tanδ), the linear viscoelastic region (LVR), zero shear (η_0), infinite shear viscosity (η_∞) and the yield stress [4].

The viscosity of a topical product measured at low shear can be related to the excipient composition of a product. The increase in molecular collisions with increasing shear leads to higher viscosities, whereas when the product contains polymers and other excipients that entangle the molecules, shear-induced changes in the network of entanglements lead to molecular disentanglement and their orientation in the shear direction, leading to a relative decrease in viscosity. The overall product viscosity, however, may appear to be the same. This situation is often termed as a plateau value of zero shear viscosity (η_0). With further increases in shear rate, there is a reduction in the resistance to shear and viscosity. After a certain shear rate value is reached, an additional increase of shear rate will not decrease the viscosity. This is termed as the plateau value of infinite shear viscosity (η_∞). As the zero-shear plateau region viscosity is related to the first product pickup, that is when the product first encounters the user, and the infinite shear plateau region is significant of the in-use product behaviour, that is while rubbing the product into the skin and just before it disappears. Relationships between storage modulus (G'), viscosity (η) and tan delta (tanδ) provide a good understanding of the stiffness, spreadability, consistency and absorption properties of the material. The difference between the storage and loss modulus of a product is indicative of the time for absorption of the product in the skin. Yield stress can be considered a good measure to understand the spreadability of topical products and ease of application and thus may be important in explaining patient compliance. Yield stress, along with zero shear viscosity (η_0), can provide a good understanding of a product's inherent strength and microstructure [2, 33, 34, 44–46].

Park and Song [47] observed that the smaller the yield stress, the easier the product can be spread on the human body in a thinner film. Once the system starts to flow, the product will rub onto the applied area, and this is manifested as shear thinning and thixotropic behaviour, which can be measured with the hysteresis area value. Differences in zero shear viscosity (η_0) between products of similar or different microstructures would suggest differences in resistance to flow and deformation possibly relating to sensorial differences in texture, stiffness and spreading. Similar findings have been reported in the literature by Parente et al. [48] and Lukic et al. [49]. They found that sensory attributes relating to difficulty of spreading, stickiness and slipperiness were related to the viscosity of the products. Wegener [38] also reported positive correlations between sensory tackiness, stickiness and viscosity quantified at high shear rates (10,000 s^{-1}) and negative relations between sensory spreadability, oiliness and viscosity.

- *Tribology*

Significantly, tribology (the science of friction) is also a useful tool to measure skin feel and investigate microstructural differences between reference and generic topical products and between products of different microstructures. Tribology essentially is the scientific study of interactions between contacting surfaces that are in relative motion. This multidisciplinary branch incorporates three dimensions: friction (physics and mechanical engineering), lubrication (mechanical engineering and chemistry) and wear (materials science) [50]. A tribometer can be used to characterize the lubrication properties of the compositions made by varying the amount and type of emollients and sensorial emulsifiers.

The "Stribeck curve", a classical product friction response to the speed of rotation, is normally used to characterize the tribology of a product. Different regimes of lubrication can be determined via analysis of Stribeck curves where the friction coefficient versus the sliding speed (speed/load) or film thickness (film thickness/roughness height) are depicted [51]. Different regions in the Stribeck curve can also relate to perceptive sensory responses. For example, a longer transition region may correlate with the perception of the smoothness of a product. The length of the thin film region may also be directly proportional to the time taken for the product to completely disappear on rubbing, and an increase in the rise of the hydrodynamic region could correspond to an increase in friction perceived by

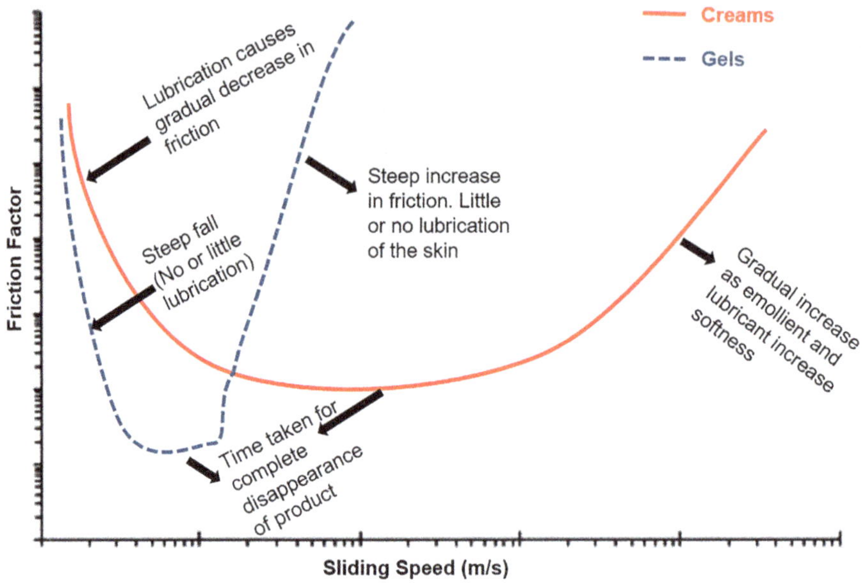

Fig. 9.3 Friction curves for creams and gels depicting differences in lubrication properties on application to skin

the subjects. Figure 9.3 illustrates two carefully chosen examples to demonstrate how this can be achieved. At low sliding speeds, the cream example (solid red line) is associated with a steady decrease in friction factor as the sliding speed (this is analogous to shear rate) increases on a tribometer. Here, the product is being forced into motion. On the other hand, in this region, the friction factor in gels (dashed blue line) shows a much steeper decrease. It is apparent that it is therefore much easier to spread the gel. In addition, there is negligible lubrication from the product itself. The middle region of the sliding curves corresponds to the bottom of the curves, where friction initially decreases and then starts to rise again. The extent of this region is directly proportional to the amount of time taken for the product to be completely rubbed in and then disappear. Due to better lubricant properties provided by the oils and other ingredients, such as surfactants and silicones, present in the cream, this region is often larger in creams than in gels. Hence, in these examples, it will take longer for the cream to completely disappear on rubbing than the gel. The last phase in the curve is the region of increasing friction after the product has been completely rubbed in and there is rubbing of skin against the skin with no product in between. Creams show a slow and gradual increase as the product has adequately lubricated the skin. This is what contributes to the silky-smooth feel of a product. On the other hand, gels experience rapid water and volatile loss. As they will usually contain no other lubricants such as oils, the friction measured by the tribometer increases rapidly with increasing sliding speed.

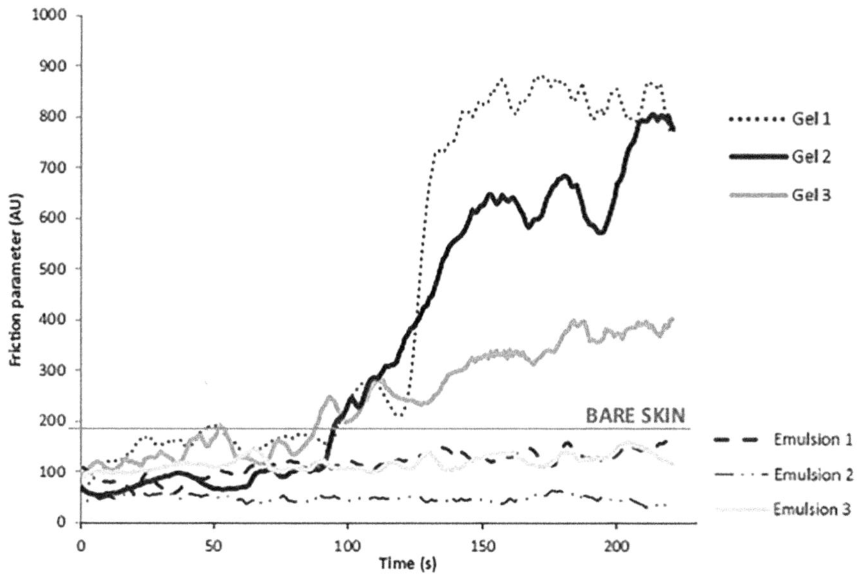

Fig. 9.4 Plot of friction parameter with time of different skin products [46]

Friction parameters can be directly related to the slipperiness, spreading behaviour and formation of a residual film after application of the product on the skin. Figure 9.4 illustrates the results of a study by Savary et al. [46], who compared the evolution of the friction curve with time in a series of gels and emulsions. A rapid increase in the friction parameter was seen for the gels, compared to the emulsion formulations. The friction parameter was stable initially because of the thickness of the film applied on the skin, but due to water and volatile loss and penetration into the skin, the friction drastically increased, as shown by the steep increases in the curves for the gels.

Products with different microstructures may have different friction parameters during spreading and under in-use conditions. Emulsions may exhibit higher lubrication due to the presence of oil and emollients, whereas gels may not result in a residual film. The perception of smoothness or roughness is predicted by the friction coefficient values. A low friction coefficient value refers to the ability of lubricant/fat to create a thin film layer in the contact zone and reduce the friction coefficient during rubbing. The extension of the boundary regime region is due to the coalescence of fat droplets adsorbed on the substrate to create bigger droplets and finally merge together to make a continuous thin film in the contact zone [52].

A convenient distinction may be made between the rheological and tribological aspects of semisolid products. Rheology is the study of bulk flow properties of fluids, and tribology evaluates the lubricating effects of fluids when present in a thin layer between sliding surfaces [40]. The viscosity of the product (bulk rheological properties) affects the hydrodynamic regime of the tribological curve. The increase

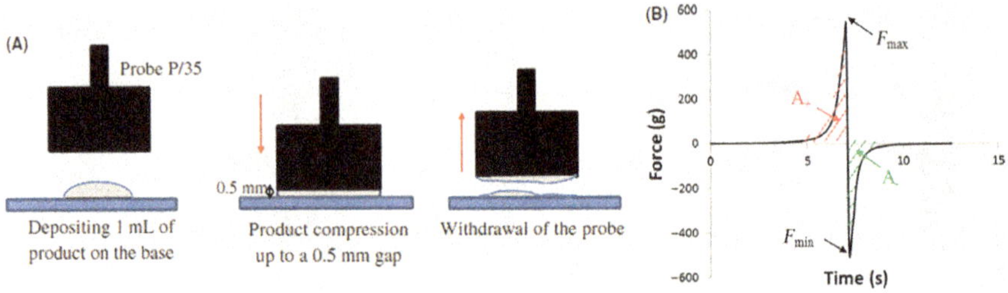

Fig. 9.5 Diagram of a compression test (**a**) and representative texture curve describing the position of the texture probe at different sampling points (**b**) [46]

of friction in the hydrodynamic regime may be governed by fluid viscosity and closely related to the gel structure and viscosity of the product [53, 54]. Correlations of the rheological data can be made to the tribological properties of the product so as to be able to make a link with the sensory aspects.

- *Texture Analysis*

Texture analysis to characterize the different forces holding the semisolid structure together can be conducted using a texture analyser. Using compression, a texture analyser can characterize different texture parameters of the semisolid sample by collecting data on force versus time [55]. On lowering the probe into the sample, the forces measured are utilized to determine the firmness/hardness and consistency of the samples; on withdrawal of the probe from the samples, the index of viscosity and adhesiveness is recorded. Figure 9.5 is a representation of a typical texture profile depicted by Savary et al. [46]. Various compression parameters could be obtained from the curves force = f (time). The maximum and minimum compression forces and the positive area correspond to the total work of compression, while the negative area under the curve corresponds to the work necessary for the texture probe to return to its initial position [46]. Texture parameters most relevant to topical dermatological formulations include adhesiveness, cohesiveness, hardness and consistency. These properties affect the applicability of the formulation at the administration site and patient adherence. The texture parameter of hardness relates to the need to exert force to apply the product. The total positive portion of the area under the curve can correlate with consistency, and adhesiveness relates to the stickiness that a patient may experience when applying the product [55].

Multiple compression allows for the topical product to be analysed under conditions like patient application when the product is repeatedly being rubbed into the site of application. Semisolids may behave differently after repeated applications of force. The texture parameter, cohesiveness, measured using double compression, can relate to how well the product holds its structure when the user applies forces during application. The lotions, gels and creams could display less cohesiveness than the ointments, meaning that the formulations' internal bonds are weaker than in the ointments. A patient might experience this as the ointment is less pliable and is difficult to spread on the area of application. Texture parameters can also be correlated with rheology data to understand the hardness, adhesiveness, consistency and spreading behaviour of the different compositions and dosage forms. It has been reported that viscosity, firmness and adhesiveness vary in a proportional manner, and textural properties could be regarded as a manifestation of the rheological properties of a product [56].

- *Measurement of water loss/rate of drying, water activity and hydration*

The evaporation of water and volatiles can cause changes in the microstructure of topical products and crystallization of actives [57], which may affect the feel and sensorial characteristics of the products. The readings of water loss from the different formulations, in vitro after the application of products to ex vivo skin and in vivo on volunteers, demonstrating collated with water activity measurements, can create a correlation between instrumental measurements and test subjects' cooling sensation. Infrared cameras and thermometers can be used to investigate cooling (reduction in skin surface temperature) induced by drugs and excipients. Inactive ingredients such as emulsifiers/surfactants, oils, humectants, solvents/solubilizers, emollients, polymers and other materials incorporated in topical dosage forms can affect evaporation of water and volatiles, frictional properties and sensory aspects of the product [2, 3].

- *Particle size and geometry/globule size*

Particle properties such as size and shape in a topical product can have a significant impact on rheological properties and hence on sensorial parameters. Regularly shaped particles are resistant to deform at higher stress due to the ease of gliding over one another, whereas irregularly sized particles struggle to glide over each other at higher shear stresses, leading to changes in viscosity as the stress increases [58]. The relatively orderly arrangement of globules within a gel formulation is related to the common incorporation of polymer gel networks within gels [27]. Lotions display coarser globules, which may possibly be linked to their water evaporation. The mean globule diameters of the ointments and gels were smaller than those of lotions and creams, but the ointments and gels exhibited higher viscosity readings [59]. Uniformly sized globules pack more tightly, creating less slip and have a higher resistance to flow, which resulted in the samples' higher viscosity readings. This results in a reduction in resistance to flow under stress, which was seen in the samples' lower viscosity readings [60]. By varying the globule sizes by either using specific excipients such as emulsifiers and oils or by modifying the processing parameters such as speed and time of homogenization, variants with different globule sizes can be manufactured, followed by systematic physical attribute characterization and connected with sensory testing.

Fig. 9.6 Schematic representation of the potential relationship between quality attributes and sensorial properties

Figure 9.6 is a schematic of the quality tests that can be performed on the different formulations, the parameters to be measured and the relationships that could exist between a sensorial attribute and quality test parameter. In general, to undertake correlations between the different parameters measured by product quality tests and to relate them to sensory perception, sensitivity analysis and correlation matrix analysis should be performed to understand the inter-dependency and individuality of each product quality test parameter on sensorial attributes, as shown in Fig. 9.6. It is also valuable to understand how far different topical products incorporating excipients can lead to differences in CQAs (e.g. rheological, textural and tribological) behaviour that can affect human sensory perceptions.

9.5 Conclusion

Human sensorial panel tests for topical drug products, like any human subject study, can have challenges in both training human subjects and the subjective nature of findings. Understanding in depth the characterization of product quality attributes in vitro is a good measure to predict functional and sensorial properties of topical products and should be very beneficial to the industry, as it would allow predictions to be made about generic product sensory attributes (during pick up, spreading and rub-in) from instrumental data only, thus saving time and money, particularly in the context of consumer perception assessments. These outcomes are not only important in designing in vitro characterizing products that are predictive of sensorial performance but also in developing regulatory guidelines that include both therapeutic and sensory equivalence.

References

1. Ghosh, P., Raney, S. G., & Luke, M. C. (2022). How Does the Food and Drug Administration Approve Topical Generic Drugs Applied to the Skin? *Dermatologic Clinics, 40*(3), 279–287. doi:https://doi.org/10.1016/j.det.2022.02.003
2. Namjoshi, S., Dabbaghi, M., Roberts, M. S., Grice, J. E., & Mohammed, Y. (2020). Quality by Design: Development of the Quality Target Product Profile (QTPP) for Semisolid Topical Products. *Pharmaceutics, 12*(3). doi:https://doi.org/10.3390/pharmaceutics12030287
3. Chang, R. K., Raw, A., Lionberger, R., & Yu, L. (2013). Generic development of topical dermatologic products, Part II: quality by design for topical semisolid products. *The AAPS Journal, 15*(3), 674–683. doi:https://doi.org/10.1208/s12248-013-9472-8
4. Dabbaghi, M., Namjoshi, S., Panchal, B., Grice, J. E., Prakash, S., Roberts, M. S., & Mohammed, Y. (2021). Viscoelastic and deformation characteristics of structurally different commercial topical systems. *Pharmaceutics, 13*(9). doi:https://doi.org/10.3390/pharmaceutics13091351
5. Jin, X., Imran, M., & Mohammed, Y. (2022). Topical Semisolid Products-Understanding the Impact of Metamorphosis on Skin Penetration and Physicochemical Properties. *Pharmaceutics, 14*(11). doi:https://doi.org/10.3390/pharmaceutics14112487
6. Barry, B., & Grace, A. (1972). Sensory testing of spreadability: investigation of rheological conditions operative during application of topical preparations. *Journal of pharmaceutical sciences, 61*(3), 335–341.
7. Brummer, R. (2006). *Rheology essentials of cosmetic and food emulsions*: Springer Science & Business Media.
8. Marriott, R. (1961). Rheological Measurements in the Cosmetic Industry. *Journal of the Society of Cosmetic Chemists, 12*, 89.
9. Cohen, T. F. (2014). The Power of Drug Color. Retrieved from https://www.theatlantic.com/health/archive/2014/10/the-power-of-drug-color/381156/
10. Rovensky, J., Micekova, D., Gubzova, Z., Fimmers, R., Lenhard, G., Vogtle-Junkert, U., & Schreyger, F. (2001). Treatment of knee osteoarthritis with a topical non-steroidal antiinflammatory drug. Results of a randomized, double-blind, placebo-controlled study on the efficacy and safety of a 5% ibuprofen cream. *Drugs Under Experimental and Clinical Research, 27*(5-6), 209–221. Retrieved from https://www.ncbi.nlm.nih.gov/pubmed/11951579
11. Dreno, B., Bettoli, V., Ochsendorf, F., Layton, A. M., Perez, M., Dakovic, R., & Gollnick, H. (2014). Efficacy and safety of clindamycin phosphate 1.2%/tretinoin 0.025% formulation for

the treatment of acne vulgaris: pooled analysis of data from three randomised, double-blind, parallel-group, phase III studies. *European Journal of Dermatology, 24*(2), 201–209. doi:https://doi.org/10.1684/ejd.2014.2293

12. Darragh, M., Chang, J. W., Booth, R. J., & Consedine, N. S. (2015). The placebo effect in inflammatory skin reactions: the influence of verbal suggestion on itch and weal size. *Journal of Psychosomatic Research, 78*(5), 489–494. doi:https://doi.org/10.1016/j.jpsychores.2015.01.011
13. Mennella, J. A., Spector, A. C., Reed, D. R., & Coldwell, S. E. (2013). The bad taste of medicines: overview of basic research on bitter taste. *Clinical Therapeutics, 35*(8), 1225–1246. doi:https://doi.org/10.1016/j.clinthera.2013.06.007
14. Kemp, S. E. (2009). *Sensory evaluation: a practical handbook*: Oxford: Wiley-Blackwell.
15. Kimball, M. (2017). Manufacturing Topical Formulations: Scale-up from Lab to Pilot Production. In D. N. (Ed.), *Handbook of Formulating Dermal Applications: A Defi nitive Practical Guide*: Wiley: Hoboken, NJ, USA.
16. Shrank, W. H., Liberman, J. N., Fischer, M. A., Girdish, C., Brennan, T. A., & Choudhry, N. K. (2011). Physician perceptions about generic drugs. *Ann Pharmacother, 45*(1), 31–38. doi:https://doi.org/10.1345/aph.1P389
17. Rathe, J., Larsen, P., Andersen, M., Paulsen, M., Jarbol, D., Thomsen, J., & Soendergaard, J. (2013). Associations between generic substitution and patients' attitudes, beliefs and experiences. *European Journal of Clinical Pharmacology, 69*(10), 1827–1836. doi:https://doi.org/10.1007/s00228-013-1539-z
18. Andersson, K., Sonesson, C., Petzold, M., Carlsten, A., & Lonnroth, K. (2005). What are the obstacles to generic substitution? An assessment of the behaviour of prescribers, patients and pharmacies during the first year of generic substitution in Sweden. *Pharmacoepidemiol Drug Safety, 14*(5), 341–348. doi:https://doi.org/10.1002/pds.1055
19. Mohammed, Y., Holmes, A., Kwok, P. C. L., Kumeria, T., Namjoshi, S., Imran, M., Matteucci, L., Ali, M., Tai, W., Benson, H. A. E., & Roberts, M. S. (2022). Advances and future perspectives in epithelial drug delivery. *Adv Drug Deliv Rev, 186*, 114293. https://doi.org/10.1016/j.addr.2022.11429359
20. Dunne, S. S., & Dunne, C. P. (2015). What do people really think of generic medicines? A systematic review and critical appraisal of literature on stakeholder perceptions of generic drugs. *BMC Medicine, 13*, 173. doi:https://doi.org/10.1186/s12916-015-0415-3
21. Poulsen, B. J., Young, E., Coquilla, V., & Katz, M. (1968). Effect of topical vehicle composition on the in vitro release of fluocinolone acetonide and its acetate ester. *Journal of pharmaceutical sciences, 57*(6), 928–933. Retrieved from https://www.ncbi.nlm.nih.gov/pubmed/5671339
22. Roberts, M. S., Triggs, E. J., & Anderson, R. A. (1975). Permeability of Solutes through Biological-Membranes Measured by a Desorption Technique. *Nature, 257*(5523), 225–227. doi:https://doi.org/10.1038/257225a0
23. Dooley, L. M., Adhikari, K., & Chambers, E. (2009). A General Lexicon for Sensory Analysis of Texture and Appearance of Lip Products. *Journal of Sensory Studies, 24*(4), 581–600. doi:https://doi.org/10.1111/j.1745-459X.2009.00227.x
24. Estanqueiro, M., Amaral, M. H., & Sousa Lobo, J. M. (2016). Comparison between sensory and instrumental characterization of topical formulations: impact of thickening agents. *International journal of cosmetic science, 38*(4), 389–398. doi:https://doi.org/10.1111/ics.12302
25. Standardization-ISO, I. O. f. (2003). Sensory analysis—Methodology—General guidance for establishing a sensory profile. In. British.
26. Meilgaard, M. C., Carr, B. T., & Civille, G. V. (1999). *Sensory evaluation techniques*: CRC press.
27. Djekic, L., Martinovic, M., Stepanovic-Petrovic, R., Micov, A., Tomic, M., & Primorac, M. (2016). Formulation of hydrogel-thickened nonionic microemulsions with enhanced percutaneous delivery of ibuprofen assessed in vivo in rats. *European Journal of Pharmaceutical Sciences, 92*, 255–265. doi:https://doi.org/10.1016/j.ejps.2016.05.005
28. Kwon, S. S., Kong, B. J., & Park, S. N. (2015). Physicochemical properties of pH-sensitive hydrogels based on hydroxyethyl cellulose-hyaluronic acid and for applications as transdermal

delivery systems for skin lesions. *European Journal of Pharmaceutics and Biopharmaceutics, 92*, 146–154. doi:https://doi.org/10.1016/j.ejpb.2015.02.025
29. Salka, B. (1997). Choosing emollients: Four factors will help you decide. *Cosmetics and toiletries, 112*(10), 101-106.
30. Wiechers, J., Taelman, M., Wortel, V., Verboom, C., & Dederen, J. (2002). Emulsifiers and emollients exert their sensory impact in different phases of the sensory evaluation process but how does one demonstrate the absence of such an influence. *IFSCC Magazine, 5*(2), 99–105.
31. Benson, H. A. E. (2019). Cosmetic formulation: principles and practice. CRC Press Taylor and Francis Group.
32. Gore, E., Picard, C., & Savary, G. (2018). Spreading behavior of cosmetic emulsions: Impact of the oil phase. *Biotribology, 16*, 17–24. doi:https://doi.org/10.1016/j.biotri.2018.09.003
33. Cyriac, F., Xin Yi, T., Chow, P. S., & Macbeath, C. (2022). Tactile friction and rheological studies to objectify sensory properties of topical formulations. *Journal of Rheology, 66*(2), 305–326. doi:https://doi.org/10.1122/8.0000341
34. Calixto, L. S., Infante, V. H. P., & Campos, P. M. B. G. M. (2018). Design and characterization of topical formulations: Correlations between instrumental and sensorial measurements. *Aaps Pharmscitech, 19*(4), 1512–1519. doi:https://doi.org/10.1208/s12249-018-0960-0
35. Gilbert, L., Picard, C., Savary, G., & Grisel, M. (2013). Rheological and textural characterization of cosmetic emulsions containing natural and synthetic polymers: relationships between both data. *Colloids and Surfaces A: Physicochemical and Engineering Aspects, 421*, 150–163. doi:https://doi.org/10.1016/j.colsurfa.2013.01.003
36. Kusakari, K., Yoshida, M., Matsuzaki, F., Yanaki, T., Fukui, H., & Date, M. (2003). Evaluation of post-application rheological changes in cosmetics using a novel measuring device: Relationship to sensory evaluation. *Journal of Cosmetic Science, 54*, 321–333.
37. Godoi, F. C., Bhandari, B. R., & Prakash, S. (2017). Tribo-rheology and sensory analysis of a dairy semi-solid. *Food Hydrocolloids, 70*, 240–250. doi:https://doi.org/10.1016/j.foodhyd.2017.04.011
38. Wegener, M. R. (1997). *A psycho-rheological study of skin-feel.* (Doctoral Thesis). University of Bristol
39. Krishnaiah, Y. S., Xu, X., Rahman, Z., Yang, Y., Katragadda, U., Lionberger, R., ... Khan, M. A. (2014). Development of performance matrix for generic product equivalence of acyclovir topical creams. *International Journal of Pharmaceutics, 475*(1–2), 110–122. doi:https://doi.org/10.1016/j.ijpharm.2014.07.034
40. Guest, S., McGlone, F., Hopkinson, A., Schendel, Z. A., Blot, K., & Essick, G. (2013). Perceptual and sensory functional consequences of skin care products. *Journal of Cosmetics, Dermatological Sciences and Applications, 3*(01), 66.
41. Henderson, N., Kostenbauder, H. B., & Meer, P. M. (1961). Approximate Rates of Shear Encountered in Some Pharmaceutical Processes. *Journal of pharmaceutical sciences, 50*(9), 788. doi:https://doi.org/10.1002/jps.2600500920
42. Langenbucher, F., & Lange, B. (1970). Prediction of the application behaviour of cosmetics from rheological measurements. *Pharmaceutica Acta Helvetiae, 45*(9), 572–582. Retrieved from https://www.ncbi.nlm.nih.gov/pubmed/5469387
43. Wibowo, C., & Ng, K. M. (2001). Product-oriented process synthesis and development: creams and pastes. *AIChE Journal, 47*(12), 2746–2767.
44. Adeyeye, M. C., Jain, A. C., Ghorab, M. K., & Reilly, W. J., Jr. (2002). Viscoelastic evaluation of topical creams containing microcrystalline cellulose/sodium carboxymethyl cellulose as stabilizer. *Aaps Pharmscitech, 3*(2), E8. doi:https://doi.org/10.1208/pt030208
45. Ozkan, S., Gillece, T. W., Senak, L., & Moore, D. J. (2012). Characterization of yield stress and slip behaviour of skin/hair care gels using steady flow and Laos measurements and their correlation with sensorial attributes. *International journal of cosmetic science, 34*(2), 193–201. doi:https://doi.org/10.1111/j.1468-2494.2012.00702.x
46. Savary, G., Gilbert, L., Grisel, M., & Picard, C. (2019). Instrumental and sensory methodologies to characterize the residual film of topical products applied to skin. *Skin Research and Technology*. doi:https://doi.org/10.1111/srt.12667

47. Park, E. K., & Song, K. W. (2010). Rheological evaluation of petroleum jelly as a base material in ointment and cream formulations: steady shear flow behavior. *Archives of Pharmacal Research, 33*(1), 141–150. doi:https://doi.org/10.1007/s12272-010-2236-4
48. Parente, M. E., Gambaro, A., & Ares, G. (2008). Sensory characterization of emollients. *Journal of Sensory Studies, 23*(2), 149–161. doi:https://doi.org/10.1111/j.1745-459X.2007.00136.x
49. Lukic, M., Jaksic, I., Krstonosic, V., Cekic, N., & Savic, S. (2012). A combined approach in characterization of an effective w/o hand cream: the influence of emollient on textural, sensorial and in vivo skin performance. *International journal of cosmetic science, 34*(2), 140–149.
50. Gohar, R., & Rahnejat, H. (2008). Fundamentals of tribology. In: Imperial College Press, London.
51. Dresselhuis, D. M., Klok, H. J., Stuart, M. A. C., de Vries, R. J., van Aken, G. A., & de Hoog, E. H. A. (2007). Tribology of o/w emulsions under mouth-like conditions: Determinants of friction. *Food Biophysics, 2*(4), 158–171. doi:https://doi.org/10.1007/s11483-007-9040-9
52. Nguyen, P. T., Bhandari, B., & Prakash, S. (2016). Tribological method to measure lubricating properties of dairy products. *Journal of Food Engineering, 168*, 27–34.
53. Nguyen, P. T., Kravchuk, O., Bhandari, B., & Prakash, S. (2017). Effect of different hydrocolloids on texture, rheology, tribology and sensory perception of texture and mouthfeel of low-fat pot-set yoghurt. *Food Hydrocolloids, 72*, 90–104.
54. Prakash, S., Tan, D. D. Y., & Chen, J. (2013). Applications of tribology in studying food oral processing and texture perception. *Food Research International, 54*(2), 1627–1635.
55. Gratieri, T., Gelfuso, G. M., Rocha, E. M., Sarmento, V. H., de Freitas, O., & Lopez, R. F. (2010). A poloxamer/chitosan in situ forming gel with prolonged retention time for ocular delivery. *European Journal of Pharmaceutics and Biopharmaceutics, 75*(2), 186–193. doi:https://doi.org/10.1016/j.ejpb.2010.02.011
56. Almeida, I. F., & Bahia, M. F. (2006). Evaluation of the physical stability of two oleogels. *International Journal of Pharmaceutics, 327*(1–2), 73–77. doi:https://doi.org/10.1016/j.ijpharm.2006.07.036
57. Jin, X., Alavi, S. E., Shafiee, A., Leite-Silva, V. R., Khosrotehrani, K., & Mohammed, Y. (2023). Metamorphosis of Topical Semisolid Products-Understanding the Role of Rheological Properties in Drug Permeation under the "in Use" Condition. *Pharmaceutics, 15*(6). doi:https://doi.org/10.3390/pharmaceutics15061707
58. Mastropietro, D. J., Nimroozi, R., & Omidian, H. (2013). Rheology in pharmaceutical formulations-A perspective. *Journal of Developing Drugs, 2*, 108–116.
59. Pal, R. (1996). Effect of droplet size on the rheology of emulsions. *AIChE Journal, 42*(11), 3181–3190. doi:https://doi.org/10.1002/aic.690421119
60. Higashi, N., & Sumita, I. (2009). Experiments on granular rheology: Effects of particle size and fluid viscosity. *Journal of Geophysical Research-Solid Earth, 114*. doi:Artn B04413 https://doi.org/10.1029/2008jb005999

Chapter 10
Novel Formulation Approaches in the Topical Delivery System

Nisha Shrestha and Ajay K. Banga

10.1 Introduction

Skin, the largest organ in the body, is a natural physical barrier that protects the body from the outer environment, microbial infections, and other hazards like UV radiation. Any injury, damage, infection, or aberrant changes in skin physiology may compromise its barrier properties and may cause redness, inflammation, itching, deformation, and pigmentation, resulting in cutaneous disorders such as acne, psoriasis, eczema, atopic dermatitis, leprosy, vitiligo, and ichthyosis. The topical administration of therapeutic agents is a practical approach to exert local action and treat such cutaneous diseases. The wide surface area of the skin provides a conventional route for many drugs, improving drug absorption and facilitating the efficient distribution of drugs into the layers of the skin [1, 2]. Topical drug delivery has several advantages, such as ease of application, minimum systemic exposure, and enhanced patient compliance over oral, parenteral, and inhalation routes. However, the primary barrier, stratum corneum, limits the permeability of active pharmaceutical ingredients (APIs) into the skin, thus posing a significant challenge in formulating topical preparations. This limitation varies based on the physicochemical characteristics of the APIs, such as their molecular weight and lipophilicity. Moreover, the changes in skin permeability and barrier integrity in cutaneous disorders are also considered in developing an effective topical formulation [3–5].

Topical preparations are available in various forms, such as gels, creams, ointments, lotions, solutions, and suspensions, classified as conventional/traditional

N. Shrestha · A. K. Banga (✉)
Center for Drug Delivery Research, Department of Pharmaceutical Sciences, College of Pharmacy, Mercer University, Atlanta, GA, USA
e-mail: banga_ak@mercer.edu

S. N. Murthy (ed.), *Topical Products and Dermal Drug Delivery*,
AAPS Introductions in the Pharmaceutical Sciences 2,
https://doi.org/10.1007/978-3-031-80525-7_10

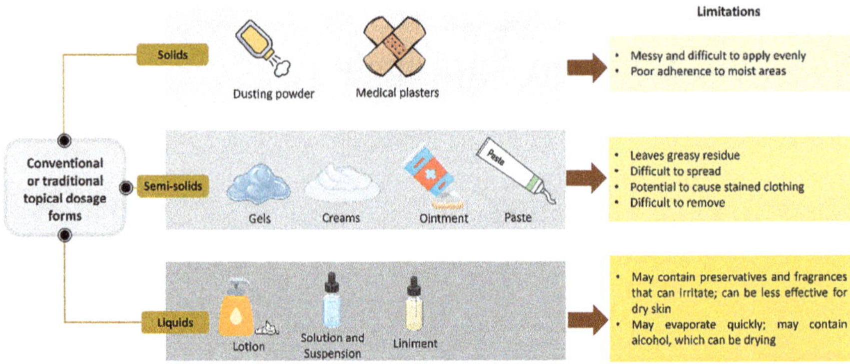

Fig. 10.1 Conventional or traditional topical dosage forms

topical dosage forms, as depicted in Fig. 10.1. The APIs intended for topical application often may not have the ideal physicochemical properties to facilitate the ease of diffusion into the skin. The formulation of topical dosage forms can be further challenging as different excipients are incorporated along with APIs to form a physically and chemically stable topical preparation. Besides the long-standing use of chemical penetration enhancers and physical enhancement techniques to bypass the stratum corneum barrier, approaches to developing novel topical formulations are gaining prominence [6]. These novel systems have different compositions and structures that enhance topical drug delivery without compromising skin barrier integrity. Further, novel delivery strategies play a pivotal role in optimizing the efficacy of therapeutic agents by enhancing the permeability of drugs, improving the stability of the APIs, and possibly reducing associated side effects, thereby improving patient compliance [5]. The underlying rationale behind applying various innovative strategies in topical delivery involves ease of penetration across the stratum corneum and utilization of the transfollicular pathway for efficient drug delivery [6].

Conventional topical dosage forms, although widely used, have certain limitations affecting their effectiveness and impact on patient outcomes. These limitations include limited drug penetration into deeper skin layers, potential skin irritation and sensitization, stability concerns including phase separation and drug instability, viscosity issues, pH changes, lack of drug-vehicle solubility, and limited duration of action [5, 7, 8]. Stability issues, such as phase separation in creams and lotions, can be mitigated through advanced delivery systems like liposomes and microemulsions [9, 10]. Similarly, viscosity challenges in topical solutions and suspensions can be overcome by employing formulations such as organogels and hydrogels, ensuring a prolonged and consistent adherence to the skin [11, 12]. Thus, developing a successful topical dermatological product is a challenging process that requires consideration of various processing parameters and critical attributes to improve its performance. Further, despite their potential advantages, other delivery strategies, such as cyclodextrins and polymeric micelles, exhibit limitations such as drug incompatibility, reduced drug loading, instability in topical formulations, and decreased release kinetics [13, 14]. Therefore, new innovative approaches are needed to mitigate these limitations and enhance the therapeutic effectiveness of medications. Various innovative topical delivery systems have been investigated over time that will be discussed in this chapter.

10.2 Novel Approaches in the Topical Delivery System

In this chapter, we have discussed the novel delivery systems for topical delivery of drugs, which include foams, organogels, hydrogels, microsponges, liposomes, microemulsions, solid lipid nanoparticles, dendrimers, and nanofibers, as shown in Fig. 10.2. The influence of the interaction of the novel topical formulations in the skin and their impact on the rate of release of the drug for optimal absorption into the skin are also discussed in this chapter of the book.

10.2.1 Foams

Pharmaceutical foams are pressurized dosage forms containing one or more APIs. API formulation is emitted as a fine dispersion of liquid or solid materials in a gaseous medium upon valve actuation. The formulation exists under pressure inside the container as one or multiple phases that can constitute a solution, suspension, or emulsion. Foams are a growing interest as topical delivery vehicles due to their apparent advantages over conventional topical dosage forms like ointments,

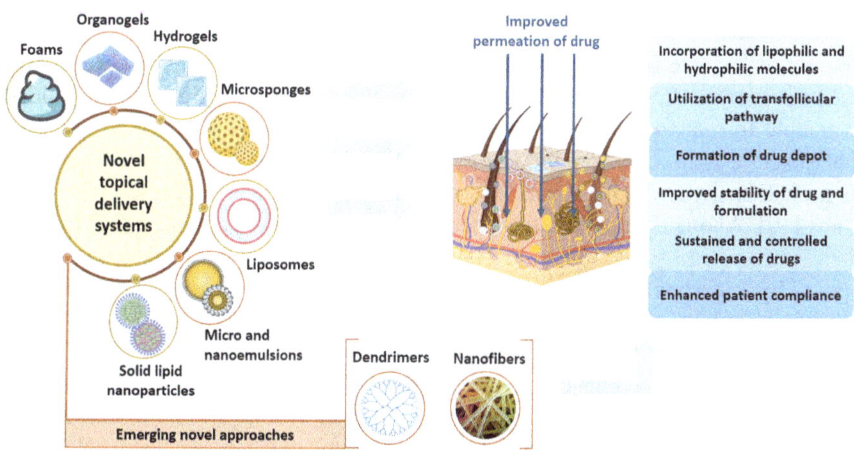

Fig. 10.2 Novel topical delivery systems

creams, and gels, as they eliminate negative sensory attributes such as greasiness or tackiness, resulting in higher cosmetic appeal. These can also be applied and spread more readily with minimal mechanical shearing, which makes them simpler to use on sensitive and inflamed skin [15, 16]. For example, patients with psoriasis adhered to calcipotriol/betamethasone dipropionate foam because it was easier to apply on the affected skin area and required less amount of formulation than ointments and gel, thus increasing the efficacy of the treatment therapy [17]. Foams are employed to deliver various therapeutic substances, including antiseptics, corticosteroids, antibiotics, antifungal medications, and anti-inflammatory agents. A few commercially available foam formulations include tazarotene, ketoconazole, azelaic acid, and clindamycin for treating skin-related conditions like psoriasis, acne vulgaris, and rosacea, respectively. Recently, the FDA approved a topical foam for roflumilast to treat seborrheic dermatitis [18]. Foams are also widely utilized in cosmetic formulation applications, particularly in formulations for skincare and haircare products [19].

Foam formulation includes propellants, which are the primary components of topical foams, as they generate a flexible matrix of liquid or semi-solid foam upon evaporation. As shown in Table 10.1, selecting a suitable propellant can be difficult despite the availability of various types of propellants. The use of chlorofluorocarbons (CFCs) in producing topical foams is prohibited because of their environmental risks. Currently, hydrochlorofluorocarbons (HFAs) are more commonly employed and investigated in foam formulation owing to their advantages, including low flammability and reduced safety risks [20]. Further, hydrocarbons such as propane and butane are frequently utilized in many commercial formulations to generate good-quality foam. However, they pose a possible safety risk due to their high flammability and explosivity. Propellant-free foam technologies are also gaining prominence due to their advantages of being environmentally safe. Moreover, they can also ensure uniform and controlled delivery, which may be challenging in aerosol foam due to the risk of drug-device combination failure [21]. We developed a propellant-free foam formulation that effectively decontaminates a significant amount of lewisite from the skin during exposure to chemical warfare agents. We used a propellant-free pump device in addition to surfactants, which primarily aided in bubble formation [22]. Airspray® pump foam dispensers are patented devices that generate foam upon mixing liquid and air without gas propellants. Compared to propellants, the foams generated with this device are less viscous with finer pores [19]. Another propellant-free approach is a "bag-in-can" system that forms a foam when a semi-solid gel is rubbed on the body. The formulation contains a low-boiling hydrocarbon like isopentane (boiling point ~28 °C), which evaporates upon application at body temperature, forming a gel. This ensures ease of application and retention to this skin, thereby increasing the efficacy of the treatment [23].

Foams are thermodynamically and mechanically unstable systems. Thus, surfactants are incorporated in foam formulation. They function as foaming agents and are essential for foam generation and stabilization when the dose is actuated. When the valve is actuated, the propellants evaporate rapidly, and the foam excipients interact with air, generating air bubbles and creating a new air/water interface.

Table 10.1 Excipients in topical foam formulations

Excipients	Examples [20, 22, 26]
Propellants	Hydrocarbons, hydrochlorofluorocarbons, dimethyl ether, and hydrofluoroalkane
Surfactants [26]	Nonionic: Polysorbates (Tween® 20/40/60/80), polyoxyethylene ethers (Brij®) Ionic: Sodium dodecyl sulfate, sodium cocoyl glutamate Amphoteric: Cocamidopropyl betaine, Capryl/capramidopropyl betaine Polymers: Poly(vinyl pyrrolidone), polyoxyethylene, poly(N-vinyl formamide), poly(vinylamine)
Viscosity modifiers	Naturally occurring polymers (xanthan gums), semi-synthetic polymers (cellulose ethers), and synthetic polymers (polyvinylpyrrolidone).
Chemical enhancers	Oleic acid, isopropyl myristate, oleyl alcohol, linoleic acid

Surfactants stabilize these air bubbles by quickly orienting and adsorbing in the new interface, ensuring adequate surface coverage once the foam is ejected from the canister [24]. Common surfactants include ionic (anionic and cationic), nonionic, and amphoteric surfactants, as shown in Table 10.1. Although ionic surfactants have higher foaming capacity, they are associated with high skin irritancy potential. Consequently, nonionic surfactants are recommended for sensitive skin, especially in cases where the administration site is infected or inflamed [25]. Further, polymers, mainly polyelectrolytes (Table 10.1), are frequently used as adjuncts to surfactants, as polymers can stabilize foams similarly to surfactants. The introduction of polymers impacts the viscosity of foaming solutions, often increasing viscosity, which contributes to foam stabilization and its collapse [26]. Additionally, foam formulation utilizes viscosity modifiers (0.5–2%), as given in Table 10.1. These contribute to enhanced texture and optimal spreading characteristics, providing better retention of the formulation in the affected skin area when compared to solutions [27].

Topical foams are classified into aqueous or nonaqueous based on the type of solvent/cosolvent incorporated. The most common one is aqueous foam, which uses water as a solvent, as it is easier to formulate than nonaqueous foam. Other excipients may be added to the foam formulation in addition to these main components, depending on the active ingredient and the desired foam properties. For example, high ethanol concentration is added in the formulation of thermolabile or temperature-sensitive foams, which rapidly break down upon contact with the skin and provide a cooling sensation. In addition, it improves the solubility of actives and their penetration into the skin [25]. Similarly, chemical enhancers, as given in Table 10.1, can also be incorporated into the foams, promoting drug molecule penetration through the skin. Topical foams may also include occlusive agents like petrolatum, promoting skin occlusion to reduce moisture loss, enhance hydration, and improve the delivery of active ingredients. Water-insoluble film-forming polymers are added to create a water-resistant film upon application. This is beneficial when there is prolonged exposure to water that might compromise the effectiveness of the formulation and is commonly employed in cosmeceuticals (e.g., sunscreens) [20].

The evaporation of volatile constituents, such as propellants or alcohol, upon application of topical foams results in supersaturation, increasing the thermodynamic activity of drugs and enhancing their delivery rate. Additionally, supersaturation may facilitate the formation of a drug reservoir within the stratum corneum, which may facilitate the controlled and sustained release of drugs. This approach is frequently employed in foams by carefully choosing the components and their composition in the formulation, resulting in a supersaturated state [28, 29]. Moreover, studies have reported incorporating nanoparticulate systems into foam formulations, offering the potential for encapsulating hydrophilic and lipophilic actives. Upon evaporation of solvents, the excipients in a formulation, such as surfactants, interact with the incorporated nanoparticles, facilitating their breakdown and promoting drug release [30]. Foam drug delivery systems enhance treatment efficacy by promoting the rapid absorption of drugs, leading to improved patient adherence [19]. The numerous advantages inherent to foams have established them as a versatile and efficacious mode of administration across diverse applications, ranging from skincare to the pharmaceutical domain.

10.2.2 Organogels

Gels are semi-solid formulations that include an external solvent phase immobilized within the existing spaces of a three-dimensional network structure formed by solid components. The nature of liquids primarily distinguishes between hydrogels and organogels. A predominantly organic or nonpolar phase characterizes organogels, whereas hydrogels have a polar continuous phase. Compared to conventional gel compositions, these systems have the advantage of forming versatile formulations that can readily incorporate organic solvents, thereby potentially resisting microbial contamination. They are biodegradable and nonimmunogenic and are suitable for developing drug depot systems for prolonged action. The functionality of organogels is also enhanced by their ability to form controlled dosage forms for both hydrophilic and lipophilic therapeutic agents, making them excellent choices for topical drug delivery [11, 31]. Different therapeutic agents have been incorporated in organogels in investigations for treating diseases such as neuropathy, hormone-dependent cancer, diabetes, inflammation, and wound healing [32].

The solid components in organogels are generally organogelators that undergo physical (hydrogen bonding, π-π stacking, Van der Waals forces) and chemical interactions, categorizing them further as physical and chemical gels, respectively. Organogelators can be categorized primarily into three groups: Low-Molecular Weight Organogelators (LMOGs), Polymeric Organic Gelators (POGs), and biomolecules [32], as illustrated in Table 10.2. The choice of organogels depends on different factors, however, primarily on their capability to establish a stable gel structure, drug compatibility, and exert suitable rheological characteristics. Organogels commonly display viscoelastic behavior, signifying a combination of both viscous and elastic properties, rendering them well-suited for easy

Table 10.2 Types of organogelators, their size, and examples

Organogelators	Size	Examples [11, 32]
LMOGs	<1 kDa	Fatty acid derivatives (stearic acid, oleic acid), anthryl derivatives, steroid derivatives (cholesterol), amino acid types (L-alanine), organometallic compounds, and sorbitan esters
POGs	>2 kDa	Poly(methyl methacrylate), copolymers of acrylic acid, polyethylene, polycarbonate, and polyesters
Biomolecules or natural organogelators	Wide range of molecular weight	Lecithin, lanolin, derivatives of vitamins, peptides, lipids, sugar

application and adherence to the skin [11]. Further, they have a unique attribute of thermoreversibility. When the temperature increases beyond critical temperature, it disrupts their three-dimensional structure, thus changing the rheological properties [33]. The incorporation of drugs in organogels also affects the rheological behavior. In a study involving sorbitan ester organogels (Span 40) with polysorbate 60, higher drug concentrations (sumatriptan) increased viscosity and the sol-to-gel temperature, attributed to a denser tubular network. In addition, the surfactant polysorbate increased the strength of organogels and improved storage stability [34]. This highlights that the addition of surfactants like polysorbate in organogel formulations can induce emulsification and can promote solubility and stability. Further, improved viscosity of the organogels can promote spreadability and retention on the skin, which is of utmost significance for dermal products, particularly when applied to inflamed or sensitive skin [11, 35]. These systems also exhibit an extended shelf life compared to conventional topical dosage forms [33].

One widely explored organogelator in topical delivery systems includes lecithin, and these are known as lecithin organogels. Lecithin is a biomolecule that exhibits the capacity to interact with and disrupt the lipids present in the skin, thereby improving the permeation of actives. Additionally, Pluronic organogels (PLO) emerged as potential topical carriers, constituting a blend of the surfactant Pluronic F127 (poloxamer). These organogels demonstrate superior stability compared to lecithin, presenting a major advantage of diminished skin sensitization in topical delivery [32]. Also, PLO gels improve hydration and moisturization of the skin, which has been proven beneficial in skin conditions such as psoriasis and atopic dermatitis [36]. A study demonstrated that the localization of halobetasol propionate was improved when using PLO organogels compared to the traditional topical dosage forms for treating atopic dermatitis [37]. Premium lecithin organogels (PrLO) and second-generation lecithin derivatives are also explored in topical formulation. These organogels exhibit superior thermostability and lack greasiness, which is commonly observed in ointments [38]. In addition, LMOGs such as sorbitan esters, isopropyl myristate, and oleic acid utilized in organogel formulation also act as penetration enhancers that effectively improve drug permeation.

Therefore, the choice of organogelators in organogel applications can significantly impact drug release and enhance permeation [32, 36]. Furthermore, the lipophilicity of organogels not only facilitates potential partitioning through the skin but also imparts an occlusive effect, effectively enhancing permeation [39]. Combinations of other delivery systems, such as microparticles [40] or microemulsions [41], have been successfully employed in conjunction with organogels to optimize their advantages and achieve sustained and prolonged release. The solvents in organogels contribute to skin hydration and promote penetration of active molecules through the skin. Nonaqueous solvents, which include hexane, alkene squalene, or vegetable oils, are more dominant in organogels and facilitate the topical delivery of lipophilic drugs. Polar solvents, including water, butanol, polyethylene glycol, and propylene glycol, are occasionally employed as cosolvents, expanding the scope to incorporate hydrophilic drugs [32, 39].

Organogels are a versatile drug carrier system for various applications in topical drug delivery, including cosmeceuticals, such as skin-aging products and sunscreens. These carriers are simple to formulate and are cost-effective, with a large potential for scale-up manufacturing [42]. However, limitations such as inadequate biocompatibility persist due to the potential toxicity of the organic solvents used in the formulation. Nevertheless, as mentioned, their attributes make them excellent matrices capable of delivering an effective drug concentration over an extended period, enhancing patient adherence to therapeutic regimens.

10.2.3 Microsponges

Microsponges are polymeric delivery systems composed of porous microspheres. These microspheres exhibit a sponge-like architecture characterized by numerous interconnected voids within a structure defined by a multitude of interconnected voids within a noncollapsible framework with a large surface area. A typical 25 μm microsponge sphere can contain up to 250,000 pores, with each pore equivalent to a length of 10 feet and a total pore volume of 1 mL/g. This provides a distinct advantage in achieving substantial drug loading or entrapment within the polymer matrix and surface of microspheres, which can be up to three times their weight. Microsponge systems consist of biologically inert and nonbiodegradable polymers that are nonirritant, nonmutagenic, nonallergenic, and nontoxic. Commonly employed polymers in microsponge formulations include polymethacrylates such as Eudragit© polymers (Eudragit RS100, Eudragit RSPO, Eudragit S100), polylactic acid, polylactide-co-glycolic acid, polyhydroxy butyrate, ethyl cellulose, and poly divinylbenzene. The combination of polymers is recognized for providing enhanced control over drug release behavior, particularly attributed to the interactions among drug, methacrylate, and polymer components. The release rate depends on factors such as surface area, pore diameter, ratio of drug to polymer, and size. Further, stimuli like moisture, pH, friction, or temperature can also affect the release rate [43–45].

Novel microsponge-based drug delivery systems have gained significant attention in topical delivery due to their potential to modify and control the release behavior of drugs, which are the major advantages over conventional topical formulations. Unlike conventional topical dosage forms (creams, gels, and ointments), where the active molecule is rapidly released upon application, forming a concentrated layer of active ingredients on the skin, microsponges exert delayed and sustained action [46]. One notable characteristic of microsponges is their inability to pass through the skin due to their size (5–300 μm). This is beneficial for gradual or delayed and sustained release of the encapsulated drugs, facilitating drug localization while simultaneously limiting the transdermal permeation. Over the years, numerous research studies have been conducted in developing microsponge-controlled delivery systems of different therapeutic agents, including anti-acne, antifungal, and anti-inflammatory agents [45]. Commercially available microsponge formulations include products such as Retin-A Micro® containing tretinoin and Carac® containing fluorouracil [18]. Additionally, various others are currently in the process of development and undergoing clinical assessment.

The particulate characteristics of microsponges make them less suitable for direct topical application. Hence, they are integrated into topical bases, including solutions, gels, creams, ointments, or emulgels. Gels are the most studied vehicle for incorporating microsponges. However, the choice of these vehicles can often be tailored to the specific skin type or skin disorder. For instance, a gel base or a low-viscosity cream is considered suitable in conditions like acne vulgaris, characterized by increased sebaceous secretion and oily skin. Similarly, a topical base providing effective occlusivity to prevent moisture loss is preferred for skin conditions leading to dryness, such as psoriasis and atopic dermatitis. The selection of the drug molecule, considering its lipophilicity and hydrophilicity, is also essential when selecting the topical base. The general mechanism of API release from microsponges includes the free diffusion of active ingredients in and out of the microsponge matrix until equilibrium is reached, saturating the vehicle. Upon application to the skin, the API in the vehicle is absorbed, shifting the equilibrium, thus making the system unsaturated. This initiates the gradual release of the active substance into the incorporating carrier and, ultimately, onto the skin. Therefore, the interaction of the drug molecule with the base matrix significantly influences the release. If the active is too soluble in the vehicle, it may not provide the desired gradual release. Nevertheless, a moderate solubility of the drug in the vehicle is accepted as it can provide the initial loading dose until release from the microsponge is activated, thus reducing the lag time. When the topical vehicle containing microsponges is dried or absorbed, microsponge particles adhere to the surface of the skin and gradually release the API. This serves as a drug depot and facilitates prolonged therapeutic effects [44, 45]. In a recent study involving microsponges of an antibacterial agent, silver sulfadiazine loaded within a gel matrix exhibited enhanced drug retention capacity, effectively functioning as a drug reservoir and reducing the dosing frequency. Significantly, the study revealed a sustained and delayed release of the drug over a period of up to 28 days, proving to be more efficacious in treating burn wounds compared to the commercially available silver sulfadiazine cream [47]. Moreover, the rheological characteristics and texture of the topical base are important. Gel-based formulations, for instance, may experience microstructure disruption during application due to an elevated shear rate caused by rubbing. Similarly, the topical base should exhibit favorable spreadability to ensure that microsponges incorporated in it spread with minimal shear. Consequently, the development of a microsponge-loaded delivery system for dermatological agents has the potential to enhance drug efficacy and mitigate local adverse effects [44].

The microsponge network protects the active ingredient from possible degradation, thus increasing the stability and reducing the potential skin irritation [43]. The commercial formulation of 5-fluorouracil, employed in treating actinic keratosis (AK) lesions and skin cancer, typically consists of solutions and creams known for their elevated skin irritation potential, manifesting as acute local inflammatory reactions and intense itching. The study involving the incorporation of microsponge-loaded 5-fluorouracil into a gel effectively reduced the skin irritancy potential of the drug. This was attributed to the efficient entrapment of the drug in the porous surface of microsponges, facilitating controlled diffusion and resulting in sustained drug release, thereby avoiding direct skin contact [48].

The quasi-emulsion solvent method is the most widely utilized preparation technique for microsponges over liquid–liquid suspension polymerization. This method requires an internal phase (solvent), which has a stronger interaction with the drug than the external phase (anti-solvent), generating a quasi-emulsion. This method provides higher drug loading and control over the microparticulate size. This approach is commonly employed when the drug exhibits sensitivity to polymerization conditions. The active ingredient in microsponge formulation should display very low water solubility, which is among the other properties required in formulating microsponges, which include inertness to monomers, preventing an increase in viscosity during formulation, and avoiding collapse of the microsponges [43, 45].

The microsponge technology is widely used in various cosmeceutical products, including sunscreens, cleansers, moisturizers, and deodorants [45]. Thus, developing a microsponge-based delivery system offers a promising strategy for optimizing drug therapy by providing controlled release, improving the therapeutic index, and enhancing patient compliance. Consequently, the development of a microsponge-loaded delivery system for dermatological agents has the potential to enhance drug efficacy and mitigate local adverse effects. This innovative drug delivery system was initially designed for the topical administration of pharmaceuticals and has gained wide prominence in tissue engineering and controlled oral delivery of drugs. Ongoing research continues to explore new formulations and applications of microsponges in topical drug delivery.

10.2.4 Hydrogels

Hydrogels are cross-linked, semi-solid, three-dimensional networks of hydrophilic polymers that can hold a large amount of water while maintaining their structural integrity. The hydrophilic domains within these polymers hydrate in an aqueous medium, resulting in the development of the three-dimensional network. Gelation is the process of formation of a large, soluble, branched polymer through the linking of macromolecular chains, influenced by the structure and conformation of the polymer used. While gels and hydrogels both undergo the gelation process, they differ in solubility. Unlike gels, the polymer constituents in hydrogels do not dissolve in water but instead swell, exhibiting the ideal characteristic of hydrogels. This swelling is facilitated by various functional groups, including amino, carboxylic acid, hydroxyl, amide, and sulfa groups present in the polymer chains, allowing for water retention or absorption. The swelling property makes them ideal for the controlled release of therapeutic agents. Hydrogels can be utilized for sustained drug delivery; thus, they are widely explored in topical delivery systems, especially for wound healing applications [49, 50]. Topical hydrogels often incorporate a range of biodegradable and biocompatible polymers, as shown in Table 10.3.

Hydrogels are tailored for topical applications and are commonly designed to be soft and elastic to ensure convenient application and removal. Significant research has been conducted on hydrogels incorporating active ingredients, such as growth factors and anti-inflammatory and antibacterial agents, establishing them as potential delivery systems for optimal wound healing. The water content in hydrogels creates a moist environment, reducing the risk of scar formation and imparting a cooling sensation to the wound. Additionally, their ability to absorb exudates and pronounced porosity contribute to effective oxygen transmission, accelerating the wound healing process. In addition to these characteristics, the change in pH in wounded skin has led to the development of pH-responsive hydrogel dressings [51]. Such hydrogels are categorized as "smart hydrogels." Responsive to various stimuli, smart hydrogels can be triggered by external factors or sensor moieties such as pH, temperature, electrical and magnetic fields, light, or the concentration of biomolecules to initiate controlled drug release. The release is primarily based on phase transitions like gel-sol reactions, swelling behavior, type of polymer, and crosslinking steps during fabrication [52]. Smart hydrogels are often employed to protect the active ingredient in the formulation and increase

Table 10.3 Types of polymers used in hydrogel preparation

Types	Examples [49, 50]
Natural polymers	Cellulose, alginate, chitosan, and hyaluronic acid
Semisynthetic polymers	Methylcellulose, hydroxypropyl cellulose, hydroxypropyl methylcellulose, carboxymethyl cellulose
Synthetic polymers	Polyacrylic acid, poly(methacrylic acid), poly(ethylene glycol), polyglycerol, poloxamers, polyvinyl alcohol

its efficacy when applied to the skin. Hydrogels rapidly release drugs to eliminate bacteria in chronic or highly infected wounds with a slightly alkaline pH (>7.3). Conversely, wounds with an acidic pH, like ulcers, result in hydrogels releasing drugs gradually and slowly [53]. In a study, a pH-responsive hydrogel dressing incorporating chitosan, hyaluronic acid, and insulin demonstrated sustained and pH-triggered drug release. It also led to a favorable microenvironment, promoting the production of growth factors for diabetic wound healing. Further, insulin loading enhanced the process by increasing neovascularization around the wound. The controlled degradation of the hydrogel is fundamental for the active ingredient to diffuse consistently, overlapping with the granulation tissue formation on wound surfaces, thus preventing any interference with the wound healing process [54]. One example of a commercially available topical hydrogel wound dressing consists of lidocaine incorporated in polyethylene glycol and is indicated for pain associated with wounds [53].

Another significant category of smart hydrogels includes temperature-responsive systems that utilize temperature differences, such as those between room and skin temperature, the thermal gradient of the skin, or temperature imbalances between healthy and injured skin. The low-critical solution temperature (LCST) is crucial in formulating temperature-responsive systems. In these systems, the polymer undergoes a sol-gel transition, initiating gelation above the LCST upon heating due to increased hydrophobic interactions among hydrophobic groups while weakening the hydrogen bonds. In wound healing applications, the formulation, when applied to the wounded area with a higher temperature, undergoes gelation, forming an overlay over the wound, facilitating ease of application and efficacy of treatment. The transitions in these systems can be characterized and tailored using techniques such as rheology, spectroscopy, and differential scanning calorimetry (DSC), depending on the chemical nature and composition of the polymer. Further, the temperature difference between the formulation, usually at room temperature, and the skin surface facilitates the quick release of the drug once applied to the skin. Poly(N-isopropyl acrylamide) (PNIPAM) is a well-known thermoresponsive polymer with LCST close to human body temperature (32 °C). However, poor mechanical properties and potential toxicity hinder its practical application in topical formulations. Copolymerization with hydrophilic and hydrophobic monomers is a common approach to mitigate this drawback, which increases and decreases LCST, respectively, thus tailoring it closer to body temperature [49, 55]. Hydrogels based on polymerization of carboxymethyl agarose with monomers and PNINAM with acrylamide led to stable hydrogels with adjusted LCST and accelerated wound healing [56]. These thermoresponsive systems also utilize the thermal gradient of skin (32–37 °C), which is generally adapted by nanosystems loaded with hydrogels. This prevents immediate drug release when applied to the skin and is released when these nanocarriers reach deeper layers of the skin [49].

Recent research has focused on hydrogels characterized by novel formulations, including integrating nanoparticles, micro/nanoemulsions, and liposomes. Incorporating nanocarriers into the polymeric network of hydrogels has made the formulation more adjustable with enhanced properties and functionalities such as enhanced percutaneous absorption, drug loading capacity, and controlled and sustained drug delivery. They are superior to traditional dosage forms like gels and creams in terms of increasing the solubilization of drugs and improving stratum corneum hydration [12]. They also contribute to increased drug-loading capacity and achieve prolonged release kinetics. In a study where fluconazole-loaded microemulsion was incorporated into hydrogels, a slight enhancement in antifungal activity was observed compared to a commercially available cream. This incorporation had a notable impact on the physical and rheological properties of the resulting formulation, thus enhancing their performance [57]. Similarly, polymeric nanoparticles loaded into hydrogels can overcome the limitations of each, such as instability and initial burst release associated with nanoparticles and difficulty in incorporating lipophilic drugs associated with hydrophilic hydrogels. Recent advancements include the development of pH- and temperature-sensitive nanocarrier systems loaded into hydrogels to enhance treatment efficacy. An example involves methotrexate encapsulated in Eudragit nanoparticles, further incorporated into a pH-sensitive chitosan polymer for treating psoriatic lesions. After equilibrating at the skin's natural pH, the cationic nature of chitosan weakens electrostatic interactions, inducing a conformational change and leading to nanoparticle release. The sustained and controlled drug release was further obtained by modulating drug diffusion from deposited nanoparticles in the skin [58]. These novel hydrogel formulations have been explored in treating a wide range of skin conditions, including psoriasis, fungal infections, inflammations, and atopic dermatitis [12].

Hydrogel formulations are also widely explored in the cosmetic industry for skin conditions such as wrinkles, cellulite, skin hydration, pigmentation, and aging, with many products available in the market [59]. Thus, hydrogel formulations have been extensively studied in topical applications due to their unique properties, such as their high water content and the ability to swell, facilitating localized delivery of drugs. Although not yet widely available, pharmaceutical applications show tremendous potential for entering clinical trials and eventually reaching the market.

10.2.5 Liposomes

Liposomes are spherical lipid vesicles composed of amphiphilic molecules in a bilayer conformation in an aqueous medium. They self-assemble and arrange in bilayers (unilamellar) and/or a concentric series of multiple bilayers (multilamellar), enclosing an equal number of aqueous compartments [60]. Phospholipids are common amphiphilic molecules that spontaneously form bilayer lamellar structures when introduced to an aqueous phase. This resulting bilayer exhibits lipophilic characteristics, while the aqueous compartment imparts hydrophilic characteristics. Cholesterol is also frequently incorporated into phospholipid membranes at high concentrations, which improves bilayer characteristics such as fluidity and stability of the membrane and increases drug entrapment efficiency [61]. Liposomes exhibit thermodynamic stability and are categorized according to their structure and lamellarity, as shown in Table 10.4.

Table 10.4 Size and lamellarity of different types of liposomes

Category [62]	Size	Lamellarity
Small unilamellar vesicles (SUVs)	20–100 nm	1
Large unilamellar vesicles (LUVs)	> 100 nm	1
Giant unilamellar vesicles (GUVs)	>1–100 μm	1
Oligolamellar vesicles (OLVs)	100–1000 nm	2–5
Multilamellar vesicles (MLVs)	>500 nm	>5
Multivesicular vesicles (MVVs)	>1 μm	Honeycomb-like

Moreover, the structural similarity of liposomes to biological membranes contributes to their biocompatibility. Research on liposome-based topical applications has been ongoing for several decades, owing to their potential advantage of enhanced skin permeation. These nanocarriers can encapsulate and improve the delivery of both hydrophilic and lipophilic active ingredients, overcoming permeability limitations imposed by the stratum corneum layer. Thus, liposome-based topical application is suitable for a wide range of active ingredients; however, most research has focused on antibiotics, antifungals, disinfectants, immunosuppressive agents, and anti-acne agents [63, 64].

Various mechanisms of penetration of liposomes through the stratum corneum have been proposed over the years. While one hypothesis suggests that conventional liposomes may not pass through the stratum corneum intact, it has also been observed that the effectiveness of penetration depends on the size of the liposomes. Specifically, small unilamellar vesicles (SUVs) have been reported to have more efficient permeation [63]. Furthermore, research findings with electron microscopy suggest that liposomes exceeding a size of 1000 nm are more likely to be retained in the stratum corneum [9]. One early study has emphasized the increased deposition of MLV loaded with triamcinolone acetonide into the epidermis and dermis when compared to conventional topical dosage forms [6]. This broadens the scope of liposomal formulations, extending their application beyond the initially developed intravenous uses. The lipids in liposomes function as penetration enhancers, modifying the lipid structure of the stratum corneum and diminishing its barrier function. Their emulsifying nature can further contribute to a moisturizing effect on the skin, promoting skin hydration and facilitating permeation. Additional mechanisms involve the fusion or adsorption of the vesicle onto the stratum corneum, resulting in the dissolution of the vesicle and an improved partitioning of the drug into the skin [63, 65]. A notable advantage of utilizing liposomes in topical delivery is their ability to facilitate the localized delivery of active compounds in the epidermis and dermis, thus forming local depots facilitating sustained release and reducing systemic absorption. Consequently, this becomes particularly significant for drugs such as glucocorticosteroids or retinoids, which are known to induce severe systemic side effects when absorbed percutaneously to a greater extent [61, 64].

Liposomal formulations offer substantial benefits for actives like dithranol, which is used in treating psoriasis, known for causing skin irritation and staining.

Studies with dithranol-loaded liposomes enhanced skin permeation and mitigated skin irritation while exhibiting no staining properties [66]. The unique bilayer structure of liposomes enables their application in degradation-preventing systems, effectively preventing the actives from degradation from the external environment and promoting chemical stability. A prominent example involves tretinoin, which is used in acne vulgaris treatment but is highly susceptible to photodegradation, which necessitates the need for photoprotected formulations. The incorporation of tretinoin into liposomes demonstrated a significant enhancement in light stability [67]. Functional liposomal formulations often face challenges of fusion/aggregation, leakage, or hydrolysis of encapsulated actives, which can be mitigated by optimization of vesicular components. These are the key factors that influence the thermodynamic activity, size, charge, lamellarity, and elasticity of the bilayer [63].

Besides pharmaceutical applications in different skin conditions such as melasma, vitiligo, alopecia, hirsutism, psoriasis, and atopic dermatitis [61], liposomes have also gained extensive significance in the field of cosmeceuticals. They are utilized in various products, including moisturizers, sunscreens, anti-aging, depigmenting agents, and eye creams, with many such formulations readily available commercially, thus resulting in a revolutionary uplift of the cosmetic industry [65]. Several modified liposome vesicles have also gained importance over some years in topical delivery, as summarized in Table 10.5, which are known to further enhance the delivery of active agents.

Thus, liposomal delivery systems are versatile nanocarriers extensively utilized in pharmaceutical and cosmeceutical applications. The research landscape over the past few decades has extensively explored the numerous advantageous features of both conventional and modified liposomes. These features encompass their ability to enhance skin hydration, enable controlled release, facilitate targeted skin permeation, and prevent systemic uptake.

10.2.6 Microemulsions and Nanoemulsions

Microemulsions (MEs) and nanoemulsions (NEs) are single-phase systems containing an oil phase, a water phase, and a surfactant, frequently in combination with a cosurfactant. These are two distinct types of colloidal dispersion systems, which are transparent or translucent and are optically isotropic solutions. Often, there has been considerable confusion about the use of the respective terms in the scientific literature. Nevertheless, they are primarily differentiated by factors such as particle size, stability, and the method of preparation. Both ME and NE share some similarities with conventional semisolid emulsions in terms of the structure and the ability to form diverse compositions (w/o, o/w, o/w/o, or w/o/w); however, they exhibit unique physical characteristics, which include transparency, viscosity, particle size, and polydispersity index [73]. A comprehensive summary differentiating between the physical chemistry of conventional emulsion, ME, and NE is given in Table 10.6.

Table 10.5 Modified vesicular systems

Modified vesicular systems	Description [63, 68–72]	
	Composition	Mechanisms
Ultra-deformable liposomes or transfersomes	Phospholipids and edge activators, typically surfactants (sodium cholate or deoxycholate, spans, and tweens) or bile salts	The presence of an edge activator destabilizes the lipids, allowing them to deform and penetrate across the stratum corneum through small channels present on it. These flexible carriers penetrate skin layers without occlusion, allowing for excess water evaporation from the formulation, maintaining the hydration gradient of the skin, and thus influencing permeation.
Ethosomes	Water, phospholipids, and high concentrations of ethanol	These are highly deformable vesicles that interact with the lipids in the stratum corneum and maintain their structure when crossing it. Further, ethanol interacts with lipids of the stratum corneum, increasing fluidity and, therefore, improving the permeation.
Transethosomes	Water, phospholipids, edge activator, and ethanol	These are the most deformable vesicles, easily changing their conformation and allowing a higher membrane passage. They are designed to synergize the advantages of transfersomes and ethosomes; thus, they include ethanol, water, phospholipids, and an edge activator.
Niosomes	Nonionic surfactants and cholesterol	These carriers improve the delivery of drugs topically by enhancing drug stability, promoting penetration into the skin, and facilitating controlled release at the target site. They also offer advantages such as uniformity, high purity, low cost, and ease of storage.
Coated liposomes	Phospholipids, cholesterol, and polymers (chitosan, polyvinyl alcohol, and polyethylene glycol)	These liposomes improve stability, prolong residence time, and promote penetration into deeper skin layers, enabling targeted delivery and reducing irritation.
Penetration enhancers containing vesicles	Phospholipids, penetration enhancers like propylene glycol, diethylene glycol, monoethyl ether, glycerol, oleic acid, and cineole	Incorporating penetration influences the permeation of disrupting lipid bilayers, altering intercellular lipids, solubilizing keratin, and enhancing drug partitioning. For example, the utilization of glycerol results in structures termed glycerosomes, which exhibit superior fluidity compared to conventional liposomes, leading to enhanced deposition in the skin.
Stimuli-responsive liposomes	Phospholipids, stimuli-responsive polymers (stimuli such as pH, temperature, or enzyme activity)	When exposed to different stimuli, the liposomes undergo structural modifications, leading to controlled drug release, enhanced skin penetration, or targeted delivery to specific skin regions.
Invasomes	Unsaturated phospholipids, a small amount of ethanol (usually 3.3% w/v), and terpenes	They are soft, flexible liposome vesicles with very high membrane fluidity. The presence of ethanol and terpenes plays the role of penetration enhancers.
Achaesomes	Lipids derived from *Archaea* species	The modified liposomes showcase increased stability and permeability of the membrane, reduced crystallization, increased resistance to oxidation, and greater resilience against phospholipases.

Table 10.6 Comparison of the properties of emulsions, microemulsions, and nanoemulsions

Characteristics [10, 73]	Emulsion	Microemulsion	Nanoemulsion
Particle size	>500 nm	100–400 nm	1–100 nm
Polydispersity	High	Low	Low
Stability	Thermodynamically unstable	Thermodynamically stable	Thermodynamically unstable but kinetically stable
Preparation speed	High	Low	Low/high
Composition: surfactant to oil ratio	Low (2–10%)	High (up to 80%)	Moderate
Physical appearance	Creamy	Transparent/ translucent	Transparent/ translucent
State	Semi-solid	Liquid	Liquid
Viscosity	High	Low	Low

The presence of cosurfactants such as short- or medium-chain alcohols or polyglycerol derivatives in combination with the primary surfactants results in low interfacial tension and small particle size in ME and NE. The balance between such components, along with the oil and water phase utilized, is carried out by constructing a ternary phase diagram, which effectively describes the behavior of various components. This process combines theoretical considerations with experimental data to refine and confirm the boundaries of the microemulsion region, which is identified by the convergence point of three primary phases [74]. While ME is formed spontaneously with gentle agitation, NE generally requires high-pressure homogenization but sometimes can also be formed with minimal energy expenditure [73]. ME and NE enhance the solubility of hydrophilic and lipophilic drugs, presenting significant benefits in topical delivery systems. These well-dispersed emulsion systems of oil and water closely resemble the surface of the skin (lipophilic sebum layer and a small fraction of water) and, consequently, are known to be compatible with the skin. Additionally, these systems offer appealing benefits by avoiding occlusion and greasiness, exhibiting cosmetic elegance, and distinguishing them from conventional forms like ointments. Furthermore, compared to liquid formulations, the viscosity of ME and NE can be readily adjusted to enhance formulation adhesion to the skin surface [74]. A commonly employed approach entails gelating these systems utilizing viscosity enhancers like Carbopol®, Aerosil®, and gelatin, forming gels or hydrogels [10]. Extensive research has been conducted on MEs and NEs for delivering a wide range of therapeutic agents, encompassing antifungals, corticosteroids, retinoids, and anti-inflammatory agents for treating different topical diseases. A study showed a significant increase in the permeating ability of penciclovir, used for herpes simplex, with the microemulsion formulation compared to the commercial cream [75]. These systems significantly enhance drug permeation, thereby increasing therapeutic efficacy and subsequently reducing associated side effects. This enhanced penetration is attributed to the combined effects of the components on the skin, as well as their unique phase structure and particle size [76].

Surfactants, in combination with co-surfactants, are integral components that exhibit a dual function of increasing the solubility of compounds with diverse physicochemical characteristics and enhancing the physical stability of the formulation. This synergistic interaction essentially reduces the interfacial tension between the water and oil phases, thereby increasing the overall stability of the formulation. These stable interfaces can facilitate interactions with the skin surface, facilitating the spread of formulation and influencing drug penetration. Additionally, surfactants and co-surfactants function as permeation enhancers by interacting with the lipid layers of the stratum corneum, altering their structural integrity and minimizing the barrier function [76, 77]. For instance, commonly utilized co-surfactants such as ethanol, isopropanol, isopropyl myristate, and propylene glycol are well-established permeation enhancers in skin drug delivery systems [78]. Similarly, surfactants such as oleic acid, polysorbates, sorbitan fatty acid esters, and proprietary surfactants like Cremophor®, Plurol Oleique®, and Labrasol® are frequently utilized in ME/NE formulations. These are nonionic surfactants, which are the preferred type due to their favorable characteristics, such as low toxicity, minimal skin irritability, high chemical stability, and the ability to promote skin permeability effectively [76]. Despite this, a combination of these surfactants can sometimes result in skin irritation, particularly when short-chain alcohol and isopropyl myristate are used in combination with emulsifiers [74]. Consequently, low-toxicity alternatives, such as naturally derived lecithin [78] and sugar-based surfactants [79], are also extensively investigated.

MEs exhibit high thermodynamic stability, leading to an enhanced permeation capacity through improved solubilization of active compounds in the vehicles and influencing drug partitioning. In contrast, as previously noted, NEs exhibit thermodynamic instability but maintain kinetic stability, which is attributed to the Brownian motion of the dispersed particles. Such kinetic stability is responsible for promoting the diffusion flow through the skin, providing better penetration and prolonged resistance against creaming, sedimentation, and flocculation processes, even in the presence of temperature variations. Further, the kinetic stability of NE provides prolonged resistance against creaming, sedimentation, and flocculation processes, even in the presence of temperature variations [73].

ME and NE utilize the transfollicular pathway for the percutaneous absorption of topically applied substances, effectively serving as a drug reservoir. Apart from the evident advantage of smaller particle size, sebum in hair follicles has a similar composition to that of these nanosystems, that is, oil and surfactant, leading to increased deposition to the hair follicles [10]. We have previously studied the enhancement of the permeation of adapalene-loaded ME in the stratum corneum, which exhibited optimal penetration into the hair follicles. The research further highlighted that deposition in hair follicles increased with a higher water content in the microemulsion. This underscores the significance of achieving an optimal composition in terms of the oil-to-water-to-surfactant ratio in the formulation for the effectiveness of the microemulsion system [80].

ME and NE also gained much attention as effective carriers delivering cosmeceuticals with many products available in the market, such as moisturizers, sunscreens, cleansers, and anti-aging products [73]. Extensive research on these carrier systems has progressed, resulting in production methods that achieve a proper balance between components and are tailored to the physicochemical properties of the actives. Moreover, the emphasis on minimizing energy requirements in the formulation process has resulted in extensive resource utilization, rendering it highly compatible with large-scale production.

10.2.7 Lipid Nanoparticles

Lipid nanoparticles (LPNs) represent heterogeneous colloidal systems characterized by an internal lipid phase and an external aqueous phase, typically stabilized by incorporating one or two surfactants. The size of LPN varies between 50 and 100 nm. This category encompasses two fundamental subclasses: solid lipid nanoparticles (SLN) and nanostructured lipid carriers (NLC). They are distinguished from nanoemulsions in that they have a solid lipid phase and are composed of solid lipids at room temperature, thus imparting structural rigidity and resistance to coalescence. The thermodynamic stability of these dispersions is further ensured by including surfactants. The distinction between SLN and NLC lies in their lipid composition. SLN exclusively consists of solid lipids, whereas NLC comprises a blend of solid and liquid lipids, with a predominant proportion of solid lipids. NLCs overcome certain limitations associated with SLN, such as lower drug loading and the risk of drug expulsion during storage, attributed to the unique nanostructure they form [81]. One of the most promising applications suggested for SLN and NLC is their topical use, particularly for therapeutic and cosmetic purposes. These nanocarriers are superior to other nanoparticle carriers in topical applications in terms of improved chemical stability and modulation of the release profile of entrapped drugs. The selection of lipids is closely linked to chemical stability, particularly for highly sensitive molecules such as retinol. For example, lipids with unfavorable characteristics, such as excessive acidity, may result in less effective stabilization [82]. These carriers are biodegradable, and physiological lipids with low systemic and cytotoxic effects are often employed. Commonly used solid lipids in SLN include glycerol palmitostearate, glycerol behenate, and cetyl palmitate, while liquid lipids in NLC include medium-chain triglycerides, castor oil, and oleic acid. Additionally, the stability of these nanoparticles also depends upon the surfactant employed. Lecithin, polysorbate 80, poloxamer 188, sodium cocoamphoacetate, and saccharose fatty acid ester are commonly employed as surfactants in the formulation of these carriers [83]. The dermal application of SLN and NLC effectively minimizes skin irritation, a primary concern in cutaneous delivery. These carriers incorporate biodegradable, well-tolerated, and physiological lipids, with typical surfactants used in topical delivery in concentrations ranging from 0.5% to 5% [84].

Following topical application of SLN and NLC, a monolayered nanoparticle film is formed on the skin, creating an occlusive effect that enhances skin hydration and adhesiveness, further influencing the percutaneous absorption of the incorporated actives [84]. In a study involving clotrimazole-loaded SLN and NLC, findings indicated that SLN exhibited a greater occlusive factor than NLC despite having the same lipid content. This can be attributed to the higher solid lipid matrix of SLN, which acts as a barrier against water evaporation, in contrast to NLC [85]. This occlusive property is more pronounced in LPNs with smaller particle sizes, leading to increased adhesiveness and a more rapid film formation. A precursor oil-in-water nanoemulsion is first formed to achieve a smaller particle size and a narrow polydispersity index, subjected to high-shear homogenization, high-pressure homogenization, and ultrasonication. The particle size of SLN and NLC is influenced by different formulation factors, including lipids, the ratio of oil to lipids, surfactants, and the drug concentration used in their preparation. The particle size of both SLN and NLC is influenced by different formulation factors, including lipids, the ratio of oil to lipid, surfactants, and the concentration of the drug used in their preparation [86, 87]. It has been reported that a lower surfactant (Chremophore® EL) concentration (1–2% w/v) resulted in the formation of smaller NLC particles; however, NLCs exhibited larger particle sizes compared to SLNs when surfactant concentration increased to 4% w/v. Additionally, increased particle size was observed with higher concentrations of clotrimazole [88]. Similarly, in another study, the impact of surfactant choice on particle size was emphasized. Notably, Tween 80 resulted in a smaller particle size than Pluronic F68 due to the steric stability provided by Tween 80, contributing to a reduction in particle size [89]. Often, the stability of the particle size of the LPNs during the storage period represents the overall physical stability of the particles, which is generally predicted by zeta potential [87]. In a size stability study, it was observed that the particle size and polydispersity index (PDI) of SLN and NLC remained unchanged over 12 months at different storage temperatures (4 °C, room temperature, and 40 °C), indicating long-term stability of the particles [90].

The smaller particle size of LPNs has contributed to the increased interest in potential transfollicular pathways. The lipids used in LPN formulation resemble the composition of sebum, contributing to an increase in the high sebum/water partition coefficient, thus facilitating their transport into hair follicles. Both SLN and NLC have been reported to enhance not only the transfollicular penetration when compared to conventional topical dosage forms but can also be less sensitizing to the skin [91]. Thus, LPNs serve as promising carriers for drug targeting in the skin, offering advantages such as enhanced permeability and compatibility with skin lipids. Research has explored the epidermal targeting of podophyllotoxin [92] and naproxen [93] using SLN and NLC, respectively, with preferential localization of LPNs on the skin surface. Both lipophilic and hydrophilic drugs can be incorporated into the LPNs. The incorporated drug gradually releases from the internal phase to the external phase and then to the skin, contributing to sustained and controlled release systems. The solid matrix structure of LPNs facilitates an extended release of the drug, effectively minimizing the adverse side effects associated with the encapsulated drug molecule [94]. Over the years, different therapeutic agents have been incorporated into LPNs in research, presenting novel strategies for treating acne, fungal infections, wound healing, atopic dermatitis, and psoriasis. These LPNs are commonly formulated in two ways: (i) as conventional semi-solid systems such as gels, ointments, and creams, or (ii) as aqueous dispersions. In the latter approach, the consistency of the dispersion is increased either by adding viscosity enhancers to the aqueous phase or by producing systems with a high concentration of lipid nanoparticles. The second approach is particularly promising as it avoids significant dilutions of lipid nanoparticle dispersions, which could lead to formulations with inadequate concentrations of drugs or active ingredients. This method involves incorporating hydrophilic gelling agents (e.g., carbomers and cellulose derivatives) into the aqueous phase of lipid nanoparticle dispersions, resulting in the formation of nanoemulgel-like systems characterized by a typical hydrogel network that entraps and stabilizes lipid nanoparticles [81].

SLN and NLC are classified as new-generation carrier systems. However, further research on skin lipid interaction and the influence of particle composition on API distribution within the skin strata is required. Nevertheless, LPNs present a highly promising and versatile platform in pharmaceutical and cosmeceutical research due to their features, including their small particle size and solid lipid-based composition.

10.3 Emerging Approaches in Topical Delivery Systems

10.3.1 Dendrimers

Dendrimers are unique macromolecules with a central core from which tree-like repeated branches protrude, ending with polyvalent peripheral functional groups. The dendrimer structure generally demonstrates the capacity to control its size based on the number of generations. The generation of a dendrimer refers to the number of layers branching from the core. Each successive generation adds more branches and increases the overall size of the dendrimer. Dendrimers are typically nanoscale structures. These nanocarriers are evolving as the next-generation nanocarriers for topical drug delivery, primarily owing to their compact size, exceptional molecular uniformity, and highly functional surfaces. The cavities and surface groups present in the dendrimers enable the encapsulation of a diverse array of drugs with varying physicochemical properties, including high molecular weight and lipophilic/hydrophilic molecules. Additionally, the peripheral functional groups contribute to distinctive characteristics, including enhanced aqueous solubility for improving the solubility of lipophilic molecules and stimuli responsiveness, such as sensitivity to pH and temperature, making it an efficient delivery system [95, 96]. The encapsulation of drugs within dendrimers occurs through mechanisms like hydrogen bonding, electrostatic interactions, and molecular encapsulation. This entrapment prevents direct contact between the drug and the solvent, offering protection against the overall degradation of the active compounds. Dendrimers exist in diverse categories, encompassing polyamidoamine (PAMAM), liquid crystalline, chiral, tecto, peptide, hybrid, poly(amidoamine-organosilicon), and glycodendrimers. PAMAM has emerged as the most extensively investigated dendrimer type for topical drug delivery applications [97]. The toxicity of dendrimers is minimal upon topical application; nonetheless, significant elimination of toxicity can be achieved by employing biodegradable dendrimers, such as peptide dendrimers [95]. While the literature on dendritic systems in topical delivery is limited, it spans various applications, including acne, psoriasis, and inflammation [98]. They are also used in the cosmetic industry to deliver creams or other personal care products such as shampoos, conditioners, and deodorants [99].

Dendrimers improve drug delivery through the skin by acting as release modifiers. In this context, dendrimers themselves do not traverse the stratum corneum but significantly improve drug delivery upon release of drug from the dendrimers to the incorporated vehicle. This maintains the thermodynamics of the vehicle and increases the driving force of permeation. Additionally, the drug release kinetics from the dendrimer-drug complex depend upon the dendrimer concentration, resulting in its functioning as a release modifier similar to cyclodextrin [95]. However, a study reported the penetration of PAMAM dendrimers through the stratum corneum into deeper layers of the skin. This phenomenon was attributed to the smaller size of the dendrimers (<10 nm) and the interactive forces between cationic dendrimers and the negatively charged intercellular domain of the skin [100]. The skin affinity of cationic dendrimers is superior to that of neutral and anionic dendrimers, which is attributed to the negatively charged nature of the skin, which enhances their affinity [101, 102]. The skin affinity of cationic dendrimers is superior to anionic and nonionic dendrimers. Further, incorporating vehicles like isopropyl myristate in dendrimers significantly contributes to penetration enhancement. This effect is particularly prominent in low-generation dendrimers, where these vehicles have the potential to disrupt the lipids in the stratum corneum, consequently resulting in a synergistic enhancement in the permeation of the encapsulated drug. A prevalent strategy within dendritic systems involves pre-treatment of the skin with placebo dendrimers that serve as reservoirs. These reservoirs or depots are established through interactions with lipids in the stratum corneum or with water molecules in the skin, especially when the dendrimers exhibit hydrophilic characteristics. When a formulation containing a drug is applied to placebo-treated skin, it is known to enhance the solubility of the drug and improve its partitioning through the skin. Dendrimers can utilize the transappendageal route for delivery, which can be achieved by controlling their size to enable entry into the hair follicles. The size of the dendrimers can be largely controlled by the types of functional groups attached to the periphery and their generation, which represents the branching [95, 103]. In a recent investigation, second generation (G2) PAMAM dendrimers of adapalene were formulated into an aqueous solution and gel with comparatively lower drug loading than the commercially available adapalene gel. The dendrimer-aqueous solution exhibited the highest efficiency in penetrating hair follicles and the skin, as compared to the commercial gel. Moreover, a delay in skin penetration was noted in the dendrimer-gel formulation compared to the dendrimer-aqueous solution. This delay may be attributed to the gel's viscous nature, which restricted penetration by diminishing thermodynamic activity [102]. Consequently, selecting the topical vehicle should align with the specific disease state and its requirements.

The type of surface functional groups in dendrimers plays an essential role in governing the penetration of encapsulated drugs. A specific investigation demonstrated that dendrimers terminated with amine (-NH_2) groups exhibited enhanced penetration in comparison to counterparts with carboxyl (-COOH) and hydroxy (-OH) functional groups [101]. The penetration enhancement by dendrimers is primarily affected by factors such as molecular size, surface charge, and concentration. In addition, these factors also affect the viscosity of dendrimers in topical applications. Typically, dendrimers exhibit low viscosity at lower concentrations and molecular weights. As these increase, viscosity also increases. Thus, tailoring these parameters results in the formulation of optimal rheological properties and spreadability, ensuring effective topical use [99].

The research on topical dendrimer delivery is limited; however, they show great potential to efficiently deliver the drug across the skin. Their well-defined structures, modifiable properties, and versatile surface functionalities enhance drug solubility and targeted delivery. While the field of dendrimer research advances, further exploration and refinement of these nanoscale structures hold significant potential for effective topical therapies and addressing specific challenges in skin-related treatments.

10.3.2 Nanofibers

Nanofibers have recently gained significant attention for their role in topical applications. These nanostructured carriers are characterized by a fibrous shape and a diameter of less than 100 nm. They are predominantly composed of natural (e.g., chitosan, gelatin, collagen) and synthetic polymers (polyglycolic acid, polylactic-co-glycolic acid, polylactic acid, polyvinyl alcohol, polyvinyl pyrrolidone, polyurethane). These polymers are employed individually or in combination to utilize the distinctive attributes of each type. Natural polymers contribute to high porosity, water solubility, and inherent homogeneity. Meanwhile, synthetic polymers bring forth characteristics like elevated chemical and thermal stability, superior degradation time, particle uniformity, and an extended storage lifespan. Various techniques are used to produce nanofibers, including electrospinning, drawing techniques, spinneret-based tunable engineered parameter methods, phase separation, freeze-drying synthesis, template synthesis, and interfacial polymerization. Among these, electrospinning is a commonly employed method due to its cost-effectiveness, capacity for large-volume production, scalability, and versatility. The method yields various types of nanofibers, including net fibers, hybrid, composite, blended fibers, and core-shell fibers, which can be further accumulated into fiber mats or sheetlike structures. The attributes of nanofibers are advantageous in topical delivery applications, with a notable focus on wound healing—an extensively studied area in nanofiber research [104, 105]. Nanofiber scaffolds efficiently absorb exudate from wounds and maintain an optimal moisture level, thereby expediting the overall wound-healing process. Further, smart wound dressings are also emerging approaches that utilize different polymers that are sensitive or resistant to external stimuli [106]. In a recent investigation, nanofibers were fabricated with a shape memory polyurethane (SMPU) polymer in combination with complementary polymers like chitosan and gelatin. The hydrophobic nature of SMPU plays a pivotal role in maintaining the stability of the nanofiber dressing upon exudate absorption. Thus, unlike conventional wound dressings, these nanofibers do not undergo expansion or contraction, retaining their shape [107]. In addition to wound healing, nanofibers have been investigated across diverse indications, including atopic dermatitis, acne, skin cancer, and fungal infections [104]. A study demonstrated that chitin nanofibers not only inhibited the expression of inflammatory molecules in atopic dermatitis but also suppressed the progression of atopic dermatitis in its initial phase, highlighting the significance of these nanocarriers [108].

Due to their porous nature, nanofibers facilitate high drug loading for both hydrophilic and lipophilic molecules, which is achieved through hydrophilic and lipophilic polymers, respectively [104]. The release of encapsulated drugs from nanofibers depends on the polymer degradation rate, necessitating a careful selection of the polymer to achieve the desired therapeutic effect [109]. The drug released from polymeric nanofibers is classified into two types: burst release, characterized by rapid drug release primarily attributed to the large surface area of the nanofibers, and sustained release, indicating a continuous release of a constant drug concentration over a specific duration. The burst release is due to the drug present on the nanofibers' surface, which easily diffuses into the medium and causes an initial, quick burst. However, in cases where burst release is not required, it can be modified by adjusting the surface characteristics of the scaffold, the dispersibility of the drug in the matrix, the pore size, and the pore density. Surface functionalization of nanofibers involves the introduction of distinct functional groups, coatings, and responsive materials onto the nanofiber surface. Such modifications can be achieved through various techniques, including plasma treatment or wet chemical methods, surface graft polymerization, and co-electrospinning. They contribute to an improved release profile by inducing swelling, enhancing adhesion, and increasing drug loading capacity. This also facilitates controlled and targeted drug delivery in topical applications, providing several advantages in therapeutic efficacy, patient compliance, and minimizing side effects. Moreover, the modulation of drug release from nanofibers is influenced by several factors, such as the ratio of drug to polymer, fiber diameter, morphology, and/or porosity [104, 110].

Nanofiber scaffolds ensure the chemical stability of drugs, especially those that are prone to degradation. An investigation demonstrated that nanofibers loaded with tretinoin, susceptible to decomposition by light and air, maintained stability throughout a 45-day storage period [111]. Further, nanofibers are often loaded into other topical bases, such as gels, as a common approach to overcome the challenge associated with nanofibers alone, including spreading and retention. In a specific investigation, fluconazole incorporated into Eudragit nanofibers was integrated into a topical gel, demonstrating an enhanced drug release profile compared to using nanofibers alone [112]. Nanofibers have also been explored in the cosmeceutical industry, particularly in formulations aimed at anti-aging treatments and facial masks [104].

Thus, nanofibers present a range of notable advantages, including large surface area, enhanced drug entrapment, improved drug stability, and ease of manufacturing processes. These attributes make nanofibers promising carriers in topical delivery systems for sustained and controlled drug delivery. While the literature on nanofibers produced through electrospinning for topical applications is limited, there remains a need for more comprehensive research on formulating parameters and polymer combinations. Such extensive research will contribute to advancing drug delivery efficiency through nanofiber-based systems.

10.4 Conclusion

Novel drug delivery systems present a compelling rationale for topical delivery, addressing the drawbacks of conventional topical therapies such as limited penetration into deeper skin layers, drug instability, and a lack of sustained and controlled therapeutic delivery. The novel advances discussed in this chapter are promising approaches to enhance the delivery of drugs with varying physicochemical properties by utilization of various excipients such as surfactants and lipids. These systems have facilitated the development of target-specific therapies that are simple and noninvasive, thereby improving patient compliance, safety, and efficacy. Extensive research has been conducted to explore innovative methods in topical delivery, contributing to an exponential growth in available literature and publications. Despite these advancements, FDA-approved topical drugs utilizing novel delivery systems remain limited, possibly due to the substantial developmental efforts required for regulatory approval. Nonetheless, the continuous research and progress in this area represent a pivotal milestone in delivering different therapeutic agents, therefore resulting in improved therapeutic outcomes and patient care.

References

1. Tomar Y, Pandit N, Priya S, Singhvi G. Evolving Trends in Nanofibers for Topical Delivery of Therapeutics in Skin Disorders. ACS Omega. 2023;8(21):18340–57.
2. Hani Z. Asfour, Nabil A. Alhakamy, Md Shoaib Alam MWA-R& SM. Design of Experiment Navigated Methodical Development of Neem Oil Nanoemulsion Containing Tea Tree Oil for Dual Effect Against Dermal Illness: Ex Vivo Dermatokinetic and In Vivo. J Clust Sci. 2023;34:1311–1323.
3. Singh Malik D, Mital N, Kaur G. Topical drug delivery systems: A patent review. Expert Opin Ther Pat. 2016;26(2):213–28.

4. Tadwee IK, Gore S, Giradkar P. Advances in Topical Drug Delivery System: A Review. Int J Pharm Res Allied Sci. 2011;1:14–23.
5. Date AA, Naik B, Nagarsenker MS. Novel drug delivery systems: Potential in improving topical delivery of antiacne agents. Skin Pharmacol Physiol. 2005;19(1):2–16.
6. Schlich M, Musazzi UM, Campani V, Biondi M, Franzé S, Lai F, et al. Design and development of topical liposomal formulations in a regulatory perspective. Drug Deliv Transl Res. 2022;12(8):1811–28.
7. Güngör S, Erdal MS, Aksu B. New Formulation Strategies in Topical Antifungal Therapy. J Cosmet Dermatological Sci Appl. 2013;03(01):56–65.
8. Touitou E. Drug delivery across the skin. Expert Opin Biol Ther. 2002;2(7):723–33.
9. El Maghraby GMM, Williams AC, Barry BW. Can drug-bearing liposomes penetrate intact skin? J Pharm Pharmacol. 2010;58(4):415–29.
10. Nastiti CMRR, Ponto T, Abd E, Grice JE, Benson HAE, Roberts MS. Topical nano and microemulsions for skin delivery. Pharmaceutics. 2017;9(4):1–25.
11. Sagiri SS, Behera B, Rafanan RR, Bhattacharya C, Pal K, Banerjee I, et al. Organogels as matrices for controlled drug delivery: A review on the current state. Soft Mater. 2014;12(1):47–72.
12. Almoshari YH. Novel Hydrogels for Topical Applications: An Updated Comprehensive Review Based on Source. Gels. 2022;8(3).
13. Makhmalzade BS, Chavoshy F. Polymeric micelles as cutaneous drug delivery system in normal skin and dermatological disorders. J Adv Pharm Technol Res. 2018;9(1):2–8.
14. Loftsson T, Masson M. Cyclodextrins in topical drug formulations: Theory and practice. Int J Pharm. 2001;225(1–2):15–30.
15. Hoc D, Haznar-Garbacz D. Foams as unique drug delivery systems. Eur J Pharm Biopharm. 2021;167:73–82.
16. Purdon CH, Haigh JM, Surber C, Smith EW. Foam Drug Delivery in Dermatology. Am J Drug Deliv. 2003;1(1):71–5.
17. Rudnicka L, Olszewska M, Goldust M, Waśkiel-Burnat A, Warszawik-Hendzel O, Dorożyński P, et al. Efficacy and safety of different formulations of calcipotriol/betamethasone dipropionate in psoriasis: Gel, foam, and ointment. J Clin Med. 2021;10(23).
18. U.S. Food and Drug Administration. Orange Book: Approved Drug Products with Therapeutic Equivalence Evaluations. [cited 2023 Nov 28]. Available from: https://www.accessdata.fda.gov/scripts/cder/ob/search_product.cfm
19. Arzhavitina A, Steckel H. Foams for pharmaceutical and cosmetic application. Int J Pharm. 2010;394(1–2):1–17.
20. Zhao Y, Jones SA, Brown MB. Dynamic foams in topical drug delivery. J Pharm Pharmacol. 2010;62(6):678–84.
21. Farkas D, Kállai-Szabó N, Sárádi-Kesztyűs Á, Lengyel M, Magramane S, Kiss É, et al. Investigation of propellant-free aqueous foams as pharmaceutical carrier systems. Pharm Dev Technol. 2020;
22. Vora D, Dandekar AA, Srivastava RK, Athar M, Banga AK. Development and Evaluation of a Topical Foam Formulation for Decontamination of Warfare Agents. Mol Pharm. 2022;19(12):4644–53.
23. Gennari CGM, Selmin F, Minghetti P, Cilurzo F. Medicated Foams and Film Forming Dosage Forms as Tools to Improve the Thermodynamic Activity of Drugs to be Administered Through the Skin. Curr Drug Deliv. 2019;16(5):461–71.
24. Zhao Y, Brown MB, Jones SA. Pharmaceutical foams: are they the answer to the dilemma of topical nanoparticles? Nanomedicine Nanotechnology, Biol Med. 2010;6(2):227–36.
25. Parsa M, Trybala A, Malik DJ, Starov V. Foam in pharmaceutical and medical applications. Curr Opin Colloid Interface Sci. 2019;44:153–67.
26. Bureiko A, Trybala A, Kovalchuk N, Starov V. Current applications of foams formed from mixed surfactant-polymer solutions. Adv Colloid Interface Sci. 2015;222:670–7.

27. Shinde NG, Aloorkar NH, Bangar BN, Deshmukh SM, Shirke M V, Kale BB. Pharmaceutical Foam Drug Delivery System: General Considerations. Indo Am J Pharm Res. 2013;3(12):1322–7.
28. Lind M, Nielsen KT, Schefe LH, Nørremark K, Eriksson AH, Norsgaard H, et al. Supersaturation of Calcipotriene and Betamethasone Dipropionate in a Novel Aerosol Foam Formulation for Topical Treatment of Psoriasis Provides Enhanced Bioavailability of the Active Ingredients. Dermatol Ther (Heidelb). 2016;6(3):413–25.
29. Huang X, Tanojo H, Lenn J, Deng CH, Krochmal L. A novel foam vehicle for delivery of topical corticosteroids. J Am Acad Dermatol. 2005;53(1 SUPPL.):26–38.
30. Zhao Y, Moddaresi M, Jones SA, Brown MB. A dynamic topical hydrofluoroalkane foam to induce nanoparticle modification and drug release in situ. Eur J Pharm Biopharm. 2009;72(3):521–8.
31. Uzan S, Barış D, Çolak M, Aydın H, Hoşgören H. Organogels as novel carriers for dermal and topical drug delivery vehicles. Tetrahedron. 2016;72(47):7517–25.
32. Esposito CL, Kirilov P, Roullin VG. Organogels, promising drug delivery systems: an update of state-of-the-art and recent applications. J Control Release. 2018;271(October 2017):1–20.
33. Sahoo S, Kumar N, Bhattacharya C, Sagiri SS, Jain K, Pal K, et al. Organogels: Properties and applications in drug delivery. Des Monomers Polym. 2011;14(2):95–108.
34. Upadhyay KK, Tiwari C, Khopade AJ, Bohidar HB, Jain SK. Sorbitan ester organogels for transdermal delivery of sumatriptan. Drug Dev Ind Pharm. 2007;33(6):617–25.
35. Pénzes T, Csóka I, Eros I. Rheological analysis of the structural properties effecting the percutaneous absorption and stability in pharmaceutical organogels. Rheol Acta. 2004;43(5):457–63.
36. Mady FM, Essa H, El-Ammawi T, Abdelkader H, Hussein AK. Formulation and clinical evaluation of silymarin pluronic-lecithin organogels for treatment of atopic dermatitis. Drug Des Devel Ther. 2016;10:1101–10.
37. Lalan MS, Khode SS, Shah KS, Patel PC. Preliminary Development Studies of Halobetasol Propionate Organogel for Management of Atopic Dermatitis. Int J Pharm Sci Res. 2017;8(2):1000–9. Available from: www.ijpsr.com
38. Mujawar NK, Ghatage SL, Yeligar VC. Organogel: Factors And Its Importance. Int J Biol Chem Sci. 2014;4(3):758–73.
39. Sharma C, Agrawal D, Goyal R, Sharma AK, Khandelwal M. Organogel: A new approach in topical drug delivery system. Eur J Pharm Med Res. 2021;8(11):304–7.
40. Simsolo EE, Eroğlu İ, Tanrıverdi ST, Özer Ö. Formulation and Evaluation of Organogels Containing Hyaluronan Microparticles for Topical Delivery of Caffeine. AAPS PharmSciTech. 2018;19(3):1367–76.
41. Hamed R, Farhan A, Abu-Huwaij R, Mahmoud NN, Kamal A. Lidocaine Microemulsion-Laden Organogels as Lipid-Based Systems for Topical Delivery. J Pharm Innov. 2020;15(4):521–34.
42. Martinez RM, Rosado C, Velasco MVR, Lannes SCS, Baby AR. Main features and applications of organogels in cosmetics. Int J Cosmet Sci. 2019;41(2):109–17.
43. Ingale DJ, Aloorkar NH, KulkarnI AS, Patil RAP. Microsponges as Innovative Drug Delivery Systems. Int J Pharm Sci Nanotechnol. 2012;5(1):1597–606.
44. Mahant S, Kumar S, Nanda S, Rao R. Microsponges for dermatological applications: Perspectives and challenges. Asian J Pharm Sci. 2020;15(3):273–91.
45. Kaity S, Maiti S, Ghosh AK, Pal D, Ghosh A, Banerjee S. Microsponges: A novel strategy for drug delivery system. J Adv Pharm Technol Res. 2010;1(3):283–90.
46. Junqueira MV, Bruschi ML. A Review About the Drug Delivery from Microsponges. AAPS PharmSciTech. 2018;19(4):1501–11.
47. Kumar PM, Ghosh A. Development and evaluation of silver sulfadiazine loaded microsponge based gel for partial thickness (second degree) burn wounds. Eur J Pharm Sci. 2017;96:243–54.
48. Jain SK, Kaur M, Kalyani P, Mehra A, Kaur N, Panchal N. Microsponges enriched gel for enhanced topical delivery of 5-fluorouracil. J Microencapsul. 2019;36(7):677–91.

49. Van Gheluwe L, Chourpa I, Gaigne C, Munnier E. Polymer-based smart drug delivery systems for skin application and demonstration of stimuli-responsiveness. Polymers (Basel). 2021;13(8).
50. Ahsan A, Tian WX, Farooq MA, Khan DH. An overview of hydrogels and their role in transdermal drug delivery. Int J Polym Mater Polym Biomater. 2021;70(8):574–84.
51. Zhang X, Qin M, Xu M, Miao F, Merzougui C, Zhang X, et al. The fabrication of antibacterial hydrogels for wound healing. Eur Polym J. 2021;146(January):110268.
52. Mantha S, Pillai S, Khayambashi P, Upadhyay A, Zhang Y. Smart Hydrogels in Tissue Engineering and. Materials (Basel). 2019;12(3323):33.
53. Cascone S, Lamberti G. Hydrogel-based commercial products for biomedical applications: A review. Int J Pharm. 2020;573(November 2019):118803.
54. Li Z, Zhao Y, Liu H, Ren M, Wang Z, Wang X, et al. pH-responsive hydrogel loaded with insulin as a bioactive dressing for enhancing diabetic wound healing. Mater Des. 2021;210:110104.
55. Huang H, Qi X, Chen Y, Wu Z. Thermo-sensitive hydrogels for delivering biotherapeutic molecules: A review. Saudi Pharm J. 2019;27(7):990–9.
56. Cai Y, Xin L, Sun P, Li H, Liu C, Fang L. Temperature-sensitive multifunctional intelligent responsive hydrogel based on carboxymethyl agarose and N-isopropylacrylamide: Controlled drug release and accelerated wound healing. Carbohydr Polym. 2023;322(June):121327.
57. Coneac G, Vlaia V, Olariu I, Muţ AM, Anghel DF, Ilie C, et al. Development and Evaluation of New Microemulsion-Based Hydrogel Formulations for Topical Delivery of Fluconazole. AAPS PharmSciTech. 2015;16(4):889–904.
58. Asad MI, Khan D, Rehman AU, Elaissari A, Ahmed N. Development and in vitro/in vivo evaluation of ph-sensitive polymeric nanoparticles loaded hydrogel for the management of psoriasis. Nanomaterials. 2021;11(12).
59. Aswathy SH, Narendrakumar U, Manjubala I. Commercial hydrogels for biomedical applications. Heliyon. 2020;6(4):e03719.
60. Mdanda S, Ubanako P, Kondiah PPD, Kumar P, Choonara YE. Recent advances in microneedle platforms for transdermal drug delivery technologies. Polymers (Basel). 2021;13(15).
61. Pierre MBR, Dos Santos Miranda Costa I. Liposomal systems as drug delivery vehicles for dermal and transdermal applications. Arch Dermatol Res. 2011;303(9):607–21.
62. Liu P, Chen G, Zhang J. A Review of Liposomes as a Drug Delivery System: Current. Molecules. 2022;27(4):1372.
63. Carita AC, Eloy JO, Chorilli M, Lee RJ, Leonardi GR. Recent Advances and Perspectives in Liposomes for Cutaneous Drug Delivery. Curr Med Chem. 2017;25(5):606–35.
64. Prajapati BG. Topical Liposomes in Drug Delivery: A Review. 2016;(January):1–7.
65. Soni V, Chandel S, Jain P, Asati S. Role of liposomal drug-delivery system in cosmetics. Nanobiomaterials in Galenic Formulations and Cosmetics: Applications of Nanobiomaterials. Elsevier Inc.; 2016. 93–108 p.
66. Myśliwiec H, Myśliwiec P, Baran A, Flisiak I. Dithranol treatment of plaque-type psoriasis increases serum TNF-like weak inducer of apoptosis (TWEAK). Adv Med Sci. 2016;61(2):207–11.
67. Ioele G, De Luca M, Garofalo A, Ragno G. Photosensitive drugs: a review on their photoprotection by liposomes and cyclodextrins. Drug Deliv. 2017;24(0):33–44.
68. Ascenso A, Raposo S, Batista C, Cardoso P, Mendes T, Praça FG, et al. Development, characterization, and skin delivery studies of related ultradeformable vesicles: Transfersomes, ethosomes, and transethosomes. Int J Nanomedicine. 2015;10:5837–51. Available from: https://doi.org/10.2147/IJN.S86186
69. Hamishehkar H, Rahimpour Y, Kouhsoltani M. Niosomes as a propitious carrier for topical drug delivery. Expert Opin Drug Deliv. 2013;10(2):261–72.
70. Lee EH, Lim SJ, Lee MK. Chitosan-coated liposomes to stabilize and enhance transdermal delivery of indocyanine green for photodynamic therapy of melanoma. Carbohydr Polym. 2019;224(May).

71. Amnuaikit T, Limsuwan T, Khongkow P, Boonme P. Vesicular carriers containing phenylethyl resorcinol for topical delivery system; liposomes, transfersomes and invasomes. Asian J Pharm Sci. 2018;13(5):472–84. Available from: https://doi.org/10.1016/j.ajps.2018.02.004
72. Yang Q, Luo S, Tong S, Zhao H, Chen R, Song B, et al. Advances in the Study of Liposomes Gel with Stimulus Responsiveness in Disease Treatment. J Clust Sci. 2023;(0123456789). Available from: https://doi.org/10.1007/s10876-023-02510-y
73. Souto EB, Cano A, Martins-Gomes C, Coutinho TE, Zielińska A, Silva AM. Microemulsions and Nanoemulsions in Skin Drug Delivery. Bioengineering. 2022;9(4):1–22.
74. Lu GW, Gao P. Emulsions and Microemulsions for Topical and Transdermal Drug Delivery. Handbook of Non-Invasive Drug Delivery Systems. 2010. 59–94 p.
75. Zhu W, Yu A, Wang W, Dong R, Wu J, Zhai G. Formulation design of microemulsion for dermal delivery of penciclovir. Int J Pharm. 2008;360(1–2):184–90.
76. Alves LP, da Silva Oliveira K, da Paixão Santos JA, da Silva Leite JM, Rocha BP, de Lucena Nogueira P, et al. A review on developments and prospects of anti-inflammatory in microemulsions. J Drug Deliv Sci Technol. 2020;60(March).
77. Peltola S, Saarinen-Savolainen P, Kiesvaara J, Suhonen TM, Urtti A. Microemulsions for topical delivery of estradiol. Int J Pharm. 2003;254(2):99–107.
78. Hoppel M, Ettl H, Holper E, Valenta C. Influence of the composition of monoacyl phosphatidylcholine based microemulsions on the dermal delivery of flufenamic acid. Int J Pharm. 2014;475(1–2):156–62.
79. Todosijević MN, Savić MM, Batinić BB, Marković BD, Gašperlin M, Randnelović D V., et al. Biocompatible microemulsions of a model NSAID for skin delivery: A decisive role of surfactants in skin penetration/irritation profiles and pharmacokinetic performance. Int J Pharm. 2015;496(2):931–41.
80. Bhatia G, Zhou Y, Banga AK. Adapalene microemulsion for transfollicular drug delivery. J Pharm Sci. 2013;102(8):2622–31.
81. Garcês A, Amaral MH, Sousa Lobo JM, Silva AC. Formulations based on solid lipid nanoparticles (SLN) and nanostructured lipid carriers (NLC) for cutaneous use: A review. Eur J Pharm Sci. 2018;112(November 2017):159–67.
82. Müller RH, Radtke M, Wissing SA. Solid lipid nanoparticles (SLN) and nanostructured lipid carriers (NLC) in cosmetic and dermatological preparations. Adv Drug Deliv Rev. 2002;54(SUPPL.):131–55.
83. Schäfer-Korting M, Mehnert W, Korting HC. Lipid nanoparticles for improved topical application of drugs for skin diseases. Adv Drug Deliv Rev. 2007;59(6):427–43.
84. Souto EB, Almeida AJ, Müller RH. Lipid nanoparticles (SLN®, NLC®) for cutaneous drug delivery: Structure, protection and skin effects. J Biomed Nanotechnol. 2007;3(4):317–31.
85. Souto EB, Wissing SA, Barbosa CM, Müller RH. Development of a controlled release formulation based on SLN and NLC for topical clotrimazole delivery. Int J Pharm. 2004;278(1):71–7.
86. Puglia C, Bonina F. Lipid nanoparticles as novel delivery systems for cosmetics and dermal pharmaceuticals. Expert Opin Drug Deliv. 2012;9(4):429–41.
87. Bahari LAS, Hamishehkar H. The impact of variables on particle size of solid lipid nanoparticles and nanostructured lipid carriers; A comparative literature review. Adv Pharm Bull. 2016;6(2):143–51.
88. Das S, Ng WK, Tan RBH. Are nanostructured lipid carriers (NLCs) better than solid lipid nanoparticles (SLNs): Development, characterizations and comparative evaluations of clotrimazole-loaded SLNs and NLCs? Eur J Pharm Sci. 2012;47(1):139–51.
89. Luo Z, Liu C, Quan P, Yang D, Zhao H, Wan X, et al. Mechanistic insights of the controlled release capacity of polar functional group in transdermal drug delivery system: the relationship of hydrogen bonding strength and controlled release capacity. Acta Pharm Sin B. 2020;10(5):928–45.
90. Saupe A, Wissing SA, Lenk A, Schmidt C MR. Solid lipid nanoparticles (SLN) and nanostructured lipid carriers (NLC)—Structural investigations on two different carrier systems. Biomed Mater Eng. 2005;15(5):393–402.

91. Lauterbach A, Müller-Goymann CC. Applications and limitations of lipid nanoparticles in dermal and transdermal drug delivery via the follicular route. Eur J Pharm Biopharm. 2015;97:152–63.
92. Chen H, Chang X, Du D, Liu W, Liu J, Weng T, et al. Podophyllotoxin-loaded solid lipid nanoparticles for epidermal targeting. J Control Release. 2006;110(2):296–306.
93. Puglia C, Blasi P, Rizza L, Schoubben A, Bonina F, Rossi C, et al. Lipid nanoparticles for prolonged topical delivery: An in vitro and in vivo investigation. Int J Pharm. 2008;357(1–2):295–304.
94. Fang JY, Fang CL, Liu CH, Su YH. Lipid nanoparticles as vehicles for topical psoralen delivery: Solid lipid nanoparticles (SLN) versus nanostructured lipid carriers (NLC). Eur J Pharm Biopharm. 2008;70(2):633–40.
95. Sun M, Fan A, Wang Z, Zhao Y. Dendrimer-mediated drug delivery to the skin. Soft Matter. 2012;8(16):4301–5.
96. Pentek T, Newenhouse E, O'Brien B, Singh Chauhan A. Development of a Topical Resveratrol Formulation for Commercial Applications Using Dendrimer Nanotechnology. Molecules. 2017;22(1).
97. Nanjwade BK, Bechra HM, Derkar GK, Manvi F V., Nanjwade VK. Dendrimers: Emerging polymers for drug-delivery systems. Eur J Pharm Sci. 2009;38(3):185–96.
98. Tolia G, Choi H. The role of Dendrimers in Topical Delivery. Pharamceutical Technol. 2008;32(11).
99. Ammala A. Biodegradable polymers as encapsulation materials for cosmetics and personal care markets. Int J Cosmet Sci. 2013;35(2):113–24.
100. Venuganti VVK, Perumal OP. Effect of poly(amidoamine) (PAMAM) dendrimer on skin permeation of 5-fluorouracil. Int J Pharm. 2008;361(1–2):230–8.
101. Venuganti VV, Sahdev P, Hildreth M, Guan X, Perumal O. Structure-skin permeability relationship of dendrimers. Pharm Res. 2011;28(9):2246–60.
102. Gökçe BB, Boran T, Emlik Çalık F, Özhan G, Sanyal R, Güngör S. Dermal delivery and follicular targeting of adapalene using PAMAM dendrimers. Drug Deliv Transl Res. 2021;11(2):626–46.
103. Mignani S, El Kazzouli S, Bousmina M, Majoral JP. Expand classical drug administration ways by emerging routes using dendrimer drug delivery systems: A concise overview. Adv Drug Deliv Rev. 2013;65(10):1316–30.
104. Kamble P, Sadarani B, Majumdar A, Bhullar S. Nanofiber based drug delivery systems for skin: A promising therapeutic approach. J Drug Deliv Sci Technol. 2017;41:124–33.
105. Talebi N, Lopes D, Lopes J, Macário-Soares A, Dan AK, Ghanbari R, et al. Natural polymeric nanofibers in transdermal drug delivery. Appl Mater Today. 2023;30.
106. Ambekar RS, Kandasubramanian B. Advancements in nanofibers for wound dressing: A review. Eur Polym J. 2019;117(April):304–36.
107. Tan L, Hu J, Huang H, Han J, Hu H. Study of multi-functional electrospun composite nanofibrous mats for smart wound healing. Int J Biol Macromol. 2015;79:469–76.
108. Izumi R, Azuma K, Izawa H, Morimoto M, Nagashima M, Osaki T, et al. Chitin nanofibrils suppress skin inflammation in atopic dermatitis-like skin lesions in NC/Nga mice. Carbohydr Polym. 2016;146:320–7.
109. Joshi M, Butola BS, Saha K. Advances in topical drug delivery system: Micro to nanofibrous structures. J Nanosci Nanotechnol. 2014;14(1):853–67.
110. Yoo HS, Kim TG, Park TG. Surface-functionalized electrospun nanofibers for tissue engineering and drug delivery. Adv Drug Deliv Rev. 2009;61(12):1033–42.
111. Khoshbakht S, Asghari-Sana F, Fathi-Azarbayjani A, Sharifi Y. Fabrication and characterization of tretinoin-loaded nanofiber for topical skin delivery. Biomater Res. 2020;24(1):1–7.
112. Arunachalam A, Karthikeyan M, Kumar DV, Prathap M, Sethuraman S, Ashutoshkumar S. Review Article Current Pharma Research Transdermal Drug Delivery System: A Review. Micro. 2010;1(1).

Chapter 11
Approaches to Minimize the Dermal Absorption of Actives

Sesha Rajeswari Talluri, Vinam Puri, Romit Jani, Nirali Dholaria, Mittal Bhadiyadra, and Bozena Michniak-Kohn

11.1 Introduction

The skin, covering the body's entire external surface, is one of its largest organs. It is composed of three layers—the epidermis, dermis, and hypodermis—each with distinct anatomical structures and functions [1]. It also possesses associated structures termed cutaneous appendages—hair follicles, sebaceous glands, and sweat glands (Fig. 11.1), and these are illustrated below.

The epidermis [2–6] is the primary layer of the skin (between 50 μm and 200 μm thick) and provides the first barrier of protection from the invasion of substances into the body. The epidermis consists of five layers: stratum corneum (SC), stratum granulosum, stratum lucidum, stratum spinosum, and stratum basale. Among these, the stratum corneum (horny layer) is the outermost and the most important layer limiting the absorption of the drug from the formulation/skin product. The stratum corneum contains a 10- to 15-mm thick layer of flat, partially desiccated, dead epidermal cells. This horny layer is composed of approximately 40% protein and 40% water, with the remainder being lipids, cholesterol, free phospholipids, and triglycerides. This layer plays a significant role in determining the permeability of

S. R. Talluri · N. Dholaria · M. Bhadiyadra · B. Michniak-Kohn (✉)
Center for Dermal Research, Rutgers, The State University of New Jersey, Piscataway, NJ, USA

Dept. of Pharmaceutics, Ernest Mario School of Pharmacy, Rutgers, The State University of New Jersey, Piscataway, NJ, USA
e-mail: michniak@pharmacy.rutgers.edu

V. Puri
Bristol Myers Squibb, New Brunswick, NJ, USA

R. Jani
Solaris Pharma Corporation, Somerset, NJ, USA

S. N. Murthy (ed.), *Topical Products and Dermal Drug Delivery*,
AAPS Introductions in the Pharmaceutical Sciences 2,
https://doi.org/10.1007/978-3-031-80525-7_11

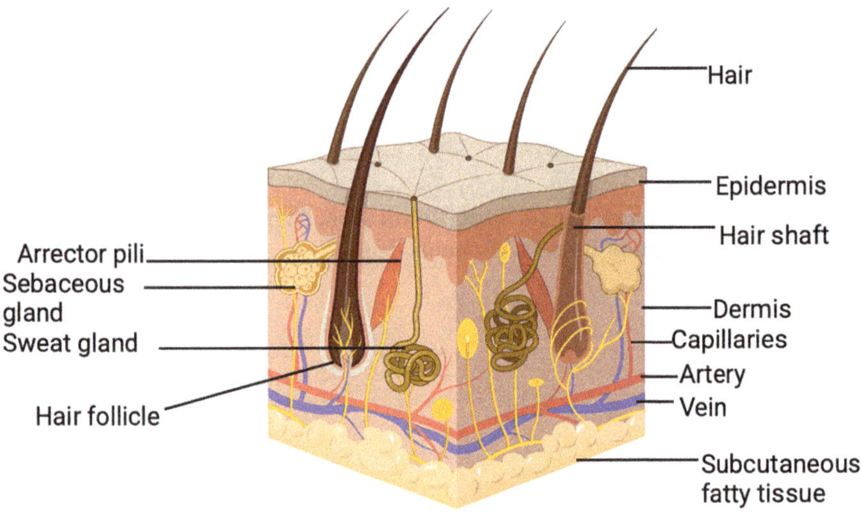

Fig. 11.1 Layers of human skin and associated features. (Image from Human Biology by D. Wilkin and J. Brainard)

drug substances across the skin. Some drug molecules of small molecular weight can enter by passive diffusion. The dermis [2, 5–9] is located below the epidermis and has a thickness of about 3–5 mm. This layer is mechanically strong and elastic, supplies nutrients to the area, and removes toxins and waste products. The blood supply to the dermis helps maintain a low concentration of actives/drugs and other substances within the dermis, which creates a concentration gradient necessary for permeation through the epidermal layer above. The hypodermis [3, 4], or subcutaneous fat tissue, supports the dermis and epidermis, serving as a fat storage area. This layer helps regulate temperature, provides nutritional support, and offers additional mechanical protection. It contains major blood vessels and nerves that supply the skin and may also house sensory pressure organs.

Overall, the skin's structure forms a complex network that acts as the body's first line of defense against pathogens, UV light, chemicals, and mechanical injury. The cell envelope establishes the epidermal water barrier with a layer of insoluble proteins on the inner surface of the plasma membrane, formed by the cross-linking of small proline-rich proteins. Larger proteins like cystatin, desmoplakin, and filaggrin enhance the barrier's mechanical strength. The lipid envelope, a hydrophobic layer, is attached to the outer surface of the plasma membrane [10, 11].

There are three primary pathways for substances to penetrate the skin: the intercellular route, the intracellular (transcellular) route, which traverses through successive layers and dead cells, and the route via skin appendages, such as hair follicles and sweat glands, forming shunt pathways through the intact epidermis (Fig. 11.2). The route via skin appendages is generally considered insignificant,

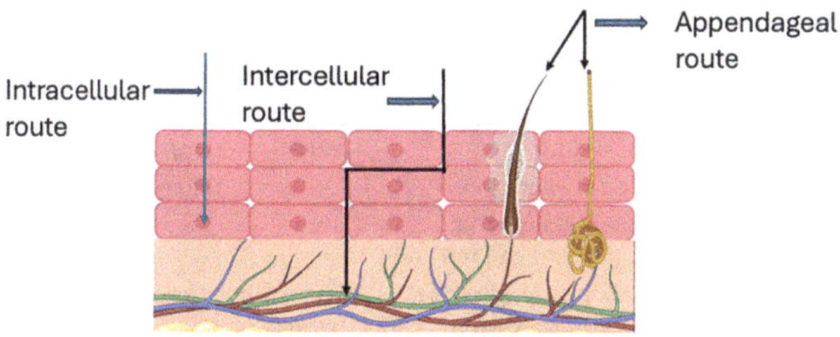

Fig. 11.2 Permeation routes via skin

as only 0.1% of the total human skin is occupied by these structures [12]. The intracellular route is the most direct but involves transport through densely packed, keratin-filled corneocytes and multiple transfers between corneocytes and lipid-filled intercellular areas, presenting significant resistance to permeation [13]. Consequently, the intercellular passage between corneocytes is the more common route for drug permeation [14].

Optimizing drug delivery to the precise skin compartment and biophase where it is needed is crucial for a) enhancing effectiveness and b) reducing systemic side effects. Certain drugs and chemicals are intended to be retained in the skin; some do not even pass into skin layers in high concentrations or cross the skin and enter the systemic circulation for either of the applications [15]. Examples of such substances and drugs are provided below in Table 11.1.

11.2 Pharmacokinetics of Dermal Absorption

The study of pharmacokinetics and prediction of dermal absorption is critical for the pharmaceutical and cosmetics industries, as well as environmental regulatory agencies concerned with the occupational safety of workers exposed to chemicals through skin contact. Understanding the absorption rate following dermal administration is crucial for determining the concentration of the interest molecule, which can be either an active pharmaceutical ingredient, a cosmetic product, or xenobiotics in systemic circulation [37]. Dermal absorption occurs in three main steps. The first is penetration, which involves the initial entry of a substance into a specific skin layer. Molecules may either accumulate within the stratum corneum or pass into the lower skin and enter the bloodstream. The second step is permeation/diffusion, which involves the movement of chemicals from one layer to another through hydrophilic/lipophilic channels. The final step is resorption, which requires the uptake of molecules into the vascular system, including lymph and blood vessels, which serve as the central compartment [38]. The dermal absorption is the cumulative result of these three processes. The absorption process is either a transit of the molecules through the stratum corneum alone or a transit through

Table 11.1 Examples of drugs and other substances that should not pass through the skin or be absorbed in high concentrations by the systemic circulation

Active ingredient/chemical	Use/condition	Category	Reference
Avobenzone, octocrylene	UV filters	Sunscreens/cosmetic[1]	[16]
Hyaluronic acid/hyaluronic acid-conjugated drugs	Skin hydration, anti-aging, drug delivery, and retention	Cosmetic, therapeutic	[17–19]
Clobetasol propionate	Skin itching, redness, swelling, rashes	Therapeutic (psoriasis, eczema)	[20]
Antifungal drugs	Topical dermatophyte infections	Therapeutic	[21]
Crisaborole	Atopic dermatitis	Therapeutic (non-steroidal)	[22]
Paromomycin	Cutaneous leishmaniasis	Therapeutic (antibiotic)	[23]
Anticancer drugs	Skin cancer	Therapeutic (chemotherapy)	[24]
Isotretinoin	Acne	Therapeutic	[25]
Analgesics/Anesthetics	Pain	Therapeutic	[26]
Curcumin	Wound healing	Therapeutic	[27, 28]
Nanosilver and nanogold	Antifungal, antibacterial	Cosmetic	[29]
Per- and polyfluoroalkyl substances (PFAS)	Sunblock, water repellant	Cosmetic/fabric	[30]
Para-phenylenediamine (PPD)	Permanent hair dye	Cosmetic	[31]
Lidocaine	Topical anesthetic	Therapeutic	[32]
Phthalates	Plasticizers in erasers	Nontherapeutic/educational supplies	[33]
Inorganic nanoparticles—titanium, zinc oxides	Sunscreens	Cosmetics	[34]
Dimethoate	Pesticide	Non-therapeutic	[35]
Brimonidine	Topical ophthalmic alpha-2 adrenergic agonist to treat glaucoma	Therapeutic	[36]

the entire skin, including the epidermis and dermis. The parameters of flux across the stratum corneum, representing actual absorption, adhere to Fick's law. These include steady-state absorption flux per unit skin area (J), permeability coefficient (P), solubility ratio in stratum corneum to vehicle (K), diffusion coefficient (P), solubility ratio in the stratum corneum to vehicle (K), diffusion coefficient (D), lag time (τ), concentration gradient (C), thickness of skin membrane (h), and steady-state concentration (Cs). The relationship between these parameters and their equations is provided in Table 11.2. An understanding of these parameters

Table 11.2 Some equations governing skin pharmacokinetics (for the explanation of parameters used, please see text)

Parameter	Equation
Fick's first law	$J = -D\ (dC/dx)$
Fick's second law	$\partial C/\partial t = D\ \partial 2C/\partial x2$
Lag time (τ)	$(\tau) = h^2/6D$
Steady-state flux	$Jss = (D * K * Cs) / h$
Permeability coefficient	$Kp = D{*}K/h$

is essential for accurate interpretation of absorption as well as the application to different situations [39].

Different pharmacokinetic (PK) models are used to predict and analyze the dermal absorption of other molecules. One-compartment models treat the skin as either a single compartment representing the stratum corneum, viable epidermis, or epidermis, or a combination of epidermis and dermis. Two-compartment models typically represent the stratum corneum and viable epidermis separately. These models simplify the complex structure of the skin into more manageable mathematical forms, which can be easily combined with existing PK models of the body. Physiological-based PK models are also used to estimate dermal absorption for both explanatory and predictive purposes in skin toxicology and pharmacology. These models simulate the absorption of molecules through animal/human skin using Franz cells, considering various physiochemical factors such as diffusivity and mass transfer factors. This type of model also incorporates parameters related to skin structure, solvent evaporation rates, and absorption dynamics. The physiologically based pharmacokinetic models can also be validated using dermal absorption studies such as tape-stripping data. The flux/time curve obtained from the model can be used to predict dermal absorption [40].

11.3 Factors Affecting Dermal Absorption

The factors affecting dermal absorption can be divided into three main categories: biological factors, physico-chemical factors, and formulation factors (Fig. 11.3). Below, these factors and their effects will be discussed in more detail.

11.3.1 Biological Factors

11.3.1.1 Skin pH

Skin surface is usually acidic (4–6), causing a gradient between the surface and underlying epidermis and dermis. This acidity was historically thought to be only a defense mechanism against invading organisms, but recent studies show it also impacts key enzymes involved in the synthesis and maintenance of a competent

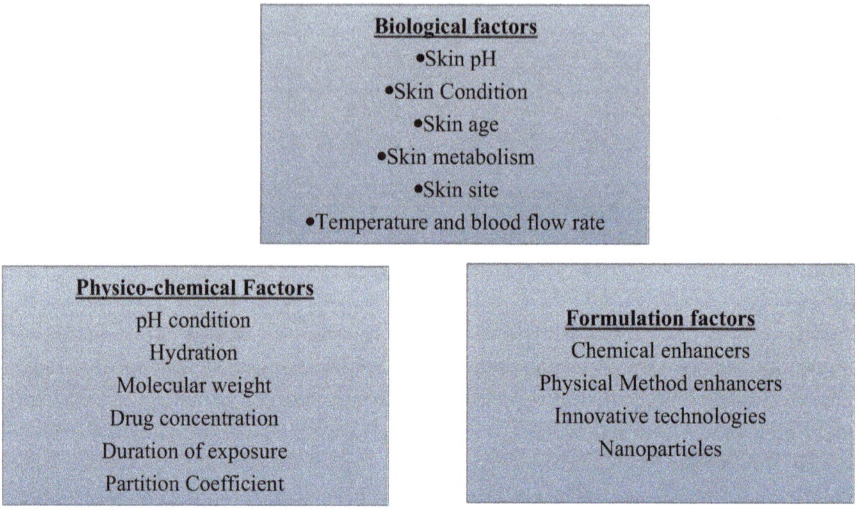

Fig. 11.3 Some factors affecting dermal absorption

skin barrier and is also very important in maintaining a healthy skin microbiome [41].

11.3.1.2 Skin Condition

The stratum corneum often limits how quickly compounds are absorbed through the skin. However, damage to the skin, such as cuts, wounds, abrasions, or sunburn, can increase this absorption rate. Skin conditions that weaken the stratum corneum's ability to act as an effective barrier, such as psoriasis, eczema, rashes, or dermatitis, can also lead to higher absorption in individuals with these conditions [42].

11.3.1.3 Skin Age

Normal functioning (barrier function) of the skin, and so the drug absorption rate changes with age. There are numerous reports in the literature of toxic effects in infants following the application of various topical drugs or other pharmaceutical agents. These effects are most likely due to the increased surface-to-volume ratio in infants, which leads to greater total absorption of these compounds, rather than an increased permeability of their skin compared to adults. Research has shown that full-term infants have a fully functional stratum corneum with excellent barrier properties [43]. However, a study conducted by Kearns et al. revealed that young children's skin is generally more permeable than that of older individuals, making them more sensitive to the skin absorption of toxins. Therefore, the age of the skin is an important factor affecting drug penetration [44]. A study conducted by Potts et al. demonstrated that alterations in aged skin encompass increased dryness of the stratum corneum, a decrease in sebaceous gland activity leading to reduced skin surface lipids, flattening of the dermal-epidermal junction, and atrophy of the skin capillary network, which gradually diminishes the blood supply to the viable epidermis [45]. A study conducted by Furue et al. demonstrated the toxic effects of topically applied steroids were higher in an infantile group than that of adults [46]. Skin, being a dynamic organ, changes structurally and functionally as individuals age [47]. The aging process affects the skin in multiple ways. With aging, the skin thins progressively at an accelerated rate. The thickness of the epidermis decreases with age by about 6.4% per decade, and water content in aged skin decreases, making the skin more prone to dermal-related conditions or diseases [48, 49].

11.3.1.4 Skin Metabolism

Its levels are generally lower when compared to the main metabolism site of the body, that is, the liver. Certain enzymes, like cytochrome P 450, in the skin are less abundant and may not significantly impact the delivery of substances throughout the body. However, local metabolism can transform chemicals into active forms that bind directly to skin DNA, for example, benzopyrene metabolism. Rapidly diffusing chemicals are generally less metabolized than slow-diffusing chemicals [50].

11.3.1.5 Skin Site

The thickness of the skin varies by body location. The eyelid skin is around 0.05 cm thick, while the skin on the palms and soles is about 0.4 cm thick. The thickness of the skin, the characteristics of the stratum corneum, and the density of appendages differ from one body location to another. These factors significantly influence how substances penetrate the skin. For instance, substances can be absorbed up to 50 times more effectively through highly permeable areas like scrotal skin compared to thicker, less permeable skin on the legs and abdomen [51, 52].

11.3.1.6 Temperature and Blood Flow Rate

Skin temperature influences the rate at which chemicals penetrate the skin in two key ways. First, higher skin temperatures directly increase chemical diffusion within the skin. Additionally, temperature changes can alter the structure of the stratum corneum, specifically the crystalline structure of the lipid bilayers, leading to greater permeability. Second, temperature affects blood flow to the skin. Increased blood flow primarily impacts the absorption of lipophilic chemicals, as their removal by the bloodstream often limits their penetration rate. This process also depends on the skin's physiology. Warmer skin boosts cutaneous blood flow, potentially accelerating chemical penetration. In summary, blood clearance is especially important for small, moderately lipophilic chemicals [51].

11.3.2 Physico-Chemical Factors

11.3.2.1 pH Condition

For weakly acidic and weakly basic drugs, skin permeability is pH dependent. Ionized drugs are known to permeate poorly when compared with non-ionized drugs. By changing the pH of the drug delivery system/formulation, the permeability of nonionized drugs may be enhanced [53].

11.3.2.2 Hydration

Water is the most widely used and safest method to enhance the absorption of substances into the skin, whether they are hydrophilic or lipophilic. By increasing skin hydration, the solubility of these substances can change, which may improve their absorption from the carrier into the skin. Additionally, hydration can swell and open the structure of the stratum corneum (SC), further boosting penetration. Methods such as using occlusion, moisturizers, paraffins, and oils from emulsions can help improve skin hydration and thereby increase the penetration rate [54].

11.3.2.3 Molecular Weight

The molecular weight of a substance significantly influences its ability to pass through the skin. The skin acts as a barrier, blocking large molecules such as peptides and proteins from entering the skin layers easily. Besides the molecular weight, the molecular structure also affects how well the substance binds to skin components, impacting its bioavailability. Ideally, a drug's molecular weight should be under 500 Daltons to ensure effective passive delivery, although larger molecules can sometimes be delivered efficiently if the required dose is very low [55].

A recently reported study by Chen et al., 2024, where a large molecular size compound—recombinant human growth hormone (rhGH)—was delivered transdermally using a liposomal delivery system [56]. Banerjee et al. (2017) reported higher penetration amounts of fluorescein isothiocyanate (FITC)-labeled bovine serum albumin (BSA; molecular weight $\approx$66 kDa), ovalbumin (OVA; molecular weight $\approx$45 kDa), and insulin (INS; molecular weight 5.8 kDa) in the presence of deep eutectic solvent CAGE (Choline and Geranate) compared to phosphate buffered saline (PBS) into porcine skin [57]. Many studies demonstrated that macromolecules (peptides, proteins) do not penetrate the skin without external support of physical or chemical methods. Various methods like electroporation, iontophoresis, and ultrasound are promising methods that successfully enable high molecular weight compounds to travel through or along the skin. Such an option of using external methods for sufficient drug penetration helps achieve active dermal delivery of the macromolecules [58].

11.3.2.4 Polarity

The ability of a substance to penetrate the skin depends partly on its polarity, which refers to the uneven distribution of electron density in a molecule. Polar compounds tend to be poorly absorbed through the skin, while nonpolar compounds are absorbed more easily. Polarity can be quantified by the dipole moment, which depends on the size of the partial charges in the molecule and the distance between these charges. The degree of polarity in a molecule is influenced by the presence and placement of electronegative atoms like nitrogen, oxygen, and fluorine, especially if they can ionize. Conversely, non-electronegative atoms such as hydrogen and carbon reduce polarity. Adding an electronegative functional group to a nonpolar compound increases its polarity. However, the overall molecular size and structure also play a significant role in determining the compound's polarity. Generally, the higher the polarity of a compound, the lower its lipophilicity [42].

11.3.2.5 Drug Concentration

The absorption of drugs through the skin occurs via passive diffusion, moving from areas of high concentration to low concentration. Therefore, the concentration of the drug in the formulation applied to the skin determines its diffusion rate. Higher concentrations result in greater permeation through the skin [59].

11.3.2.6 Duration of Exposure

Several factors complicate the effective measurement of the applied chemical load on the skin. For instance, a volatile chemical can both penetrate the skin and evaporate from its surface. Evaporation reduces the amount absorbed through the skin, but it is unclear how to quantitatively assess the competing processes of absorption and evaporation. The duration of exposure, or how long the skin remains in contact with the chemical, significantly affects the amount absorbed and the fractional absorption factor. An eight-hour exposure might be most relevant for workplace scenarios, whereas low-level environmental exposures, such as those from indoor air pollutants, could persist indefinitely. When designing experiments and evaluating finite-dose dermal absorption data, it is crucial to consider the relationship between the absorption kinetics of a chemical and the duration of exposure. This helps determine whether the calculated fractional absorption indicates a steady state or if further absorption would occur with longer exposure [60].

11.3.2.7 Partition Coefficient

In order to achieve high levels of penetrants in the first layer of skin, the stratum corneum, the penetrants need to readily leave the vehicle and move into the skin. This tendency is measured by the partition coefficient (Kp) of the penetrant, which is the ratio of its solubility in the stratum corneum to its solubility in the vehicle. The partition coefficient depends on the chemical structure of the penetrant. Drugs suitable for transdermal and topical drug delivery typically are recommended with a partition coefficient between 1 and 4 and an effective daily dose of less than 10 mg. For a specific drug, adjusting the nature or composition of the vehicle to increase the tendency of the penetrant to leave the vehicle can raise the Kp [61].

11.3.3 Formulation Factors

11.3.3.1 Chemical Enhancers [62, 63]

These are also referred to as chemical permeation enhancers. These are substances used to enhance drug absorption by altering the interaction of the active with the skin barrier, changing the extent of the skin hydration levels, adjusting the partition coefficient, impacting the cellular pathway, and modifying the solubility of the drug both in the formulation as well as in the skin. Various chemical enhancers have been used for many years, for example, deep eutectic solvents, ionic, peptides, nanoparticles, and dimethyl sulfoxide (DMSO). Using these enhancement technologies would significantly increase the number of drugs suitable for dermal drug delivery.

11.3.3.2 Physical Enhancement [63]

In this approach, various physical methods are used to enhance the drug delivery through skin. Methods such as electroporation (application of high-voltage pulses to induce skin perturbation), iontophoresis (application of a low-level electric current, either directly to the skin or indirectly via the dosage form), ultrasound, laser radiation and photomechanical waves, magnetophoresis (application of a magnetic field that acts as an external driving force to enhance the diffusion of a diamagnetic solute across the skin), and microneedle-based devices (microneedles of length 50 to 110 mm will penetrate the stratum corneum and epidermis to deliver the drug from the reservoir). The above methods can be used alone or in combination or even with chemical penetration enhancers to increase drug absorption.

11.3.3.3 Innovative Technologies

Stratum corneum is the main barrier in determining the permeability of drug substances across the skin. Techniques for bypassing the stratum corneum have been extensively investigated over the years. One approach involves using the transfollicular (via hair follicles) route for drug delivery [64]. As previously explained, absorption of drugs through this route is of minor importance because of its relatively small area of absorption, approximately 0.1% of the skin surface area; however, for some drugs, nanosized particles, and enhancement methods (iontophoresis), this route becomes the predominant one, and studies in all these areas are in progress.

11.3.3.4 Nanoparticles

Nanoparticles of various types, including liposomes, polymeric nanoparticles, dendrimers, solid lipid nanoparticles (SLNs), nanostructured lipid carriers (NLCs), and inorganic nanoparticles, have been explored for dermal drug delivery. The characteristics of nanoparticles, such as size, surface charge, and surface modification, can significantly affect their rate of dermal penetration as well as drug release [65].

11.4 Detection Method of Actives in Skin

An important aspect of delivering actives into the skin with the objective of skin layer retention is the ability to detect the applied actives in the skin. There are various techniques that may be applied for such detection and may be selected based on factors relating to the activity being analyzed. These techniques range from simply extracting the active from the skin and quantifying it using standard analytical techniques such as high performance liquid chromatography (HPLC) or liquid chromatography-mass spectrometry (LC/MS) to applying more advanced technology to identifying penetration depths and skin imaging. Described below are some of the techniques used for detecting actives within skin layers.

11.4.1 Extraction of Active from Skin

The most straightforward in vitro method of detection of actives meant for retention in the skin is the extraction of the active into suitable extraction solvents and quantification using a validated analytical technique. After an in vitro skin permeation experiment, extraction of the active can be performed on the entire skin tissue used for the permeation experiment, or the epidermis can be separated from the dermis, and extraction can be performed on each of the layers [66]. Tape stripping of the

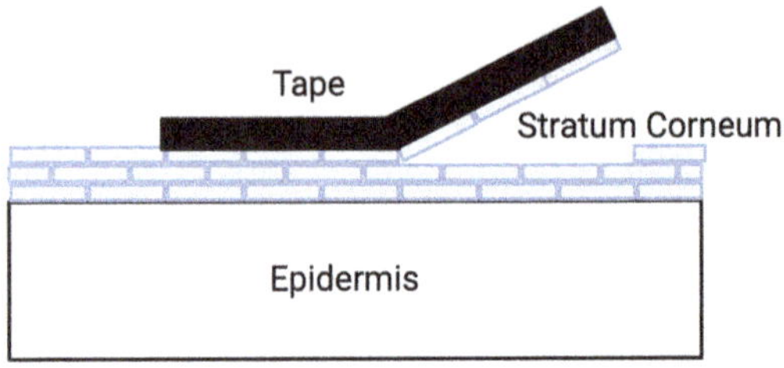

Fig. 11.4 Diagrammatic representation of tape stripping for extraction of active from skin layers

skin can also be employed for progressive removal of stratum corneum with each tape strip and determination of activity in each tape, which results in a penetration depth profile [67] (Fig. 11.4).

11.4.2 Fluorescence Microscopy

Fluorescence imaging can be applied to visualize actives that fluoresce, or fluorescent dyes with similar properties to the active can be used in place of actives. Actives can also be tagged with fluorescent dyes, but this process often makes the molecule larger and may alter the overall permeation behavior of the combination. This technique provides a visual representation of the penetration into the skin layers but may not be an accurate representation of the permeant.

11.4.3 ATR-FTIR Spectroscopy

The concentration of actives in deeper skin layers can also be determined using spectroscopic techniques such as attenuated total reflectance Fourier transform infrared spectroscopy (ATR-FTIR) [66, 67]. The stratum corneum, or epidermis, is placed in close contact with the ATR crystal, and the formulation is applied to the skin. Diffusion of the active in the skin layers is monitored by the active-specific infrared spectroscopy (IR) absorbance as a function of time [68, 69]. This is a non-destructive technique of analysis requiring minimal sample preparation capable of generating results at a fast rate [69] (Fig. 11.5).

11.4.4 Confocal Raman Spectroscopy and Microscopy

Another advanced non-invasive technology that has been used for real-time profiling of diffusion of actives into the skin is confocal Raman spectroscopy (CRS). This technique also carries the advantage of not needing any pretreatment or labeling [70]. Confocal Raman microscopy (CRM) has been identified as a more attractive

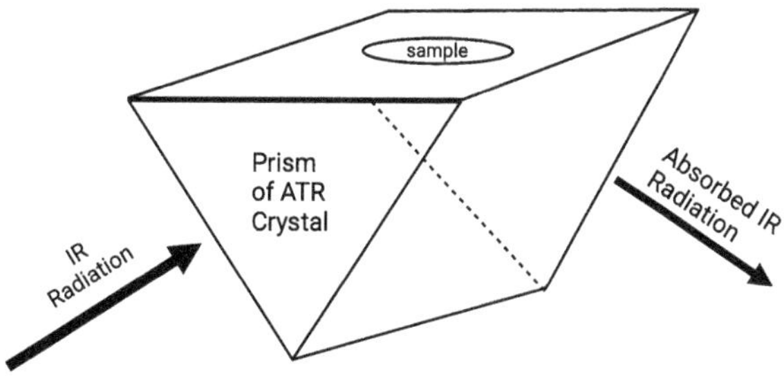

Fig. 11.5 Schematic representation of experimental setup for ATR-FTIR-based detection of skin permeation of actives

alternative to time- and labor-intensive techniques such as tape stripping that lack spatial resolution [71]. Iliopoulos et al. developed an approach to draw an in vitro-in vivo correlation by studying the permeation of niacinamide in vitro using conventional Franz diffusion cells and in vivo using a quantitative CRS method [72]. Caspers et al. demonstrated a method to detect and simultaneously quantify actives and flux through the stratum corneum in vivo following topical application using CRS [73]. Zarmpi et al. utilized CRS to characterize dermatopharmacokinetics ex vivo using 4-cyanophenol (CP) as a model skin penetrant [74]. This work showed that CRS can robustly follow an active's delivery into the deeper layers of skin. In another work, Zarmpi et al. tracked the drug disposition of the same model drug, CP, in the skin using stimulated Raman scattering (SRS) microscopy, which is said to be more precise than confocal Raman microscopy [75]. This allows us to visualize the localization of an active ingredient and study its preference for the stratum corneum lipids or corneocytes.

Deeper penetration may also be measured using spatially offset Raman spectroscopy (SORS), which can enable even stronger signals than CRS [71].

11.4.5 Other Techniques

There are other techniques that are used less frequently or may have been more prevalent in the past. Radiolabeling and radiography have been utilized, but these can be slow and expensive methods [66]. Photothermal spectroscopy involves using a modulated UV beam on the skin, and the resulting increase in temperature causing pressure oscillations can be read using a sensitive microphone. This method would only apply to UV-absorbing actives. Alternative skin models, such as parallel artificial membrane permeability assay (PAMPA), have been used to visualize the permeation behavior of actives in the skin and as a tool for screening enhancers [76]. This does not completely substitute for the use of skin in permeation studies but could reduce its utilization.

11.5 Methods of Reduction of Dermal Absorption

11.5.1 Encapsulation in Vesicles

Ultra-flexible liposomes: An ideal formulation should increase the concentration of active drugs in the target tissue (the skin) without enhancing penetration to the extent that the drugs enter the systemic circulation in toxic amounts. For diseases where the target tissue is the skin and the active drugs are highly toxic to other organs, ultra-flexible liposomes would be the optimal choice of formulation [77]. A study conducted by Peralta et al. showed that the ultra-flexible liposomes loaded with amphotericin B could be an alternative source of treatment for the treatment of cutaneous leishmaniasis. The ultra flexibility of liposomes is due to the ratio of soy phosphatidylcholine and sodium cholate [78]. Encapsulated drugs in the form of transferosomes, ethosomes, and transethosomes have shown increased skin retention and enhanced therapeutic effects in the treatment of various skin disorders. For example, transferosomal and ethosomal formulations were proven effective in the treatment of topical fungal diseases [79, 80]. These ultra-flexible liposomes were used for the encapsulation of non-steroidal anti-inflammatory drugs for the treatment of rheumatoid arthritis [81, 82]. Ethosomes have been shown to be effective carriers for delivering erythromycin to bacteria located within the deep skin layers to eradicate staphylococcal infections [83]. A recent study conducted on curcumin loaded in hyaluronic acid-coupled propylene glycol-based ethosomes effectively treated psoriasis by localizing the drug in the skin layers [84].

11.5.2 Polymeric Complexation

Complexation of drugs with cyclodextrin, hydroxypropyl-β-cyclodextrin, or poloxamer hydrogels could be an alternative method for increasing dermal retention. These poloxamers are composed of hydrophilic (polyethylene oxide) and hydrophobic (polypropylene oxide) blocks that form micelles at certain temperatures, alter the solubility of the molecules, and restrict them in the skin layers. A recent study conducted by Padula et al. revealed that the complexation of budesonide with cyclodextrin accumulated the drug more in the epidermis than dermis and hence restricted the drug from entering deeper skin layers [85].

11.5.3 Permeation Retardants

Chemicals such as perfluoroalkyl polyether and polytetrafluoroethylene are sometimes used as permeation retardants. These chemicals were formulated as creams to provide protection against chemical warfare agents such as sulfur mustard, malathion, and parathion. The addition of a carboxylic or phenolic group to the penetration enhancers reversed its action to retardation. The addition of high molecular weight chemical moieties also causes skin retardation. The chemical bonding of a bulky polymer to the permeant also inhibits dermal uptake. The reduction in dermal uptake can be attributed to the reduced thermodynamic activity, reduced solubility of a permeating agent, or increasing the micro-viscosity of skin lipids [85].

11.5.4 Regeneration of Stratum Corneum

The stratum corneum is a semipermeable barrier that protects against external agents like bacteria, fungi, and chemicals, while also preventing the loss of fluids and solutes from the body. It is thickest on the palms and soles. Desquamation is the process of shedding cells from the surface of the stratum corneum, balancing the proliferating keratinocytes formed in the stratum basale [86]. These cells migrate through the epidermis to the surface in a journey that takes about fourteen days. Ceramide is a crucial component in the formation of the stratum corneum, which has a brickwork-like structure. The study conducted by Jung SM et al. supported the effect of a ceramide nanoformulation as an effective skin regeneration treatment for atopic dermatitis [87].

11.6 Conclusion

In this chapter, the multifaceted topic of dermal drug/active retention was reviewed, underscoring its importance in dermatology and therapeutic treatments. Initially, the need was highlighted for effective dermal retention to ensure that active ingredients remain within the target skin layer site, enhancing therapeutic outcomes and minimizing systemic exposure.

The chapter then explored the pharmacokinetics of dermal absorption, detailing the processes by which substances penetrate and are absorbed into and through the skin. The intricate balance required to optimize the delivery and retention of actives within the dermal layers was discussed, ensuring maximum efficacy of the applied compound.

A comprehensive analysis of the factors affecting dermal absorption was provided, illustrating how variables such as skin condition, formulation characteristics, and external environmental factors can influence the absorption rate and extent. This section emphasized the complexity of achieving consistent and controlled dermal absorption.

We also reviewed the detection methods of actives in the skin, highlighting advanced techniques used to measure and monitor the presence and concentration of therapeutic agents within the dermal layers. These methods are crucial for both research and clinical practice, providing insights into the effectiveness and safety of dermal treatments.

Finally, we examined various methods to reduce dermal absorption when necessary, such as in cases where minimizing systemic exposure is critical. Techniques to modulate and control absorption rates were discussed, offering strategies to tailor treatments to specific therapeutic needs. Through this comprehensive exploration, we have laid a solid foundation for understanding the mechanisms and challenges associated with dermal retention and absorption. The insights gained here are essential for developing effective dermatological therapies and enhancing patient outcomes. By integrating the principles discussed in this chapter with cutting-edge research and clinical practices, we aim to push the boundaries of dermatological science, paving the way for improved therapeutic interventions and patient care.

References

1. Yousef, Hani, Mandy Alhajj, and Sandeep Sharma. "Anatomy, skin (integument), epidermis." (2017).
2. Waugh, Anne, and Allison Grant. *Ross & Wilson Anatomy and physiology in health and illness E-book*. Elsevier Health Sciences, 2014.
3. Tanwar, Himanshi, and Ruchika Sachdeva. "Transdermal drug delivery system: A review." *International journal of pharmaceutical sciences and research* 7.6 (2016): 2274.
4. Sharma, N., Agarwal, G., Rana, A. C., Bhat, Z. A., & Kumar, D. (2011). A review: Transdermal drug delivery system: A tool for novel drug delivery system. *International Journal of Drug Development & Research, 3*(3), 70–84.
5. Cevc, Gregor. "Drug delivery across the skin." *Expert opinion on investigational drugs* 6.12 (1997): 1887–1937.
6. Allen, Loyd, and Howard C. Ansel. *Ansel's pharmaceutical dosage forms and drug delivery systems*. Lippincott Williams & Wilkins, 2013.
7. Tanwar, Himanshi, and Ruchika Sachdeva. "Transdermal drug delivery system: A review." *International journal of pharmaceutical sciences and research* 7.6 (2016): 2274.
8. Sharma, N., Agarwal, G., Rana, A. C., Bhat, Z. A., & Kumar, D. (2011). A review: Transdermal drug delivery system: A tool for novel drug delivery system. *International Journal of Drug Development & Research, 3*(3), 70–84.
9. Allen, Loyd, and Howard C. Ansel. *Ansel's pharmaceutical dosage forms and drug delivery systems*. Lippincott Williams & Wilkins, 2013.
10. O'Connell, Rachel L., and Jennifer E. Rusby. "Anatomy relevant to conservative mastectomy." *Gland Surgery* 4.6 (2015): 476.
11. Andersson, Tilde, et al. "Common skin bacteria protect their host from oxidative stress through secreted antioxidant RoxP." *Scientific reports* 9.1 (2019): 3596.
12. Förster, Matthias, et al. "Topical delivery of cosmetics and drugs. Molecular aspects of percutaneous absorption and delivery." *European Journal of Dermatology* 19.4 (2009): 309–323.
13. Singh, S., and J. Singh. "Transdermal drug delivery by passive diffusion and iontophoresis: a review." *Medicinal Research Reviews* 13.5 (1993): 569–621.
14. Talreja, Priva S., et al. "Visualization of the lipid barrier and measurement of lipid pathlength in human stratum corneum." *Aaps PharmSci* 3 (2001): 48–56.
15. Gorzelanny, Christian, et al. "Skin barriers in dermal drug delivery: which barriers have to be overcome and how can we measure them?" *Pharmaceutics* 12.7 (2020): 684.

16. Cozzi, Arianna C., Paola Perugini, and Samuel Gourion-Arsiquaud. "Comparative behavior between sunscreens based on free or encapsulated UV filters in term of skin penetration, retention and photo-stability." *European Journal of Pharmaceutical Sciences* 121 (2018): 309–318.
17. Brown MB, Jones SA. Hyaluronic acid: a unique topical vehicle for the localized delivery of drugs to the skin. Journal of the European Academy of Dermatology and Venereology. 2005;19(3):308–18.
18. Ni, Chang, et al. "Hyaluronic acid and HA-modified cationic liposomes for promoting skin penetration and retention." *Journal of Controlled Release* 357 (2023): 432–443.
19. Zhu, Jieyu, et al. "Applications and delivery mechanisms of hyaluronic acid used for topical/transdermal delivery–a review." *International journal of pharmaceutics* 578 (2020): 119127.
20. de Andrade, Diego Fontana, et al. "Nanoencapsulation of clobetasol propionate decreases its penetration to skin layers without changing its relative skin distribution." *Journal of Nanoscience and Nanotechnology* 15.1 (2015): 875–879.
21. Sahni, Kanika, Sanjay Singh, and Sunil Dogra. "Newer topical treatments in skin and nail dermatophyte infections." *Indian dermatology online journal* 9.3 (2018): 149–158.
22. Fantini, Adriana, et al. "In vitro skin retention of crisaborole after topical application." *Pharmaceutics* 12.6 (2020): 491.
23. Ferreira, Luciana S., et al. "In vitro skin permeation and retention of paromomycin from liposomes for topical treatment of the cutaneous leishmaniasis." *Drug development and industrial pharmacy* 30.3 (2004): 289–296.
24. Slavkova, Marta, et al. "Gel formulations for topical treatment of skin cancer: a review." *Gels* 9.5 (2023): 352.
25. Patel, Mrunali R., et al. "Novel isotretinoin microemulsion-based gel for targeted topical therapy of acne: Formulation consideration, skin retention and skin irritation studies." *Applied nanoscience* 6.4 (2016): 539–553.
26. Leppert, Wojciech, et al. "Transdermal and topical drug administration in the treatment of pain." *Molecules* 23.3 (2018): 681.
27. Mohanty, Chandana, and Sanjeeb K. Sahoo. "Curcumin and its topical formulations for wound healing applications." *Drug discovery today* 22.10 (2017): 1582–1592.
28. Kulac, Mustafa, et al. "The effects of topical treatment with curcumin on burn wound healing in rats." *Journal of molecular histology* 44 (2013): 83–90.
29. Pulit-Prociak, Jolanta, et al. "Safety of the application of nanosilver and nanogold in topical cosmetic preparations." *Colloids and Surfaces B: Biointerfaces* 183 (2019): 110416.
30. Harris, Keegan James. *Occurrence of Per-and Polyfluoroalkyl substances in Cosmetics and Personal Care Products.* Diss. Carleton University, 2022.
31. More, Sharlee L., et al. "Dermal exposure and hair dye: Assessing potential bladder cancer risk from permanent hair dye." *Regulatory Toxicology and Pharmacology* 138 (2023): 105331.
32. Bacon, Brandon, et al. "Local anesthetic systemic toxicity induced cardiac arrest after topicalization for transesophageal echocardiography and subsequent treatment with extracorporeal cardiopulmonary resuscitation." *Journal of cardiothoracic and vascular anesthesia* 33.1 (2019): 162–165.
33. Wu, Chen-Chou, et al. "Particle transfer mediates dermal exposure of consumers to plasticizers in eraser and pen accessories." *Environment International* 180 (2023): 108191.
34. Saweres-Argüelles, Clara, et al. "Skin absorption of inorganic nanoparticles and their toxicity: A review." *European Journal of Pharmaceutics and Biopharmaceutics* 182 (2023): 128–140.
35. Pedon de Araujo Cardoso, Thalita, et al. "Pesticide dermal absorption: Case study x in vitro study." *Environmental Toxicology and Pharmacology* 75 (2020): 103313.
36. Tobarran, Natasha, et al. "Systemic toxicity from subcutaneous brimonidine injection successfully treated with naloxone." *The American Journal of Emergency Medicine* 59 (2022): 218–e5.
37. Polak, Sebastian, et al. "Prediction of concentration–time profile and its inter-individual variability following the dermal drug absorption." *Journal of pharmaceutical sciences* 101.7 (2012): 2584–2595.

38. Singh, Ashok K. *Engineered nanoparticles: structure, properties and mechanisms of toxicity*. Academic Press, 2015.
39. Lboutounne, Y. and P. Muret. *Skin Absorption in Man In Vivo*.Springer International Publishing. (2015):1–24.
40. Van der Merwe, D., et al. "A physiologically based pharmacokinetic model of organophosphate dermal absorption." *Toxicological Sciences* 89.1 (2006): 188–204.
41. Ali, Saba M., and Gil Yosipovitch. "Skin pH: from basic science to basic skin care." *Acta dermato-venereologica* 93.3 (2013).
42. Exposure Assessment Group, Office of Health and Environmental Assessment, U.S. Environmental Protection Agency. (January 1992). *Dermal Exposure Assessment: Principles and Applications* (EPA/600/8-91/011B) (pp. 2-1 to 2-35).
43. Shankar, Patole Bhushan, Shinkar Dattatraya Manohar, and Saudagar Ravindra Bhanudas. "Patches: A Novel approach for development of topical drug delivery system." *Journal of Advanced Pharmacy Education and Research* 3.4-2013 (2013): 347–358.
44. Kearns, Gregory L., et al. "Developmental pharmacology—drug disposition, action, and therapy in infants and children." *New England Journal of Medicine* 349.12 (2003): 1157–1167.
45. Potts, Russell O., Edmund M. Buras Jr, and Dan A. Chrisman Jr. "Changes with age in the moisture content of human skin." *Journal of investigative dermatology* 82.1 (1984): 97–100.
46. Furue, M., et al. "Clinical dose and adverse effects of topical steroids in daily management of atopic dermatitis." *British Journal of Dermatology* 148.1 (2003): 128–133.
47. Lu, Hong, and Sara Rosenbaum. "Developmental pharmacokinetics in pediatric populations." *The Journal of Pediatric Pharmacology and Therapeutics* 19.4 (2014): 262–276.
48. Farage, Miranda A., et al. "Clinical implications of aging skin: cutaneous disorders in the elderly." *American journal of clinical dermatology* 10 (2009): 73–86.
49. Farage, Miranda A., et al. "Characteristics of the aging skin." *Advances in wound care* 2.1 (2013): 5–10.
50. Williams, F. M. "Potential for metabolism locally in the skin of dermally absorbed compounds." *Human & experimental toxicology* 27.4 (2008): 277–280.
51. Kielhorn, J., Melching-Kollmu, S. and Mangelsdorf, I. Environmental Health Criteria (EHC) 235: Dermal Absorption. World Health Organization, (2006):102, 6–36.
52. Semple, Sean. "Dermal exposure to chemicals in the workplace: just how important is skin absorption?" *Occupational and environmental medicine* 61.4 (2004): 376–382.
53. Akula, Pranitha, and Lakshmi PK. "Effect of pH on weakly acidic and basic model drugs and determination of their ex vivo transdermal permeation routes." *Brazilian Journal of Pharmaceutical Sciences* 54.02 (2018): e00070.
54. Mbah, C. J., Philip F. Uzor, and E. O. Omeje. "Perspectives on transdermal drug delivery." *Journal of Chemical and Pharmaceutical Research* 3.3 (2011): 680–700.
55. Bos, Jan D., and Marcus MHM Meinardi. "The 500 Dalton rule for the skin penetration of chemical compounds and drugs." *Experimental Dermatology: Viewpoint* 9.3 (2000): 165–169.
56. Chen, Jiayi, et al. "Transdermal Delivery of Recombinant Human Growth Hormone by Liposomal Gel for Skin Photoaging Therapy." *ACS Applied Biomaterials* (2024).
57. Banerjee, Amrita, et al. "Transdermal protein delivery using choline and geranate (CAGE) deep eutectic solvent." *Advanced healthcare materials* 6.15 (2017): 1601411.
58. Münch, S., J. Wohlrab, and R. H. H. Neubert. "Dermal and transdermal delivery of pharmaceutically relevant macromolecules." *European journal of Pharmaceutics and Biopharmaceutics* 119 (2017): 235–242.
59. Jhawat, Vikas Chander, et al. "Transdermal drug delivery systems: approaches and advancements in drug absorption through skin." *Int J Pharm Sci Rev Res* 20.1 (2013): 47–56.
60. Frasch, H. Frederick, et al. "Analysis of finite dose dermal absorption data: implications for dermal exposure assessment." *Journal of exposure science & environmental epidemiology* 24.1 (2014): 65–73.
61. Bhowmick, Mithun, Tamizharasi Sengodan, and Sivakumar Thangavel. "TRANSDERMAL THERAPEUTIC SYSTEMS: A BRIEF OVERVIEW OF FACTORS AFFECTING PENETRATION AND PERMEATION." (2013).

62. Sugumar, Vaisnevee, et al. "Current development of chemical penetration enhancers for transdermal insulin delivery." *Biomedicines* 11.3 (2023): 664.
63. Ramadon, Delly, et al. "Enhancement strategies for transdermal drug delivery systems: Current trends and applications." *Drug delivery and translational research* (2021): 1–34.
64. Salimi, Anayatollah, and Saeed Mohammad Soleymani. "Transfollicular drug delivery systems." *Jundishapur Journal of Natural Pharmaceutical Products* 13.3 (2018).
65. Mohit, Martolia, J., & Singh, N. P. A review on nanoparticle technology for dermal drug delivery. European Chemical Bulletin, Mohit, Martolia, J., & Singh, N. P. (2023). A review on nanoparticle technology for dermal drug delivery. *European Chemical Bulletin,* Mohit, Martolia, J., & Singh, N. P. (2023). A review on nanoparticle technology for dermal drug delivery. *European Chemical Bulletin, (2023): 12*(S3), 5745–5761.
66. Touitou, Elka, Victor M. Meidan, and Ehud Horwitz. "Methods for quantitative determination of drug localized in the skin." *Journal of controlled release* 56.1-3 (1998): 7–21.
67. Moser, Katrin, et al. "Passive skin penetration enhancement and its quantification in vitro." *European journal of pharmaceutics and Biopharmaceutics* 52.2 (2001): 103–112.
68. McAuley, William James, et al. "The effects of esterified solvents on the diffusion of a model compound across human skin: An ATR-FTIR spectroscopic study." *International journal of pharmaceutics* 447.1–2 (2013): 1–6.
69. Tetteh, J., et al. "Local examination of skin diffusion using FTIR spectroscopic imaging and multivariate target factor analysis." *Analytica chimica acta* 642.1–2 (2009): 246–256.
70. Ameen, Dina, et al. "An overview of bioequivalence requirements for generic transdermal products." *Time-Proof Perspectives on Bioequivalence* (2023): 197–220.
71. Krombholz, Richard, and Dominique Lunter. "A new method for in-situ skin penetration analysis by confocal Raman microscopy." *Molecules* 25.18 (2020): 4222.
72. Iliopoulos, Fotis, et al. "Franz cell diffusion testing and quantitative confocal Raman spectroscopy: In vitro-in vivo correlation." *Pharmaceutics* 12.9 (2020): 887.
73. Caspers, Peter J., et al. "Method to quantify the in vivo skin penetration of topically applied materials based on confocal Raman spectroscopy." *Translational Biophotonics* 1.1–2 (2019): e201900004.
74. Zarmpi, Panagiota, et al. "Confocal Raman Spectroscopic Characterization of Dermatopharmacokinetics Ex Vivo." *Molecular Pharmaceutics* 20.11 (2023): 5910–5920.
75. Zarmpi, Panagiota, et al. "Evaluation of chemical disposition in skin by stimulated Raman scattering microscopy." *Journal of Controlled Release* 368 (2024): 797–807.
76. Rahma, Annisa, Majella E. Lane, and Bálint Sinkó. "A comparative study of the in vitro permeation of 2-phenoxyethanol in the skin PAMPA model and mammalian skin." *International Journal of Pharmaceutics* 635 (2023): 122692.
77. Sudhakar, Kalvatala, et al. "Ultraflexible liposome nanocargo as a dermal and transdermal drug delivery system." *Nanomaterials* 11.10 (2021): 2557.
78. Peralta, Ma F., et al. "Liposomes can both enhance or reduce drugs penetration through the skin." *Scientific reports* 8.1 (2018): 13253.
79. Agrawal, Rumjhum, et al. "Development and evaluation of curcumin-loaded elastic vesicles as an effective topical anti-inflammatory formulation." *Aaps Pharmscitech* 16 (2015): 364–374.
80. Abd El-Alim, Sameh Hosam, et al. "Comparative study of liposomes, ethosomes and transfersomes as carriers for enhancing the transdermal delivery of diflunisal: in vitro and in vivo evaluation." *International journal of pharmaceutics* 563 (2019): 293–303.
81. Agrawal, Rumjhum, et al. "Development and evaluation of curcumin-loaded elastic vesicles as an effective topical anti-inflammatory formulation." *Aaps Pharmscitech* 16 (2015): 364–374.
82. Abd El-Alim, Sameh Hosam, et al. "Comparative study of liposomes, ethosomes and transfersomes as carriers for enhancing the transdermal delivery of diflunisal: in vitro and in vivo evaluation." *International journal of pharmaceutics* 563 (2019): 293–303.
83. González-Rodríguez, M. L., et al. "Deformability properties of timolol-loaded transfersomes based on the extrusion mechanism. Statistical optimization of the process." *Drug development and industrial pharmacy* 42.10 (2016): 1683–1694.

84. Kaur, P. R. A. B. H. J. O. T., et al. "Formulation, systematic optimization, in vitro, ex vivo and stability assessment of transethosome based gel of curcumin." *Asian J Pharm Clin Res* 11.2 (2018): 41–7.
85. Hadgraft, Jonathan, et al. "Mechanisms of action of skin penetration enhancers/retarders: Azone and analogues." *International Journal of Pharmaceutics* 141.1–2 (1996): 17–25.
86. Menon, Gopinathan K., Gary W. Cleary, and Majella E. Lane. "The structure and function of the stratum corneum." *International journal of pharmaceutics* 435.1 (2012): 3–9.
87. Jung, Sang-Myung, et al. "Thermodynamic insights and conceptual design of skin-sensitive chitosan coated ceramide/PLGA nanodrug for regeneration of stratum corneum on atopic dermatitis." *Scientific reports* 5.1 (2015): 18089.

Chapter 12
The Maximum Usage Trial: From Inception to Standardization

Chinmay Shukla and Edward Bashaw

12.1 Overview

The Maximum Usage Trial, or MUsT, has been a key component of the safety evaluation of topically applied/locally acting dermatologic products since the mid-1990s in the United States. Developed by the US Food and Drug Administration (FDA), its goal is to assess the absorption potential under the highest labeled conditions of use. In this chapter, we will present a historical overview of what led to the development of this methodology, its implementation, and its evolution over the last 30 years, and finally discuss future potential alternatives.

12.2 The MUsT Beginnings: The Paleo Regulatory Era

Up until the early 1990s, the assessment of in vivo plasma levels as part of the drug development program for topically applied products was an exception and not the rule. Many reasons accounted for this lack of evaluation, but it was primarily due to the lack of suitable bioanalytical methods. When such studies were done, they were done under "less than optimal or less than maximal" conditions, which had little or no clinical relevance. That is, studies were done not in patients with diseased skin (where skin permeability could be different) but with healthy volunteers and often done over a very small circumscribed area with no relationship to the actual clinical

C. Shukla (✉)
Lead Pharmacokineticist, DIIP, OCP, OTS, CDER, FDA, Silver Spring, MD, USA
e-mail: Chinmay.Shukla@fda.hhs.gov

E. Bashaw
Bashaw Consulting LLC, Monroe, NJ, USA

S. N. Murthy (ed.), *Topical Products and Dermal Drug Delivery*,
AAPS Introductions in the Pharmaceutical Sciences 2,
https://doi.org/10.1007/978-3-031-80525-7_12

presentation and clinical use. It was not uncommon to have the study done as a single application over a small area for example, 10 cm^2 area. Unsurprisingly, even when analytical methods did exist that were sufficiently sensitive to enable assessment of in vivo blood levels, the design conditions of the study were such that the potential for any drug absorption was minimized because of the less-than-optimal design of the pharmacokinetic (PK) studies.

Oftentimes, it was assumed that since topical dermatological products are meant to be locally acting, they are not designed to be absorbed systemically. As noted above, even when studies were conducted, their less-than-optimal design produced results that reinforced this position in the minds of the drug developers. Prior to the adoption of the MUsT paradigm, the only class of drugs that had an in vivo assessment were the topical corticosteroids that utilized the Stoughton-McKenzie Vasoconstrictor [1] assay. This was primarily utilized, in a regulatory sense, in the evaluation of generic drug products [2] and in the assignment of a potency class for a new molecular entity or reformulation on the new drug side. This is, of course, a pharmacodynamic assessment of the degree of skin blanching, which is a surrogate of absorption that may be related to the potential for hypothalamic-pituitary-adrenal (HPA) axis suppression, depending on the clinical usage and duration.

This was the state of review science in dermatology up until 1994. In 1994, a standalone Division of Dermatology and Dental Products (DDDP) was established at the FDA [3]. The establishment of a standalone review division focused on dermal drugs allowed for the development of a firm scientific underpinning of dermatological review at the FDA.

12.3 Recognition of Need

During this time, it was apparent that more and more potent dermatological products were in development. Drugs such as finasteride and tacrolimus both of which have known systemic effects and could have the potential for dermal absorption following topical administration. In light of this, there was a need to assess in vivo concentrations of dermal drugs, and the only options that readily presented themselves at that time were radiolabeled studies, tape stripping, and microdialysis.

Radiolabeled studies were and are routinely used in Phase 1 drug development for drug disposition studies [4]. They are expensive to perform and needing the production of radiolabeled compound, had an emotional baggage of using radioactive substances, making subject recruitment an issue, and normally used a very small number of subjects and were thus not predictive of larger, more varied populations.

Tape stripping, that is, the removal of the stratum corneum via repeated application and removal of tape, also referred to as "dermatopharmacokinetics" or "DPK," was the subject of an FDA Draft Guidance Document [5] that was issued in 1998. The method had the advantage of being relatively easy to do and low cost. Unfortunately, the method was limited in regard to its use on diseased skin and issues surrounding reproducibility [6], resulting in the guidance being withdrawn in 2002. Since the withdrawal of DPK Guidance, additional research has been conducted with DPK, and a better understanding of the failure in this instance has been elucidated, and an improved methodology [7] has been developed. Even so, it still had limitations in that it is designed to be used on healthy skin and is more amenable to generic or bioequivalence determinations.

Microdialysis [8] technique that utilizes small probes inserted into the skin below the area of topical application was widely viewed as a contender for a regulatory tool in the assessment of dermal penetration. There was much excitement around its ability to sample the intercellular fluid in the skin and provide a direct assessment of concentrations in the dermis. Its limitations were, however, substantial, as it was primarily limited to single dosing due to patient inconvenience, required a fair degree of dexterity to place the probe properly, and had analytical system issues in the bioanalysis of microsampling and microdialysis membrane selection [9]. Still, while not deployable on a large scale, microdialysis continues to be used as a research tool and has been employed in other tissues successfully in drug development [10].

Faced with this, it was decided that a new approach would be needed and that it would need to assess and address all the recognized factors of drug absorption, and the data that was generated should be sufficiently robust to cover a range of exposures. Additionally, it was desirable to have a methodology that would be as clinically relevant as possible and one that would not require specialized devices or training to achieve reproducible results. As necessity is the mother of invention, the MUsT concept was born.

12.3.1 Development of MUsT Principles

As it is well recognized today and discussed elsewhere in this book, one can break the factors affecting dermal absorption into three broad domains, all of which would need to be considered in assessing dermal absorption.

- Delivery system technology.
- Ingredient factors (both active pharmaceutical ingredients (APIs) and excipients).
- Skin factors.

Each of these factors has been gone into by other authors in this book in sufficient detail to preclude a repetitive discussion of them here, beyond indicating that in vivo bioavailability assessment resides at the intersection of these three domains (Fig. 12.1).

Recognizing this reality, the initial assessment of the pharmacokinetics began in a subgroup of patients that represented the upper limit of patient exposure within larger clinical trials. This was a reaction to the previously described situation where, for example, subjects with psoriasis over 30% of body surface area (BSA) had their absorption evaluated following only a 10 cm^2 application area. To guide the design of the maximal use study, in the mid-1990s a "template" comment was shared with sponsors developing topical dermatological products. An example of

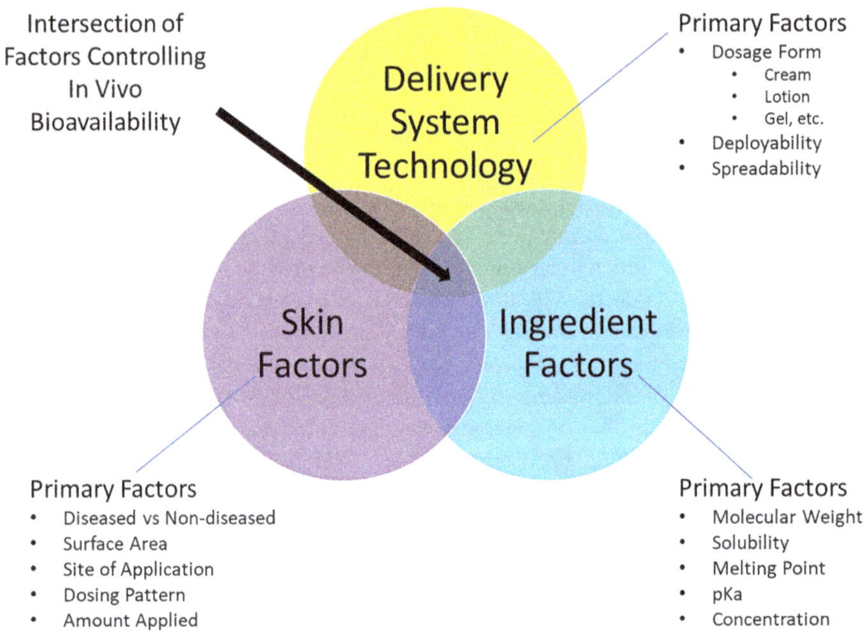

Fig. 12.1 Intersection of key factors in determining dermal bioavailability

the standard comment excerpted from the FDA's Draft 2005 "Guidance for Industry Acne Vulgaris: Developing Drugs for Treatment" [11] is shown below:

> ***B. Biopharmaceutical Considerations***
>
> *Pharmacokinetic studies to assess the degree of and/or potential for systemic absorption may be needed to fulfill the requirements of 21 CFR part 320 (Bioavailability and Bioequivalence Requirements). Under this section, a new drug application (NDA) must either contain an assessment of in vivo bioavailability or sufficient information which would allow the Agency to issue a waiver of in vivo bioavailability testing. In general, waivers of in vivo biostudies are the exception and are only granted in specific cases. Because of the variable dosing nature of topical products, the Agency recommend that studies of the in vivo assessment of systemic exposure be done under so-called maximal use conditions. The recommended elements of such a maximal use study are as follows:*
>
> - *A formulation identical to the clinically studied/to-be-marketed formulation should be used.*
> - *The study should be done in an adequate number of patients with area of involvement and disease severity index measuring toward the upper end of that in the proposed indication and anticipated clinical use.*
> - *The topical dosing used should represent the maximal dosing anticipated in both phase 3 trials and in the proposed package insert for the following:*
>
>   *(a) Frequency of dosing*
>   *(b) Duration of treatment*
>   *(c) Use of highest proposed strength*
>   *(d) Extent of involved area to be treated at one time*

(e) Amount applied per square centimeter
(f) Method of application/site preparation

- *The analytical method should be properly validated for both parent compound and metabolites.*

  The objective of this study is to maximize those elements affecting dermal penetration such that systemic absorption can be determined.

These were chosen to optimize the three identified domains of systemic availability in vivo [12]. To better convey the utility of the resulting data, we will present a short overview of some of the key MUsT elements that must be considered in trial design and are of particular importance to the "rollout" section below.

12.3.2 Subjects

As noted previously, one of the common elements seen in the early days of dermatologic research was the use of subjects with intact skin barriers, i.e., healthy skin. The skin, as denoted in other parts of this book, is an effective barrier to outside elements and regulates water balance in the body. The archetype of diseased skin used in discussing the MUsT is the psoriatic plaque, where the skin cells are laid down not in the classic "brick and mortar" model with tight joints but are laid down erratically [13]. These plaques, although thick, do not present a uniform barrier and significantly impair the epidermal barrier. In a study of healthy skin vs. psoriatic plaques conducted by Alejandro Molina-Leyva, they found "*Transepidermal Water Loss (TEWL) was significantly higher at psoriatic plaques (18.45* $g{\cdot}m^{-2}{\cdot}h^{-1}$*) than at uninvolved psoriatic skin (12.06* $g{\cdot}m^{-2}{\cdot}h^{-1}$*) and healthy skin (12.34* $g{\cdot}m^{-2}{\cdot}h^{-1}$*), while no differences were found between uninvolved psoriatic skin and healthy skin.*" [14] This and other studies using the TEWL methodology clearly indicate that the use of healthy subjects as a surrogate for diseased skin is not scientifically defensible, save in those cases where the disease being treated, such as hyperpigmentation disorders or alopecia, occurs in normal skin.

12.3.3 Age

Closely related to the issue of healthy vs. diseased skin is one of age. Skin is one of the first manifestations of aging and is one that significant amounts of money are spent on to camouflage, reverse, or arrest it. That being said, the MUsT design needs to incorporate a sensitivity to this in relation to the prevalence of the disease (and its clinical manifestations). This is especially important for drugs targeted to defined populations at both ends of the age scale, that is, pediatric and geriatric. Accordingly, for drugs that are intended to be applied to subjects from 6 months to

12 years, for example, there should be a representation of subjects at both age limits in the MusT. Studies where the bulk of the subjects are clustered at the upper age range are unacceptable as it dilutes, for example, the information provided by the 1 subject at 6 months. While the older subjects are easier to recruit and to obtain informed consent in, we cannot lose sight of the differentiation in neonatal skin and the alteration of body surface area to body volume ratios associated [15] with aging. The same holds true for the geriatric population, where a lifetime of occupational exposure can affect skin elasticity and permeability, making the "healthy adult population" of 18–45 or 65 a less relevant model for absorption in these populations.

12.4 Surface Area Factors

The surface area chosen, and its preparation, were another driver in establishing the need for a standardized study design. As noted before, prior to the MUsT rollout and implementation (to be discussed below), when in vivo studies were attempted, beyond using healthy subjects, they also used surface areas that did not align with the intended therapeutic use. There were situations where sponsors were requesting 30% body surface areas and then presented "supporting" PK studies with skin involvement that was far below the anticipated use. The lack of in vivo plasma data is being held up as evidence of systemic safety or non-absorbability, even though the studied use had no relationship to the proposed clinical use. This kind of approach is of limited to no value from a PK perspective and even less from a systemic safety perspective.

Additionally, there is also the situation, such as with acne vulgaris, where the potential sites of application are disparate and include the face, shoulders, upper back, and upper chest. As part of an acne MUsT, all affected areas should be treated as skin does vary in thickness, degree of occlusion,[a] and ease of application.

The final element in this topic is the issue of site preparation. That is, prior to applying the topical product, are any site preparation steps taken? For example, is the site washed? Or is there a need for debridement prior to drug application, as is the case in drugs for wound healing? If so, then the same cleanser and/or site preparation procedures should be provided to each subject who has been trained in the amounts and manner of use. Is the site trimmed, such as in the case of topical pre-surgical agents? No matter how carefully done, surgical clippers can induce miniature nicks in the stratum corneum that could damage the skin and result in increased permeation. We already accept the situation that everyone's skin is different for a variety of environmental, genetic, and lifestyle issues. In conducting a MUsT, we need to proactively control these factors to the best of our ability to ensure that the data that comes out of a MUsT is relevant to the clinical situation.

As noted in the MUsT template language provided above, there are a variety of factors to be considered in designing the MUsT. What has been presented in this section is a brief overview of key factors and does not represent a full discussion of the issues. The FDA has published guidance entitled "Maximal Usage Trials for

Topically Applied Active Ingredients Being Considered for Inclusion in an Over-the-Counter Monograph: Study Elements and Considerations" [15] and a "Guidance Snapshot" that provides a clearer overview of these approaches [16]. While this is directly related to the over-the-counter (OTC) situation, the design elements discussion is particularly on point and should serve as a more complete discussion of these factors.

12.5 MUsT Rollout and Implementation

At the time of inception, the MUsT concept was not well received, change never is, especially as it involved the costs associated with a new trial. Initially, MUsT studies started to be performed as part of the larger clinical trials program as an open-label subgroup. These days, MUsT is primariliy conducted as a standalone trial as part of the Phase 2 development program.

12.5.1 Elidel™ (Pimecrolimus)

Pimecrolimus was approved by the FDA in 2001 for the treatment of mild to moderate atopic dermatitis [17]. Its development program kicked off in the same time frame as the initial rollout of the MUsT paradigm, and it was one of the first NDAs submitted to the FDA that incorporated such studies into their clinical studies program. It was marketed to the pediatrician as a treatment for use in children as young as 2 years of age, as these subjects can have large surface area involvement up to 90% BSA [17, 18]. At the time of approval and during the discussion in the Joint Clinical Pharmacology-Pediatric Advisory Committee Meeting held in 2000, there was some concern about a tumor signal from some animals. So much so that, as part of the approval letter, the sponsor was required to commit to performing a series of additional studies to address this via a registry study [19]. Post approval, there were reports of lymphoma to such a degree that the FDA held a Pediatric Advisory Committee meeting in February 2005 [20], which voted to recommend that the FDA add a boxed warning to Elidel™ [21] and to the other class member Protopic™ (tacrolimus) [22].

While a full discussion of the issues surrounding lymphoma and topical calcineurin inhibitors is beyond the scope of this chapter and is still hotly debated [23, 24], the inclusion of a boxed warning had a dramatic impact on the sales and marketing [25] of Elidel™. In the defense of Elidel™, the sponsor was able to cite the results (contained in the label) of three maximal-use studies. One in adults and two trials were specifically done in children down to the age of two years and below. Both studies showed exceedingly low levels that are described in both the FDA [26] and EMA [27] approved labeling, using text describing the pediatric MUsT trials:

WARNINGS

> **WARNING**
> **Long term Safety of Topical Calcineurin Inhibitors Has Not Been Established**
> Although a causal relationship has not been established, rare cases of malignancy (e.g., skin and lymphoma) have been reported in patients treated with topical calcineurin inhibitors, including ELIDEL Cream.
>
> Therefore:
> - Continuous long-term u se of topical calcineurin inhibitors, including ELIDEL Cream, in any age group should be avoided, and application limited to areas of involvement with atopic dermatitis.
> - ELIDEL Cream is not indicated for use in children less than 2 years of age.

The systemic exposure to pimecrolimus from pimecrolimus cream, 1% was investigated in 28 pediatric subjects with atopic dermatitis (20% to 80% BSA involvement) between the ages of 8 months to 14 yrs. Following twice daily application for three weeks, blood concentrations of pimecrolimus were <2 ng/mL with 60% (96/161) of the blood samples having blood concentration below the limit of quantification (0.5 ng/mL). However, more children (25 children out of the total 28 children investigated) had at least one detectable blood level as compared to the adults (12 adults out of the total 52 adults investigated) over a 5-week treatment period. Due to the erratic nature of the blood levels observed, no correlation could be made between amount of cream, degree of BSA involvement, and blood concentrations. In general, the blood concentrations measured in adult atopic dermatitis subjects were comparable to those seen in the pediatric population.

In a second group of 30 pediatric subjects aged to 25 months with 10% to 92% BSA involvement, following twice daily application for three weeks, blood concentrations of pimecrolimus were <2.6 ng/mL with 65% (75/116) of the blood samples having blood concentration below 0.5ng/mt, and 27% (31/116) below the limit of quantification (0.1 ng/mL) for these trials.

Overall, a higher proportion of detectable blood levels was seen in the pediatric subject population as compared to adult population. This increase in the absolute number of positive blood levels may be due to the larger surface area to body mass ratio seen in these younger subjects. In addition, a higher incidence of upper respiratory symptoms/infections was also seen relative to the older age group in the pK trials. At this time, a causal relationship between these findings and pimecrolimus cream, 1% use cannot be ruled out.

While this information did not fully explain the situation with regard to lymphoma, it did put the absorption profile out there for physicians and patients and allowed them to make an informed decision. Looking at it from the MUsT implementation perspective, we can develop the following table:

Pimecrolimus data from NDA 21-302	Adult patients	Pediatric atopic dermatitis	Pediatric atopic dermatitis
N	52	28	30
Age	19–59 years	8 months–14 years	3–23 months
BSA	13–62% involvement	20–80% involvement	10–92% involvement
Dosing	BID	BID	BID
Duration	Up to 1 year	Three Weeks	Three Weeks
% samples below BLQ	91%	60%	27%
"Peak concentration"	1.4 ng/mL	<2 ng/mL	<2.6 ng/mL

BSA Body surface area, *BLQ* Below quantifiable limits, *BID* Twice daily dosing

Thus, we can see that the adult MUsT was done, as recommended at that time, as a subgroup in the Phase 3 study, while the two pediatric studies had a standalone study design utilizing a three-week duration [28]. The number of subjects was more than adequate, and the populations studied were reflective of the clinical studies program and the proposed labeling. The utility of the MUsT program was demonstrated here by being able to provide some perspective with regard to the degree of systemic exposure. Had the sponsor not been convinced to incorporate this new study design into their program, they would have had to face, as former Defense Secretary Rumsfeld said the "known unknowns", "*...known knowns; there are things we know we know. We also know there are known unknowns; that is to say we know there are some things we do not know...*" [29] In this case, knowing the blood levels following clinical usage at maximal usage conditions allowed them to put into some context the "known unknown" of the degree of systemic absorption.

The MUsT themselves were well-designed, had subjects with disease at the upper limits of exposure, included subjects at the anticipated age limits, and utilized the to-be-marketed formulation and dosing regimen for a sufficient duration. Had a lesser-designed study been conducted, the Advisory Committee and the FDA would have been faced with a challenge of making decisions without a clear understanding of patient exposure vis-a-vis the association with lymphoma. While having a boxed warning is not a desired outcome, having robust information on absorption in the target population at their disposal helped to some degree in avoiding removal from the market.

12.6 The MUsT Implementation in Varying Cases of Dermal Development

The previously described case of Elidel™ occurred at the very dawn of the implementation of the MUsT in drug development. Since then, several MUsTs have been performed in support of drug development; however, in recent years, the number of MUsTs being performed has dropped. This is more a reflection of the transition away from topical drug therapy to the area of injected biological products and oral small molecules.

Today, MUsT continues to be done and has a vital role to play in the development of new topical dermatological drugs, the transition of drugs from the prescription to OTC setting, and in assessing absorption in monograph products.

12.7 Examples of the Use of MUsT in Various Settings

12.7.1 Use in the 505(b) (2) Space

For topical NDA applications following a 505(b) (2) regulatory pathway, a relative bioavailability study with the new proposed product used under maximal use conditions provides the data to aid in the establishment of a clinical bridge with the listed drug. The comparative systemic exposure would then be used to help sponsors cross-reference previous FDA findings of safety from an FDA-approved product.

The criticality of an adequately designed MUsT should not be overlooked just because there is an oral product approved. In the case of Zorvye™ cream [30], which is approved for the treatment of psoriasis, the systemic exposure of this topically applied product exceeded the one observed following oral administration of a 500 μg dose [31]. Hence, adequately designed MUsT would become pivotal in providing the assessment of systemic safety.

Treatment with the JAK inhibitors class of drugs is gaining popularity in the treatment of dermatological diseases. According to Li. et al. [32], the efficacy of some of the topically applied drugs was more than orally administered ones. The concern of systemic safety makes PK assessment under MUsT a critical "must-have" information in support of safety. In the year 2001, Ruxolitinib cream (Opzelura™) became the first topical JAK inhibitor product to be approved for the topical treatment of atopic dermatitis in subjects 12 years of age and older, and in support of the approval, PK assessment was conducted under maximal usage conditions in subjects 13 years and older with a mean BSA involvement of 37.5% (range 25% to 90%) [33].

12.7.2 Rx to OTC Switch

Dermatologic products are often seen as a natural category for over-the-counter (OTC) use. They have the advantage of having a demonstratable therapeutic effect that is easily observable, unlike the treatment of hyperlipidemia with statins. A clear example of the role of MUsT in this situation was its use as one of the pivotal studies to inform the Rx to OTC switch of Differin™ Gel, 0.1%, for the treatment of acne vulgaris. Adapalene, the active pharmaceutical ingredient in Differin™ Gel, 0.1%, is a retinoid-like substance with teratogenicity concerns. The study design and the results of the MUsT were presented at an Advisory Committee meeting, and the safety margin assessment based on the PK assessment from MUsT played a pivotal role in the decision to switch from Rx to OTC [34].

While the above is not an exhaustive list of MUsT experience on the new drug side, the goal here is to provide an overview of how the concept of MUsT has matured over the years, and today it is a pivotal study for topically administered dermatological products with widespread applications, be it assessment of systemic safety or supporting establishment of a clinical bridge under the 505(b) (2) regulatory pathway.

12.7.3 Sunscreen

There likely has been no other use of the MUsT that has been more reported on, than the use of MUsT in the assessment of sunscreen exposure. The results of the two MUsTs performed by the FDA itself were widely reported in the news media and in social media. Sunscreen use in the US has been increasing year by year due to both a recognition of the risks of sun damage-related aging and increased rates of skin cancer. When most of the US OTC sunscreens were first marketed, they were a seasonal product and had low concentrations of sunscreens in them as they were marketed as products to "promote tanning" and provided relatively low sun protection factor (SPF) values with SPFs of 8, which blocks approximately 87.5% of UVB radiation. Over time, as recognition of the risk of sun exposure has increased, more and more topical products and cosmetics have incorporated sunscreens into their formulations and at higher concentrations to produce higher SPFs up to SPF 100, which blocks 99% of UVB rays from reaching your skin [35]. While no one issue triggered the concern over sunscreen exposure, it was the publication of articles such as the one by Schlumpf and her colleagues in 2010 that reported sunscreen residues present in 46 out of 54 post-partum breast milk samples [36] that started to raise regulatory interest. While this study was conducted in the EU and contained a number of sunscreens not allowed in the US, it still represented a concern given that the usage pattern and near-ubiquitousness of these agents in the cosmetic and OTC categories had grown since their introduction and there was not a clear idea as to the resulting absorption potential.

Due to these concerns, the FDA held a Nonprescription Drugs Advisory Committee Meeting in 2014 to discuss the issues surrounding the dermal absorption of sunscreens and associated risks [37]. One of the conclusions from the meeting as it relates to MUsT was an endorsement of the methodology in general and a call for specific MUsT to be performed on sunscreens:

> *The committee expressed the need for human maximal use studies in all cases, absorption studies in infants and children looking at neurodevelopmental and other endpoints, and clinical trials for long-term safety, in the framework. [38]*

It was this endorsement of the methodology that ultimately led to the conduct of the two in vivo absorption studies by the FDA that resulted in publications in JAMA in 2019 and 2020 [39, 40]. These studies used commercially available sunscreens, applied as directed at a rate of 2 mg/cm^2 every two hours for eight hours, to 75% BSA for four days. PK samples were collected and analyzed using state-of-the-art analytical methods at the FDA laboratories. While the actual products used, their formulation, and plasma sampling differed between each study, to provide a better survey of marketed products and study endpoints, the end result was the same; that is, each of the tested sunscreens was detectable at plasma levels in excess of the FDA's theoretical threshold for the conduct of long-term animal safety trials of 0.5 ng/mL [41].

The Fig. 12.2 below is representative data from the 2019 study, showing the absorption profile of each ingredient plotted individually across formulations as their corresponding geometric means.

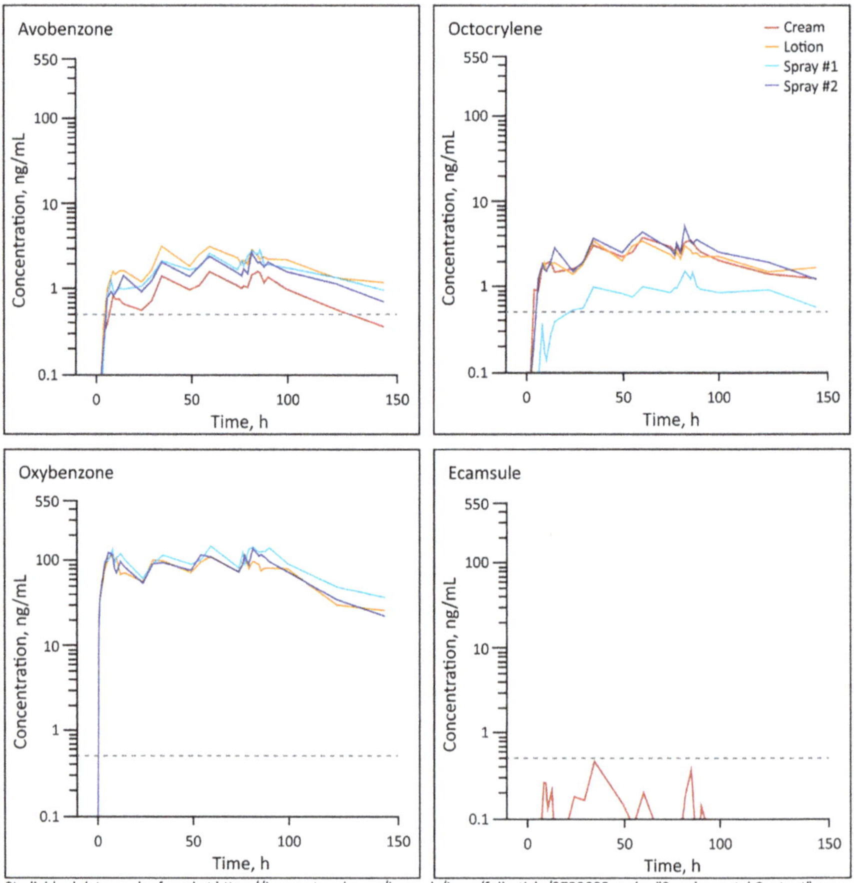

*Individual data can be found at https://jamanetwork.com/journals/jama/fullarticle/2733085 under "Supplemental Content".

Fig. 12.2 Geometric mean plasma concentrations of individual sunscreen ingredients across formulation following four daily applications across five days

The dashed horizontal lines indicate the 0.5-ng/mL plasma concentration threshold. The lower limit of quantitation (LLOQ) were 0.2 ng/mL for avobenzone, 0.4 ng/mL for oxybenzone, 0.4 ng/mL for octocrylene, and 0.2 ng/mL for ecamsule. All samples below the LLOQ were set to 0.1 ng/mL for plotting individual profiles [39].

As detailed in the publications and in the figures here, all of the tested sunscreens (except for ecamsule) and formulations tested here demonstrated significant dermal absorption following the current recommended dosing paradigm. While there is a long history of use for these sunscreens for decades prior to these studies, with the exception of ecamsule which was introduced into the US market in 2006 via the New Drug Application route, no meaningful assessment of in vivo absorption had been undertaken. Even with this history of use the actual usage patterns (amount applied), ubiquitousness of their deployment in cosmetics (formulation), and potential for extended chronic use all are factors that exceed their historical use. Even with the absorption demonstrated here the FDA in both articles, press releases, and subsequent publications has clearly stated that while:

> *The systemic absorption of sunscreen ingredients supports the need for further studies to determine the clinical significance of these findings. These results do not indicate that individuals should refrain from the use of sunscreen. [39]*

This data validated the "call to action" from the 2014 advisory committee and has kicked off more research in sunscreen absorption and increased regulatory activity within the OTC sunscreen space. This is a clear example of the MUsT being used to identify dermal absorption with systemic exposure and allowing for decisions to be made from a position of knowledge.

12.8 Future

The MUsT paradigm has been, in one form or another, in use at the FDA and in other countries since the mid-1990s. As it approaches 30 years of use, it can be seen to have served the industry and the FDA well. On the other hand, time and science march on, and one must continually assess the potential for its retirement and/or alteration. In 2014, the Pharmaceutical Quality Research Institute held a workshop in which both of the authors of this chapter participated, discussing the state-of-the-art technology with regard to topical bioavailability and bioequivalence assessment [42]. During the two-day workshop, a variety of methods were debated, including DPK (tape stripping), Raman Spectroscopy, Open Flow Microperfusion (OFM), Microdialysis, and the MUsT. One of the issues that came up during that meeting was the tendency for most researchers to look at the issue with a gaze toward developing a "one size fits all" approach, which is forcing one solution to work for all situations.

Clearly, for some elements, this is not possible, as Raman spectroscopy has limitations around fluoresce in the dermis of interfering substances. DPK has the issue of its utility in diseased skin. Microdialysis, as noted previously, due to its invasive nature, is limited to single-dose evaluation in patients and requires the development of specific semi-permeable membranes. This latter issue has been addressed in the more recent evolution of this technique into open flow microperfusion, which negates the need for specialized membranes [43], but still retains the invasive nature and the single-dose limitation in that leaving the probes in place for days is not feasible, and repeated insertion and removal of the probes leads to issues of reproducibility of probe placement and local irritation, which could affect permeation and probe sampling. Even so, overcoming this requirement for specialized membranes, the potential exists for OFM to find a home in the evaluation of generic drug products or possibly a part of a package for reformulations of existing topical drug products.

More recently, there has been increased interest in the use of computer or "in silico" methods to assess in vivo bioavailability for topical products [44]. These methods are an outgrowth of the use of pharmacometric tools for drug development, which have become an ever-increasing element in drug development. While there has been some use of these tools, and they are continually evolving, they have limitations in the differentiation of the skin and disease progression and "regression" to the healed state. One of the supplemental advantages of the MusT is the requirement that it be done under multiple-dose conditions. Thus, we are able to capture with it both accumulation (which can be readily modeled) and also the healing of the lesions over this observed timeframe (normally 2–3 weeks). That means that the skin, under active treatment, is being "remodeled" and returned to a more homeostatic nature with an improving barrier function. A computer model would have to somehow account for this progressive improvement, which may or may not be consistent over the entire site of application due to differences in the skin morphology itself. Acne again is the representative example, as disease involvement can be on the face and the trunk, two entirely different skin thicknesses and environments vis-a-vis occlusions. To date, computer simulation has not been able to demonstrate the degree of fidelity necessary to accomplish this. As methods evolve for use in transdermal simulations through healthy skin, it is possible that these tools could be adapted and may yet find a role in the future.

Even as time has passed, the final conclusion of the 2014 PQRI workshop continues on; that is

> *...workshop emphasized both a need for greater attention to quality, possibly, via a Quality-By-Design (QBD) approach and a need to develop a "whole toolkit" approach [emphasis added] towards the problem of determination of rate and extent in the assessment of topical bioavailability. [44]*

A whole toolkit approach in that we should be looking not for a "one size fits all approach" but that we should be validating multiple tools that are the "right tool" for the situation. Currently, of the methods discussed and proposed for use, only the MusT has the flexibility to be used for essentially any topically applied product and can result in relevant information when used under planned clinical conditions.

12.9 Conclusion

In this chapter, we have attempted to provide both a historical overview of the evolution of regulatory thought in the area of dermal absorption and how the Maximal Usage Trial was envisioned as a way to obtain useful information related to systemic safety. Ultimately, that is what a MusT is, a safety trial. A trial that can collect relevant in vivo exposure data and provide a pathway to link back both the preclinical data and also other data streams developed in the drug review process. It is a "provocative" study in that it operates in those subjects and conditions designed to produce the highest drug levels, in accordance with planned clinical use in the Phase 3 program and in the anticipated label, as these subjects would be the ones to have the highest anticipated absorption and thus the highest risk of systemic effects. As time marches on, it may yet be replaced with newer methods that can provide the same or better information in a more controllable manner, but for almost 30 years the MUsT has served the dermatological community well by providing a standardized drug absorption assessment technique that is relevant and achievable in all clinical situations.

Disclaimer This article reflects the views of the authors and should not be construed to represent Food and Drug Administration's views or policies.

References

1. McKENZIE AW, STOUGHTON RB. Method for Comparing Percutaneous Absorption of Steroids. *Arch Dermatol.* 1962;86(5):608–610. doi:https://doi.org/10.1001/archderm.1962.01590110044005
2. Williams RL. Bioequivalence of topical corticosteroids: a regulatory perspective. Int J Dermatol. 1992 Oct;31 Suppl 1:2–5. doi: https://doi.org/10.1111/j.1365-4362.1992.tb04004.x. PMID: 1428462.
3. https://pink.citeline.com/PS082375/FDA-TOPICAL-DRUG-REVIEW-DIVISION-WILL-BE-HEADED-BY-OHIO-STATEs-JONATHAN-WILKIN-MD
4. https://www.fda.gov/media/158178/download
5. https://wayback.archive-it.org/7993/20170405180513/; https://www.fda.gov/ohrms/dockets/ac/00/backgrd/3661b1c.pdf
6. https://wayback.archive-it.org/7993/20170405173544/; https://www.fda.gov/ohrms/dockets/ac/01/slides/3804s2_03_franz.pdf
7. N'Dri-Stempfer B, Navidi WC, Guy RH, Bunge AL. Improved bioequivalence assessment of topical dermatological drug products using dermatopharmacokinetics. Pharm Res. 2009 Feb;26(2):316–28. doi: https://doi.org/10.1007/s11095-008-9742-9. Epub 2008 Oct 22.
8. Au WL, Skinner MF, Benfeldt E, Verbeeck RK, Kanfer I. Application of dermal microdialysis for the determination of bioavailability of clobetasol propionate applied to the skin of human subjects. Skin Pharmacol Physiol. 2012;25(1):17–24. doi: https://doi.org/10.1159/000330489. Epub 2011 Aug 22. PMID: 21860254.
9. Holmgaard R, Nielsen JB, Benfeldt E. Microdialysis sampling for investigations of bioavailability and bioequivalence of topically administered drugs: current state and future perspectives. Skin Pharmacol Physiol. 2010;23(5):225–43. doi: https://doi.org/10.1159/000314698. Epub 2010 May 18. PMID: 20484965.

10. Jørgensen AR, Hansen J, Bue M, Hanberg P, Stilling M. Microdialysis as a sampling tool for the chemotherapeutic agent Doxorubicin. J Pharm Biomed Anal. 2024 Feb 15;239:115872. doi: https://doi.org/10.1016/j.jpba.2023.115872. Epub 2023 Nov 24. PMID: 38039870.
11. https://downloads.regulations.gov/FDA-1975-N-0012-0317/attachment_250.pdf
12. Bashaw ED, Tran DC, Shukla CG, Liu X. Maximal Usage Trial: An Overview of the Design of Systemic Bioavailability Trial for Topical Dermatological Products. Ther Innov Regul Sci. 2015 Jan;49(1):108–115. doi: https://doi.org/10.1177/2168479014539157. Epub 2014 Jun 27. PMID: 26634191; PMCID: PMC4663190.
13. Madison KC. Barrier function of the skin: "la raison d'être" of the epidermis. J Invest Dermatol. 2003 Aug;121(2):231–41. doi: https://doi.org/10.1046/j.1523-1747.2003.12359.x. PMID: 12880413.
14. Montero-Vilchez T, Segura-Fernández-Nogueras MV, Pérez-Rodríguez I, Soler-Gongora M, Martinez-Lopez A, Fernández-González A, Molina-Leyva A, Arias-Santiago S. Skin Barrier Function in Psoriasis and Atopic Dermatitis: Transepidermal Water Loss and Temperature as Useful Tools to Assess Disease Severity. J Clin Med. 2021 Jan 19;10(2):359. doi: https://doi.org/10.3390/jcm10020359. PMID: 33477944; PMCID: PMC7833436.
15. https://www.fda.gov/media/125080/download
16. https://www.fda.gov/media/140321/download?attachment
17. https://www.accessdata.fda.gov/drugsatfda_docs/nda/2001/21-302_ELIDEL_prntlbl.pdf
18. Eichenfield LF, Stripling S, Fung S, Cha A, O'Brien A, Schachner LA. Recent Developments and Advances in Atopic Dermatitis: A Focus on Epidemiology, Pathophysiology, and Treatment in the Pediatric Setting. Paediatr Drugs. 2022 Jul;24(4):293–305. doi: https://doi.org/10.1007/s40272-022-00499-x. Epub 2022 Jun 13. PMID: 35698002; PMCID: PMC9191759.
19. https://www.accessdata.fda.gov/drugsatfda_docs/nda/2001/21-302_ELIDEL_Approv.pdf
20. https://wayback.archive-it.org/7993/20170404062534/; https://www.fda.gov/ohrms/dockets/ac/05/agenda/2005-4089a2_01_DRAFT%20agenda.pdf
21. https://www.medpagetoday.com/dermatology/generaldermatology/2523
22. https://wayback.archive-it.org/7993/20170404062538/; https://www.fda.gov/ohrms/dockets/ac/05/transcripts/2005-4089T2.pdf
23. Huang HH, Shen D, Chan TC, Cho YT, Tang CH, Chu CY. Association Between the Use of Topical Calcineurin Inhibitors and the Risk of Cancer Among Patients with Atopic Dermatitis: A Nationwide, Population-Based, Retrospective Cohort Study. Am J Clin Dermatol. 2023 Sep;24(5):799–808. doi: https://doi.org/10.1007/s40257-023-00787-3. Epub 2023 Jun 6. PMID: 37280416.
24. Siegfried EC, Jaworski JC, Herbert AA. Topical Calcineurin Inhibitors and Lymphoma Risk: Evidence Update with Implications for Daily Practice. Am J Clin Dermatol. 2013; May 24; 14(3): 163–178
25. https://medtech.citeline.com/PS061109/Elidel-Sales-Fall-In-Q3-On-Fewer-US-Prescriptions-Novartis-Says
26. https://dailymed.nlm.nih.gov/dailymed/drugInfo.cfm?setid=ac52a235-1b6f-47a2-b3d9-8d538eeae02f
27. https://www.medicines.org.uk/emc/product/4966/smpc
28. https://www.accessdata.fda.gov/drugsatfda_docs/nda/2001/21-302_ELIDEL_biopharmr_P1.pdf
29. https://medium.com/@andreamantovani/known-knowns-known-unknowns-unknown-unknowns-leadership-367f346b0953#:~:text=%E2%80%9C%E2%80%A6,we%20don't%20know.%E2%80%9D
30. https://www.accessdata.fda.gov/drugsatfda_docs/label/2023/215985s002lbl.pdf
31. Dose-Proportional Intraindividual Single- and Repeated-Dose Pharmacokinetics of Roflumilast, an Oral, Once-Daily Phosphodiesterase 4 Inhibitor; Bethke T. D. et. al; J. Clin Pharmacol 2007; 47: 26–36
32. Efficacy and Safety of Janus Kinase Inhibitors for the Treatment of Atopic Dermatitis: A Systematic Review and Meta-Analysis; Li C. et al.; Dermatology; 2002 Jul; 238(4): 725–735.

33. https://www.accessdata.fda.gov/drugsatfda_docs/label/2022/215309Orig1s000correctedlbl.pdf
34. https://www.fda.gov/downloads/AdvisoryCommittees/CommitteesMeetingMaterials/Drugs/NonprescriptionDrugsAdvisoryCommittee/UCM495796.pdf
35. https://www.mdanderson.org/cancerwise/should-you-use-very-high-spf-sunscreen.h00-159460845.html
36. Schlumpf M, Kypke K, Wittassek M, Angerer J, Mascher H, Mascher D, Vökt C, Birchler M, Lichtensteiger W. Exposure patterns of UV filters, fragrances, parabens, phthalates, organochlorine pesticides, PBDEs, and PCBs in human milk: correlation of UV filters with use of cosmetics. Chemosphere. 2010 Nov;81(10):1171–83. doi: https://doi.org/10.1016/j.chemosphere.2010.09.079. Epub 2010 Oct 27. PMID: 21030064.
37. https://wayback.archive-it.org/7993/20170404152721/; https://www.fda.gov/downloads/AdvisoryCommittees/CommitteesMeetingMaterials/Drugs/NonprescriptionDrugsAdvisoryCommittee/UCM415403.pdf
38. https://wayback.archive-it.org/7993/20170404152726/; https://www.fda.gov/downloads/AdvisoryCommittees/CommitteesMeetingMaterials/Drugs/NonprescriptionDrugsAdvisoryCommittee/UCM421304.pdf
39. Matta MK, Zusterzeel R, Pilli NR, Patel V, Volpe DA, Florian J, Oh L, Bashaw E, Zineh I, Sanabria C, Kemp S, Godfrey A, Adah S, Coelho S, Wang J, Furlong LA, Ganley C, Michele T, Strauss DG. Effect of Sunscreen Application Under Maximal Use Conditions on Plasma Concentration of Sunscreen Active Ingredients: A Randomized Clinical Trial. JAMA. 2019 Jun 4;321(21):2082–2091. doi: https://doi.org/10.1001/jama.2019.5586. PMID: 31058986; PMCID: PMC6549296.
40. Matta MK, Florian J, Zusterzeel R, Pilli NR, Patel V, Volpe DA, Yang Y, Oh L, Bashaw E, Zineh I, Sanabria C, Kemp S, Godfrey A, Adah S, Coelho S, Wang J, Furlong LA, Ganley C, Michele T, Strauss DG. Effect of Sunscreen Application on Plasma Concentration of Sunscreen Active Ingredients: A Randomized Clinical Trial. JAMA. 2020 Jan 21;323(3):256–267. doi: https://doi.org/10.1001/jama.2019.20747. Erratum in: JAMA. 2020 Mar 17;323(11):1098. PMID: 31961417; PMCID: PMC6990686.
41. https://www.fda.gov/regulatory-information/search-fda-guidance-documents/nonprescription-sunscreen-drug-products-safety-and-effectiveness-data
42. Yacobi A, Shah VP, Bashaw ED, Benfeldt E, Davit B, Ganes D, Ghosh T, Kanfer I, Kasting GB, Katz L, Lionberger R, Lu GW, Maibach HI, Pershing LK, Rackley RJ, Raw A, Shukla CG, Thakker K, Wagner N, Zovko E, Lane ME. Current challenges in bioequivalence, quality, and novel assessment technologies for topical products. Pharm Res. 2014 Apr;31(4):837–46. doi: https://doi.org/10.1007/s11095-013-1259-1. Epub 2014 Jan 7. PMID: 24395404.
43. Birngruber T, Tiffner KI, Mautner SI, Sinner FM. Dermal open flow microperfusion for PK-based clinical bioequivalence studies of topical drug products. Front Pharmacol. 2022 Nov 22;13:1061178. doi: https://doi.org/10.3389/fphar.2022.1061178. PMID: 36483734; PMCID: PMC9723326.
44. Khotimchenko M, Antontsev V, Chakravarty K, Hou H, Varshney J. In Silico Simulation of the Systemic Drug Exposure Following the Topical Application of Opioid Analgesics in Patients with Cutaneous Lesions. Pharmaceutics. 2021 Feb 21;13(2):284. doi: https://doi.org/10.3390/pharmaceutics13020284. PMID: 33669957; PMCID: PMC7924840

Index

S. N. Murthy (ed.), *Topical Products and Dermal Drug Delivery*,
AAPS Introductions in the Pharmaceutical Sciences 2,
https://doi.org/10.1007/978-3-031-80525-7

GPSR Compliance
The European Union's (EU) General Product Safety Regulation (GPSR) is a set of rules that requires consumer products to be safe and our obligations to ensure this.

If you have any concerns about our products, you can contact us on

ProductSafety@springernature.com

In case Publisher is established outside the EU, the EU authorized representative is:

Springer Nature Customer Service Center GmbH
Europaplatz 3
69115 Heidelberg, Germany

www.ingramcontent.com/pod-product-compliance
Lightning Source LLC
Chambersburg PA
CBHW051638110825
30947CB00003B/93
* 9 7 8 3 0 3 1 8 0 5 2 4 0 *